T0318659

This book gives a comprehensive and lucid account of the science of the atmospheric boundary layer (ABL). Its purpose is to provide a moderately advanced text aimed at the experienced researcher or teacher working in the atmospheric and related sciences. Its importance lies in the breadth of the material covered, with a careful balance between mathematical description and physical interpretation.

Cambridge atmospheric and space science series

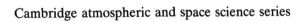

The atmospheric boundary layer

Cambridge atmospheric and space science series

Editors
John T. Houghton
Michael J. Rycroft
Alexander J. Dessler

Titles in print in this series

M. H. Rees, *Physics and chemistry of the upper atmosphere*
Roger Daley, *Atmospheric data analysis*
Ya. L. Al'pert, *Space plasma*, Volumes 1 and 2
J. R. Garratt, *The atmospheric boundary layer*
J. K. Hargreaves, *The solar–terrestrial environment*
Sergei Sahzin, *Whistler-mode waves in a hot plasma*

The atmospheric boundary layer

J. R. Garratt

*Division of Atmospheric Research, CSIRO,
Melbourne, Australia*

CAMBRIDGE
UNIVERSITY PRESS

CAMBRIDGE UNIVERSITY PRESS
Cambridge, New York, Melbourne, Madrid, Cape Town,
Singapore, São Paulo, Delhi, Tokyo, Mexico City

Cambridge University Press
The Edinburgh Building, Cambridge CB2 8RU, UK

Published in the United States of America by Cambridge University Press, New York

www.cambridge.org
Information on this title: www.cambridge.org/9780521467452

First published 1992
First paperback edition (with corrections) 1994

A catalogue record for this publication is available from the British Library

Library of Congress Cataloguing in Publication Data

Garratt, J. R.
The atmospheric boundary layer / J. R. Garratt.
p.   cm. – (Cambridge atmospheric and space science series)
Includes bibliographical references and index.
ISBN 0 521 38052 9
1. Planetary boundary layer. 2. Atmospheric physics. I. Title. II. Series.
QC880.4.B65G37 1992
551.5-dc20   91-34340 CIP

ISBN 978-0-521-38052-2 Hardback
ISBN 978-0-521-46745-2 Paperback

Contents

Preface

In the last few years there have appeared several books on the atmospheric boundary layer (ABL) and related topics that have helped fill a void existing for some time. These recent books have provided a more general introduction to the subject not found in the numerous advanced and specialist texts and research monographs that became available through the 1970s and 1980s.

Plans for the present work were well under way when the books by Stull and by Sorbjan appeared. However, the subject of the ABL covers so many areas (turbulence, dynamics, cloud physics, radiation, the physics of heat and mass transfer, soil physics, surface vegetation, numerical modelling) that no one book covers quite the right material with the necessary emphasis and attention to detail. It is hoped that the present book fills in some gaps in our knowledge, and brings to the subject an emphasis that many readers will find useful. It is aimed especially at the many researchers in the atmospheric and associated sciences who require a moderately advanced text on the ABL in which the many links between turbulence, air-surface transfer, boundary-layer structure and dynamics, and numerical modelling are discussed and elaborated upon. Within the book, I have attempted to emphasize the application of ABL ideas to numerical modelling of the climate, and it should be possible for any experienced research worker to trace the origins of specific ABL parameterization schemes, and to relate these to the fundamental principles and detailed physics of the process of interest. The book should be useful to postgraduate students and university teachers in the atmospheric (and related) sciences who require a treatment beyond the introductory level. Finally, it seemed natural that the Atmospheric and Space Science series planned by Cambridge University Press should contain one text on the ABL. It is hoped that this text will nicely complement the other contributions.

The book comprises nine chapters, four appendices (data tables, information sources, physical constants) and an extensive reference list. It is inevitably a compromise between a fundamental treatise on the one hand and a specialist research monograph on the other. Considerable effort has been made to provide a balance between mathematical description and physical interpretation, between observed behaviour and model simulation. Chapter 1 serves as an introduction, with Chapters 2 and 3 dealing with the development of mean and turbulence equations, and the many scaling laws and theories that are the cornerstone of any serious ABL treatment. Modelling of the ABL is crucially dependent for its realism on the surface boundary conditions, and Chapters 4 and 5 deal with aerodynamic and energy considerations, with attention to both dry and wet land surfaces and the sea. The structure of the clear-sky, thermally stratified ABL is treated in Chapter 6, including the convective and stable cases over homogeneous land, the marine ABL and the internal boundary layer at the coastline. Chapter 7 then extends the discussion to the cloudy ABL. This is seen as particularly relevant since the extensive stratocumulus regions over the sub-tropical oceans and stratus regions over the Arctic are now identified as key players in the climate system. Finally, Chapters 8 and 9 bring much of the book's material together in a discussion of appropriate ABL and surface parameterization schemes for the general circulation models of the atmosphere that are being used for climate simulation.

I have taken note of the many comments made by reviewers at various stages in the book's production, in particular regarding the question of references. Most books, depending upon their style, contain either few, if any, references (e.g. Rees' book on the physics and chemistry of the upper atmosphere) or references are almost too plentiful (e.g. Brutsaert's book on evaporation into the atmosphere). My choice was to list at the end of the book references to specialist material not discussed at length in the text, and to supplement this reference list with a set of brief notes and recommended further reading at the end of each chapter. It is hoped that any unnecessary duplication has been minimized.

Several colleagues read selected chapters of the manuscript, made corrections and suggested improvements in the presentation. I wish to thank Drs T. Beer, R. R. Brook, J. J. Finnigan, R. L. Hughes, D. H. Lenschow, J. L. Mcgregor, F. T. M. Nieuwstadt and M. R. Raupach and, in particular, Drs P. C. Manins and K. R. McNaughton. I am especially indebted to Dr B. L. Sawford for carefully and critically reading the greater part of the manuscript.

I gratefully acknowledge Ms Louise Carr for her enthusiastic approach to the drafting of the original artwork.

I wish to thank publishers and individual scientists who kindly gave their permission to reproduce figures from the published literature. Susan Parkinson, copy editor at CUP, provided the final polish to the manuscript, as well as ensuring consistency, continuity and correct grammar.

This undertaking began during a sabbatical leave as Assistant Professor at the Department of Atmospheric Science, Colorado State University in 1987–8. An early draft of the book served as the basis of a lecture course given in the Department during the period January to May 1988. Part of the material contained in the final draft is used in a lecture course given to honours and postgraduate students in the Department of Mathematics, Monash University.

The writing of the book depended ultimately on the enviable working environment provided in the Commonwealth Scientific and Industrial Research Organisation and on the patience and understanding of my wife Dianne.

Symbols

Vector quantities are bold
A circumflex denotes a vertical average
The subscript m denotes a mixed-layer value
The subscript 0 denotes a surface or surface-layer value in most cases
The superscript * denotes a saturation value in most cases

A	ABL similarity constant/function (velocity); IBL growth constant
a	drag-law constant (analytical form)
a_0	inverse scale height; cloud-top entrainment instability parameter; soil-moisture parameter
a_1, a_2	free convection constants; turbulence-closure constants
B	Bowen ratio; ABL similarity constant/function (velocity)
B^{-1}	interfacial sublayer parameter
b	soil index parameter
$b_M, b_H,$ b_M^*, b_H^*	drag-law constants (analytical form)
b_0	soil-moisture parameter
C	ABL similarity constant/function (temperature)
C_D, C_{D0}	surface-layer, interfacial drag coefficient
C_E	water vapour transfer coefficient
C_H	heat transfer coefficient
C_g	geostrophic drag coefficient; soil heat capacity per unit area
C_{gH}	large-scale heat-transfer coefficient
C_i	transfer coefficient for surface i
C_s	volumetric heat capacity for soil
C_v	canopy heat capacity per unit area
c	scalar concentration; drag-law constant (analytical form); ABL depth parameter
c_0	soil-moisture parameter
c_p	specific heat at constant pressure for air

c_s, c_c	specific heats for soil and canopy
c_1, c_2	turbulence-closure constants
D	damping depth
Da	Dalton number
D_η	soil-moisture diffusivity
d	zero-plane displacement
d_0, d_1, d_2, d_3	soil-layer thicknesses
E, E_P	evaporation, potential evaporation
E_L	evaporation quantity combining energy and aerodynamic terms
e	turbulent kinetic energy; vapour pressure
$\mathbf{e}$	the vector $(1, 1, 1)$ with components, e_i, unity
e^*	saturation vapour pressure
F_M, F_H, F_E	stability functions
F_w	vertical flux of soil water
f	Coriolis parameter; normalized frequency (nz/u)
f_i	normalized frequency (nh/u); fractional area
G	magnitude of the geostrophic wind; soil heat flux; stability function
g	acceleration due to gravity
H	sensible heat flux
h, h_b	boundary-layer depth, internal boundary-layer depth
h_c	height of canopy; depth of elevated cloudy mixed layer
h_d	depth of interfacial sublayer
h_e	depth of equilibrium NBL
h_i	depth of nocturnal (surface) inversion
h_r	sand diameter or height of roughness elements
h_s	depth of surface layer
h_{ss}	equilibrium or inner layer depth
i	square root of minus one
K	eddy diffusivity
K_b, K_t	bottom-up, top-down diffusivities
K_η	soil-hydraulic conductivity
k	von Karman constant
k_s	soil thermal conductivity; friction coefficient
k_T	molecular thermal conductivity of dry air
L	Monin–Obukhov length; averaging length
L_A	leaf area index (LAI)
L_x	horizontal length scale
l	mixing length

l_c	characteristic length scale; canopy length scale
l_b	blending height
l_s	characteristic length scale of the mean flow
l^i	integral length scale
M	water mass
$\mathbf{M}$	baroclinity parameter (components M_x and M_y)
M_d, M_v	mean molecular weight of dry air, water vapour
m	corrected absorber mass of gas; profile curvature index; depth of intercepted canopy water
N	Brunt–Vaisala frequency
n	natural frequency; profile curvature index
P, P_g	precipitation rate, precipitation rate under a canopy
Pr	Prandtl number
P_t	turbulent Prandtl number
p, p_R	atmospheric pressure, reference pressure
q	specific humidity (surface value q_0)
q^*	saturation specific humidity
q_l, q_t	liquid water content, total water specific humidity
R	universal gas constant; scale–height ratio; ratio of entrainment and surface scalar fluxes; runoff
Rb	bulk ABL Richardson number
Re, Re_*	Reynolds number, roughness Reynolds number
$Rf, \overline{Rf}$	flux Richardson number, layer-averaged flux Richardson number
Ri, Ri_c	gradient Richardson number, critical value of Ri
Ri_B	bulk Richardson number
Ro	surface Rossby number
R_d, R_v, R_w	gas constants for dry air, water vapour and moist air
R_E	radius of the Earth
R_j	radiative heat flux component
R_N, R_N^{lw}	net radiation, net longwave radiation
R_s, R_L	shortwave flux, longwave flux of radiation
r	radiative flux fraction at cloud top; mixing ratio
$\mathbf{r}$	vector space variable
r_a	aerodynamic resistance
r_b	excess resistance to transfer
r_b, r_d	internal canopy aerodynamic resistances
r_h	soil-surface relative humidity
r_s, r_s^+	surface resistance, unconstrained surface resistance
r_{st}, r_{sti}	bulk stomatal resistance, leaf stomatal resistance
S	stability function
Sc	Schmidt number
St	Stanton number

S_c	the solar irradiance (solar constant)
s	general property (wind component, temperature, mass concentration); $\partial q^*/\partial T$
s_m	vertically averaged value of s in the mixed layer
$\mathbf{s}_v$	nondimensional shear vector
s_θ	nondimensional temperature gradient
T	absolute temperature (surface value T_0); averaging period; NBL time scale; day length
T_e	effective temperature of Earth's system
T_f, T_{0f}	foliage temperature
T_v	virtual temperature
T_s, T_c	soil and canopy temperatures
T_*, T_{V*}	convective temperature scale, cloudy convective temperature scale
t	time
U	large-scale wind
u	wind component (u_1) in the x-direction
u_{ag}	ageostrophic wind component in the x-direction
u_f	local free convection velocity scale
u_g	geostrophic wind component in the x-direction
u_{gi}	geostrophic wind component
u_i	wind component
u_*	friction velocity (surface-value u_{*0})
$\mathbf{V}$	velocity vector
$\mathbf{V}_g$	geostrophic velocity vector
v	wind component (u_2) in the y-direction
v_{ag}	ageostrophic wind component in the y-direction
v_c	characteristic velocity scale
v_g	geostrophic wind component in the y-direction
v_s	characteristic velocity scale of the mean flow
v_η	Kolmogorov velocity scale
W_s, W_c	energy storage terms
W_*	cloudy convective velocity scale
w	vertical wind component (u_3)
w_h	vertical velocity at the ABL top
w_*	convective velocity scale
x	space variable; fetch; evaporation fraction
Y	nondimensional heat flux
y	space variable
Z	vertical distance above the surface
z	space variable; height above the zero-plane displacement
z'	vertical space coordinate in the soil; vertical distance measured

downwards from the cloud top
z_0 aerodynamic roughness length
z_T, z_q roughness scaling lengths for temperature and humidity
z_* depth of the roughness sublayer

α cross-isobar flow angle; slope angle; general albedo
α_c Charnock's constant
α_f, α_g albedo of foliage, ground
α_s surface albedo
α_w high-frequency water-wave spectrum constant
α_1 free convection constant; stable IBL parameter

β ratio of entrainment and surface heat fluxes
β_1 Monin–Obukhov profile constant
$\beta_u, \beta_\theta, \beta_q$ spectral inertial subrange constants

γ psychrometric constant (c_p/λ)
γ_c Zilitinkevich constant of the NBL
γ_1, γ_2 Monin–Obukhov profile constants; roughness constant (γ_1)
γ_θ potential temperature gradient above h
γ_θ' lapse-rate correction

$\Delta h, \Delta z$ layer thicknesses in the atmosphere
$\Delta z'$ layer thickness in the soil
Δs_I change in property concentration across the nocturnal (surface) inversion
Δs_B change in property concentration across the NBL or stable IBL
Δs_h difference in property concentration between that at $h + \varepsilon$ and the mixed-layer value
$\Delta^c s$ change in property concentration across the cloud top
δ_1 depth of viscous sublayer
$\delta_u, \delta_T, \delta_q$ depths of interfacial sublayer
δ_{ij} Kronecker delta tensor

ε rate of viscous dissipation of TKE; emissivity; height increment
$\varepsilon_f, \varepsilon_g$ emissivity of foliage, ground
ε_{ijk} alternating unit tensor
ε_s surface emissivity

ζ stability parameter, z/L

$\boldsymbol{\eta}$ unit vector (components η_j)
η_k Kolmogorov length scale
η, η_s soil volumetric moisture content, saturation value
η_w wilting value of soil moisture content

θ, θ_v potential temperature, virtual potential temperature
θ_e equivalent potential temperature

θ_s	sea-surface temperature
θ_*, θ_{v*}	turbulent temperature scales
κ	wavenumber
κ_s	soil thermal diffusivity
κ_T	molecular thermal diffusivity of dry air
κ_V	molecular diffusivity for water vapour in dry air
Λ	local Monin–Obukhov length; empirical turbulence length scale
λ	latent heat of vaporization of water; length scale; asymptotic mixing length
λ_1	roughness-element density
λ_s	mean orographic spacing
μ	stability parameter h/L; soil force-restore parameter; dynamic viscosity of air
μ_v	canopy force-restore parameter
ν	kinematic viscosity of air
ξ	nondimensional height
ρ, ρ_w	air density, density of water
ρ_s, ρ_c	soil, canopy density
σ	Stefan–Boltzmann constant; standard deviation; nondimensional stress
σ_f	fractional canopy cover
σ_h	root-mean-square orographic height variation
σ_{ij}	stress tensor
σ_t	surface tension
τ	shearing stress; time lag
τ^i	integral time scale
τ_a	transmissivity
Φ	gradient profile functions
ϕ	latitude; roughness sublayer gradient function
ϕ_{ss}	spectral density
ϕ_ε	nondimensional dissipation function
χ	rate of molecular dissipation of temperature fluctuations
Ψ	integral profile functions; solar elevation angle
ψ	moisture or matric potential
ψ_ε	nondimensional dissipation function
Ω	angular velocity of earth

Abbreviations

ABL	atmospheric boundary layer
AOD	absorption optical depth
AMS	American Meteorological Society
AMTEX	Air-mass Transformation Experiment
ATEX	Atlantic Tradewind Experiment
BATS	Biosphere–Atmosphere Transfer Scheme
BOMEX	Barbados Oceanographic and Meteorological Experiment
CAS	Commission for Atmospheric Science
CBL	convective boundary layer
COAST	Cooperative Operations with Acoustic Sounding Techniques
CSIRO	Commonwealth Scientific and Industrial Research Organisation
CTBL	cloud-topped boundary layer
FGGE	First GARP Global Experiment
FIFE	First ISLSCP Field Experiment
FIRE	First ISCCP Field Experiment
GARP	Global Atmospheric Research Programme
GATE	GARP Atlantic Tropical Experiment
GCM	general circulation model
HAPEX	Hydrologic–Atmospheric Pilot Experiment
IBL	internal boundary layer
ISCCP	International Satellite Cloud Climatology Project
ISLSCP	International Satellite Land Surface Climatology Project
JASIN	Joint Air Sea Interaction (Project)
JSC	Joint Scientific Committee
LAI	leaf area index
LCL	local condensation level
LES	large-eddy simulation
MASEX	Mesoscale Air–Sea Exchange (Project)
NBL	nocturnal boundary layer
NCMC	non-classical mesoscale circulation
SCR	surface cooling rate

SEB	surface energy balance
SiB	Simple Biosphere (Model)
TKE	turbulent kinetic energy
TIBL	thermal internal boundary layer

1

The atmospheric boundary layer

1.1 Introduction

The concept of a *boundary layer* in fluid flows can perhaps be attributed to Froude, who carried out a series of laboratory towing experiments in the early 1870s to study the frictional resistance of a thin flat plate when towed in still water. The term itself was probably first introduced into the literature by Prandtl (1905), working in the field of aerodynamics, who was concerned with the flow of a fluid of low viscosity close to a solid boundary. His work recognised the transition, through a thin aerodynamic boundary layer, from irrotational flow well away from the boundary to the condition of no-slip at the boundary.

In the atmospheric context, it has never been easy to define precisely what the boundary layer is. Nevertheless, a useful working definition identifies the boundary layer as the layer of air directly above the Earth's surface in which the effects of the surface (friction, heating and cooling) are felt directly on time scales less than a day, and in which significant fluxes of momentum, heat or matter are carried by turbulent motions on a scale of the order of the depth of the boundary layer or less.

The turbulent nature of the atmospheric boundary layer (ABL) is one of its most conspicuous and important features. However, turbulence in the lower atmosphere differs from most turbulence studied in wind tunnels in two main ways. Firstly, turbulence associated with thermal convection coexists with mechanical turbulence (by mechanical turbulence we mean turbulence generated by wind shear). Secondly, boundary-layer turbulence interacts with a mean flow that is influenced by the rotation of the Earth.

The structure of the atmospheric boundary layer shows many similarities to the two-dimensional turbulent boundary layer generated in a wind tunnel, in that both have a distinctive *inner* region and *outer region* (Fig. 1.1). In the outer region, the flow shows little dependence on the nature of the surface and, in the atmosphere, the Coriolis force due to the Earth's rotation is also important. This region is sometimes referred to as the *Ekman layer*, since Ekman (1905) first

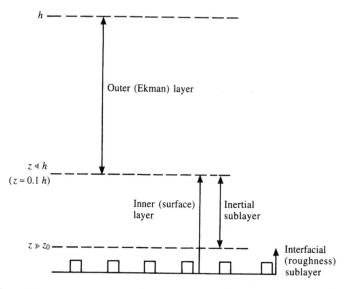

Fig. 1.1 Schematic atmospheric boundary-layer structure for aerodynamically rough flow in neutrally-stratified conditions. Several regions are identified, including the interfacial (or roughness) sublayer, the inner (or surface) layer and the outer (or Ekman) layer. The inertial sublayer is discussed in Chapter 3.

Within the roughness sublayer, the turbulence and mean profiles are strongly affected by the structure of the roughness elements. In the surface layer, wind and stress exhibit negligible rotation with height. The inertial sublayer is the region within which the velocity profile in neutrally buoyant conditions is logarithmic.

In the diagram, h is the boundary-layer depth, z is height and z_0 is the aerodynamic roughness length (see Chapter 4). The reader is referred to discussions and illustrations in Sheppard (1968), Tennekes (1973) and Brutsaert (1982).

dealt with the effects of rotation on boundary-layer flow in the ocean. In contrast with the above, flow in the *inner layer* (also termed the wall or *surface layer*) is mainly dependent on the surface characteristics and is little affected by rotation. The transition between the inner and outer layers is not abrupt, but is characterized by an overlap region. The influence of the surface is directly felt in the *interfacial sublayer*, which is the layer of air within and just above the roughness elements comprising the land or sea surface. In this layer, molecular diffusion is an important process by which heat and mass are exchanged between the surface and the air.

Over land in particular, the structure of ABL turbulence is strongly influenced by the diurnal cycle of surface heating and cooling, and by the presence of clouds. Neutral flow, in which buoyancy effects are absent, is readily produced in a wind tunnel, and may be closely approximated in the atmosphere in windy conditions with a complete cloud cover. The unstably stratified ABL, or convective boundary layer (CBL), occurs when strong surface heating (due to the sun) produces thermal instability or convection in the form of thermals and plumes, and when upside-down convection is generated by cloud-top radiative cooling. In strongly unstable conditions driven by surface heating, the outer layer in particular is dominated by convective motions and is often referred to as the *mixed layer*. In contrast, the stably stratified ABL occurs mostly (though not exclusively) at night, in response to surface cooling by longwave emission to space. The unstable ABL is characterized by a near-surface superadiabatic layer, and the stable ABL by the presence of a surface inversion.

The top of the boundary layer in convective conditions is often well defined by the existence of a stable layer (capping inversion) into which turbulent motions from beneath are generally unable to penetrate very far, though they may continually erode it, particularly where latent heat is released in rising elements of air. The height of this elevated stable layer is quite variable, but is generally below 2–3 km. The top of a convective boundary layer is well defined, in Fig. 1.2, by the sharp decrease in aerosol concentration at a height of about 1200 m, and coincides with the base of a deep and intense subsidence inversion. Over deserts in mid-summer under strong surface heating the ABL may be as much as 5 km deep, and even deeper in conditions of vigorous cumulonimbus convection. In stable conditions, in contrast to the above, the boundary layer is not so readily identified, turbulence is much weaker than in the unstable case and consequently the depth is no more than a few hundred metres at most. At night over land, under clear skies and light winds, it may be even smaller, perhaps no more than 50–100 m and strongly influenced by internal wave motions.

Over the open oceans, where low-level layer cloud (stratus and stratocumulus) is prevalent, the ABL depth may be no more than a few hundred metres and, in extratropical latitudes, may have a structure quite similar to that over land. According to results from the north-east Atlantic (the JASIN experiment; see Businger and Charnock, 1983), the stability is near neutral, with a capping stratocumulus layer and a depth of the order of 0.5 km. Such a shallow boundary layer is also found in coastal regions when warm air flows from land over a relatively cool sea. In the tropics, the mean structure is very much dependent on the season, and on whether conditions are disturbed (in the vicinity of the intertropical convergence zone) or undisturbed. In the former, developing cumulus clouds result in poor definition of the ABL top, whilst in undisturbed conditions, the ABL top is well defined by the trade-wind inversion. Under special circumstances, the ABL depth over the ocean can be comparable to that over land in the middle of the day. This can occur during intense cold-air outbreaks over the ocean, when the large "jumps" in temperature and humidity that identify the ABL top are particularly noticeable and are the result of cold, dry air flowing out from the continent over relatively warm sea.

1.2 History

The history of atmospheric turbulence and boundary-layer studies is rich and very relevant to contemporary work. The statistical theory of turbulence, related to problems of diffusion and the scale and spectrum of turbulence, owes much to G. I. Taylor in the period 1915–38. In the same period both von Karman and Prandtl were enunciating mixing-length hypotheses for direct application to the atmosphere, using eddy diffusivities and flux-gradient concepts based on analo-

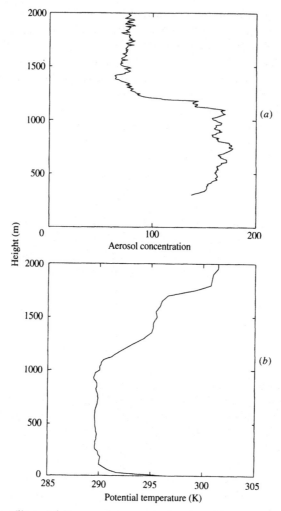

Fig. 1.2 Vertical profiles of (*a*) aerosol concentration in arbitrary units and (*b*) potential temperature observed at 1100 local time at Tarong, Queensland on 7 September 1989. The aerosol observations were made with a ground-based lidar system (data courtesy of Dr B. L. Sawford).

gies with molecular transfer. In 1941, although this was not known to western scientists until some years later, Kolmogorov was making important contributions to our understanding of the small-scale structure of turbulence and the energy transfer process from large to small scales (the "cascade" process) through his similarity theory of turbulence. In the 1950s and into the early 1960s major advances took place in our ability to interpret observations, in our understanding of the role of buoyancy in modifying the wind profile and in modifying flux-gradient relations in general. This involved the surface-layer similarity theory of Monin and Obukhov (1954) and the ABL similarity theory of Kazanski and Monin (1960, 1961). Many of the observations are associated with the major field experiments of the 1950s to 1970s. From the late 1960s to the present day, major advances in our knowledge of boundary-layer structure have taken place through the use of numerical modelling to simulate the ABL, and in the application of higher-order closure theory for representing the effects of turbulence more realistically.

1.3 Observing the ABL

Much of our knowledge of the ABL can be related to observations of turbulent flows made both in the laboratory and in the atmosphere. Investigation of the structure of the lower part of the ABL has mainly utilized sensors located on tower structures. These towers have ranged in size from relatively short masts of a few metres height to the 200–300 m tall structures that may support a very comprehensive suite of instrumentation. For studies of the whole boundary layer, early approaches had to rely on balloon-borne instrumentation – free and tethered sounding balloons, or constant-volume free balloons – until aircraft techniques and facilities became more widely available.

For measurements throughout the ABL, the aircraft is now a well-established observational platform, although balloon platforms have been used with some success. One of the major advantages of using aircraft as an atmospheric measurement system is their mobility and capacity for making extended line averages of turbulent quantities (most aircraft flights occur at speeds of about 80 m s^{-1}). Aircraft can be used both for vertical profiling well above the tower layer and for horizontal traverses. In many field experiments the two are combined. The main disadvantage of aircraft measurements is the need to measure accurately the aircraft's motion, since corrections to the velocity and temperature measurements must be made for this motion.

In the last decade or so, ABL observations have been enhanced by remote sensing techniques. These mainly involve transmitted acoustic, radio or light energy, and the detection of the scattered energy due to natural or artificial atmospheric targets (dust, salt, rain). The newest instrumentation includes, for example, sodars (acoustic sounders); acoustic radars; lidars (light radars) and Doppler radars. Remote sensing of boundary-layer variables can be done actively and passively. Active techniques involve transmission of acoustic or electromagnetic radiation to the region of interest and measuring the backscattered fraction that is returned to the instrument; sensors include lidar, radar and sodar. In contrast, passive techniques involve the measurement of radiation naturally emitted from the atmosphere, e.g. as in infrared radiometry.

Remote sensing techniques are attractive where *in situ* methods will not work or are uneconomical, and where the atmosphere interacts with the transmitted or received radiation allowing the meteorological variable of interest to be measured. Remote sensing utilizes spatial averages, usually over volumes, and so is especially suitable if this type of average is desired.

1.4 ABL modelling

In addition to observing the ABL, another important area of research involves the numerical simulation of boundary-layer structure and behaviour. This allows experimentation under carefully controlled conditions, and thus offers an advantage over real-world field experiments where no such control is possible. The approach requires numerical solutions of a set of partial differential equations which, with appropriate initial and boundary conditions, describe the conservation laws appropriate to the dynamic and thermodynamic nature of the ABL and its interaction with the underlying surface. Ideally, the effects of radiation, atmospheric composition, clouds, orography, the Earth's rotation, surface friction, gravity waves and turbulence are taken into account in order to derive realistic fields for wind, temperature, humidity and pressure.

Advanced numerical techniques are used to solve in terms of the time variable the fully coupled set of equations, with application both to limited areas of the Earth's surface (as in most mesoscale models) and to the whole globe (as in general circulation models (GCMs)). In most models, limited computer resources require calculations to be made at only a finite number of levels in the vertical (usually from 5–50), with the top level (where an upper boundary condition must be applied) usually above the mid-troposphere. In finite-grid (as opposed to spectral) models, the horizontal resolution varies from about 50 m in a large-eddy simulation (LES) model, through 1–25 km in a mesoscale model, to 500 km in a GCM. The numerical model may be integrated forward in time to determine the evolution resulting from some specified forcing, for just a few hours in the case of the LES model, through a day or so for the mesoscale model, and for periods from days to years for the GCM. In each case the time step will be set accordingly.

1.5 Applications

The atmospheric (or planetary) boundary layer plays an important role in many fields, including air pollution and the dispersal of pollutants, agricultural meteorology, hydrology, aeronautical meteorology, mesoscale meteorology, weather forecasting and climate. We can summarize just a few of the problems for which boundary-layer knowledge is important as follows.

Urban Meteorology is associated with the low-level urban environment and air pollution, including air pollution episodes involving photochemical smog and accidental releases of dangerous gases. The dispersal of smog and low-level pollutants depends strongly on meteorological conditions. Of particular importance is information on the likely growth of the shallow mixed layer resulting from surface heating, and on the factors controlling the erosion and ultimate breakdown of the surface inversion.

The control and management of air quality is closely associated with the transport and dispersal of atmospheric pollutants, including industrial plumes. Processes of concern include turbulent mixing in the ABL, particularly the role of convection, photochemistry and dry and wet deposition to the surface. In this general area, research on atmospheric turbulence has a very important practical application, and local meteorology, including the role of mesoscale circulations (sea breezes, slope winds, valley flows) and the phenomenon of decoupling of the low-level flow and the large-scale upper flow, is also of major relevance.

Aeronautical meteorology is concerned with boundary-layer phenomena such as low cloud, low-level jets and intense wind shear leading to high-intensity turbulence, of particular interest for aircraft landing and take-off. In the case of low clouds and low-level jets, factors affecting their formation, maintenance and dissipation are of great importance.

Agricultural meteorology and hydrology are concerned with processes such as the dry deposition of natural gases and pollutants to crops; evaporation, dewfall and frost formation. The last three are intimately associated with the state of the ABL, with the intensity of turbulence and with the energy balance at the surface.

Numerical weather prediction (NWP) and climate simulation based on dynamical models of the atmosphere depend on the realistic representation of the Earth's surface and the major physical processes occurring in the atmosphere. It has been said (Stewart, 1979) that no general circulation model is conceptually complete without the inclusion of boundary-layer effects, and that no prediction model can succeed without a sufficiently accurate inclusion of the influence of the boundary. The boundary layer affects both the dynamics and thermodynamics of the atmosphere. There are a variety of dynamic effects: more than a half of the atmosphere's kinetic energy loss occurs in the ABL (Palmén and Newton, 1969). Boundary-layer friction produces cross-isobar flow in the lower atmosphere, whilst boundary-layer interaction permits air masses to modify their vorticity. From the thermodynamic perspective, all water vapour entering the atmosphere by evaporation from the surface must enter through the ABL. Even the oceans are strongly influenced by the ABL, since it is through the boundary layer that they gain most of their momentum, so influencing the oceanic circulation.

From both climate and local weather perspectives, the most important ABL processes that need to be parameterized in numerical models of the atmosphere are vertical mixing and the formation, maintenance and dissipation of clouds. Those land-surface properties that are potentially crucial to accurate climate simulation include albedo, roughness, moisture content and vegetation cover.

1.6 Scope of the book

The book is primarily addressed to meteorologists, or scientists who have meteorology as a strong second subject. The overall thrust of the book is towards a balanced mathematical, dynamical and physical description of the atmospheric boundary layer, with extended material on the role of the surface and surface-related processes in determining ABL behaviour. There is a definite emphasis on models and sub-models of both surface processes and ABL phenomena, and the application of these to the parameterization of the ABL in numerical models of the atmosphere. Observations are used throughout, mainly where they lend support to a theoretical or numerical treatment or a physical process under discussion. The reader will find little discussion of observational techniques nor, for that matter, of numerical techniques. This was a deliberate decision, as was the decision to leave out any substantial discussion of the general problem of turbulence or any theoretical treatment of the turbulence closure problem.

The book is broadly divided into four parts: fundamentals and laws; surface properties; structure of the ABL; and modelling and the relationship of the ABL to climate. Chapters 2 and 3 cover basic equations (Chapter 2), scaling laws and the nature of the wind profile (Chapter 3). Chapter 4 deals with surface properties and Chapter 5 deals with the surface energy balance and evaporation. Chapters 6 and 7 discuss the structure of the thermally stratified boundary layer for both clear and cloudy conditions. In Chapter 6, the convective ABL over land and the stable ABL over land are presented, together with the marine and coastal ABL. The whole of Chapter 7 deals with the cloud-topped boundary layer from both modelling and observational perspectives, and includes discussion on radiation and turbulence fluxes. Finally, Chapters 8 and 9 deal with the parameterization problem, with emphasis on climate and climate modelling. In Chapter 8, surface schemes involving the treatment of soil and canopy processes, and turbulence closure schemes appropriate to a range of model types, are discussed. The final chapter explores the relationship between climate and the ABL, and considers current priorities in ABL research relevant to climate problems.

The book primarily deals with the ABL over horizontally homogeneous land surfaces under clear skies. However, the reader will find substantial material on boundary-layer structure in advective conditions: the internal boundary layer is discussed in the latter parts of Chapters 4 and 6; the marine boundary layer is also discussed in Chapter 6; and the cloud-topped boundary layer is discussed in Chapter 7.

1.7 Nomenclature and some definitions

A summary of nomenclature and a list of symbols are given elsewhere. Here, for the reader's convenience we present the most commonly used quantities and conventions, including the sign convention for vertical fluxes. The height z in the atmosphere and depth z' in the ground are positive and increase away from the surface. SI units are used throughout, except where specified.

1.7.1 Boundary-layer quantities and parameters

The Cartesian coordinate system is used, with the x-axis chosen appropriately. Cartesian coordinates (x, y) can be defined in the following ways (see Fig. 1.3).

  (i) Geographic coordinates, with the x-axis positive towards the east;
 (ii) Surface-layer coordinates, with the x-axis positive in the mean surface-stress or surface-wind (at a height of 10 m) direction;
(iii) Geostropic wind coordinates, with the x-axis positive in the surface geostrophic wind direction.

ABL Quantities in common use are listed below.

  (i) Mean fields: velocity components are $\bar{u}(z)$ and $\bar{v}(z)$, u and v being along the x- and y-axis respectively; potential temperature is $\bar{\theta}(z)$, virtual potential temperature is $\bar{\theta}_v(z)$ and specific humidity is $\bar{q}(z)$.

  Here, and throughout, z is the height above the zero-plane displacement, Z is the height above the surface.

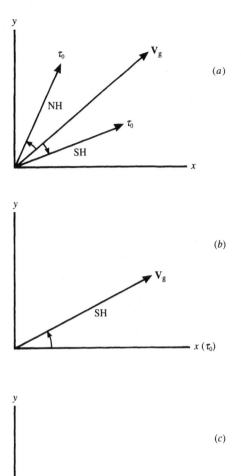

Fig. 1.3 Examples of a Cartesian coordinate system with the x-axis in (a) the east-west direction, (b) along the surface-stress or near-surface wind direction and (c) along the surface geostrophic wind direction. Surface-stress (τ_0) and geostrophic wind ($\mathbf{V}_g$) vectors are shown for flows in the northern (NH) and southern (SH) hemispheres.

(ii) Geostrophic wind components are u_g and v_g, where

$$(u_g, v_g) = (\rho f)^{-1}(-\partial \bar{p}/\partial y, \partial \bar{p}/\partial x) \tag{1.1}$$

where ρ is air density, $\bar{p}$ is the mean pressure and f is the Coriolis parameter. Note that, with $|f| = 10^{-4}\,\text{s}^{-1}$, $\rho = 1\,\text{kg}\,\text{m}^{-3}$ and $\partial \bar{p}/\partial y = 1\,\text{hPa}$ $(100\,\text{km})^{-1}$, $u_g \approx 10\,\text{m}\,\text{s}^{-1}$.

(iii) Baroclinity, or thermal wind, components are defined by

$$\partial(u_g, v_g)/\partial z = (g/f\bar{T})(-\partial \bar{T}/\partial y, \partial \bar{T}/\partial x) \tag{1.2}$$

where g is the acceleration due to gravity and $\bar{T}$ is mean absolute temperature. With $g = 9.8 \, \mathrm{m\,s^{-2}}$, $\bar{T} = 300 \, \mathrm{K}$ and $\partial T/\partial y = 1 \, \mathrm{K} \, (100 \, \mathrm{km})^{-1}$, we have $\partial u_g/\partial z \approx 3.3 \, \mathrm{m\,s^{-1}\,km^{-1}}$.

(iv) Turbulent fluxes: vertical momentum fluxes are designated $\rho \overline{u'w'}(z)$, $\rho \overline{v'w'}(z)$, with surface stress components

$$\tau_{x0} = -\rho(\overline{u'w'})_0 \quad \text{and} \quad \tau_{y0} = -\rho(\overline{v'w'})_0. \tag{1.3}$$

Vertical fluxes involving temperature and humidity include the surface sensible heat flux defined as

$$H_0 = \rho c_\mathrm{p}(\overline{w'\theta'})_0, \tag{1.4}$$

the surface buoyancy flux defined as

$$H_{v0} = \rho c_\mathrm{p}(\overline{w'\theta_\mathrm{v}'})_0 \tag{1.5}$$

and the surface latent heat flux

$$\lambda E_0 = \rho\lambda(\overline{w'q'})_0. \tag{1.6}$$

In the above E_0 is the surface moisture flux (evaporation or condensation, in units of $\mathrm{kg\,m^{-2}\,s^{-1}}$), c_p is the specific heat of air at constant pressure ($1006 \, \mathrm{J\,kg^{-1}\,K^{-1}}$) and λ is the latent heat of vaporization of water ($2.5 \times 10^6 \, \mathrm{J\,kg^{-1}}$). By taking maximum H_0 and λE_0 values ($500 \, \mathrm{W\,m^{-2}}$), we expect $w'\theta' \approx 0.4 \, \mathrm{m\,s^{-1}\,K}$ and $E_0/\rho = w'q' \approx 1.6 \times 10^{-4} \, \mathrm{m\,s^{-1}}$ (or an evaporation rate $E_0/\rho_\mathrm{w} \approx 17 \, \mathrm{mm\,day^{-1}}$, where $\rho_\mathrm{w} = 1000 \, \mathrm{kg\,m^{-3}}$ is the density of water).

(v) Turbulent scaling parameters are defined as follows.

The surface friction velocity u_{*0} is defined by

$$u_{*0}^2 = [(\overline{u'w'})_0^2 + (\overline{v'w'})_0^2]^{1/2} \tag{1.7}$$
$$= -(\overline{u'w'})_0$$

if surface-wind coordinates are used, with $\bar{v} = 0$. This follows, since we assume the surface stress is in the direction of the mean wind at $z = 0$, so that τ_{y0} (Eq. 1.3) is zero.

Temperature scales θ_{*0} and θ_{v*0} and a humidity scale q_{*0} are defined by

$$\theta_{*0} = -(\overline{w'\theta'})_0/u_{*0} \tag{1.8}$$

$$\theta_{v*0} = -(\overline{w'\theta_\mathrm{v}'})_0/u_{*0} \tag{1.9}$$

$$q_{*0} = -(\overline{w'q'})_0/u_{*0}. \tag{1.10}$$

A length scale (the Obukhov stability length) L is defined by

$$L = u_{*0}^2\overline{\theta_\mathrm{v}}/(kg\theta_{v*0}) \tag{1.11}$$

with k the von Karman constant. Taking $u_{*0} = 0.25 \, \mathrm{m\,s^{-1}}$, and H_0, λE_0 equal to $500 \, \mathrm{W\,m^{-2}}$ gives $\theta_{*0} \approx -1.5 \, \mathrm{K}$, $q_{*0} \approx -1 \, \mathrm{g\,kg^{-1}}$, and $L = -5 \, \mathrm{m}$.

Mixed-layer or free convection scales w_* for velocity and T_* for temperature are defined by

$$w_*^3 = g(\overline{w'\theta_\mathrm{v}'})_0 h/\overline{\theta_\mathrm{v}} \tag{1.12}$$

$$T_*^3 = \overline{\theta_v}(\overline{w'\theta_v'})_0^2/gh \qquad (1.13)$$

where h is the depth of the CBL. With $h = 1$ km, we find $w_* \approx 2.5 \text{ m s}^{-1}$ and $T_* \approx 0.2$ K.

1.7.2 Sign convention

A dominant characteristic of the ABL concerns the ability of turbulence to transfer mass, heat and momentum vertically, and a major task in the study of the ABL involves quantifying the associated surface fluxes and the height variation of the fluxes through the entire depth of the boundary layer. One successful approach has involved the use of similarity theories, or sophisticated dimensional analysis, in which a number of turbulent scaling parameters are defined.

The sign convention used for vertical fluxes in this book should be explained here. It is not universal but is the one found in many texts. All non-radiative fluxes (turbulent and molecular) are positive when directed away from the surface, and radiative fluxes are positive when directed towards the surface. Choice of this sign convention is also consistent with taking vertical velocities positive upwards. A simple illustration of this is shown in Fig. 1.4 in terms of a turbulent transfer concept. In each of the examples, parcels or eddies are transported from one point on the profile to another (with a positive or negative turbulent fluctuation w'). During the process, any parcel retains its original property before mixing with the local environment. In Figs. 1.4(a) and 1.4(b), where the mean gradient is positive, it readily follows that w' is anticorrelated with any other property fluctuation, so that mean covariances or fluxes are negative. For downgradient transfer they are therefore directed towards the surface. In Fig. 1.4(c), where the mean gradient is negative w' is positively correlated with other fluctuations, so the fluxes must be positive and therefore directed downgradient away from the surface. For positive diffusivities K, any flux-gradient relation involving transfer of a property s must then be defined as

$$\overline{w's'} = -K\partial\bar{s}/\partial z. \qquad (1.14)$$

1.7.3 Taylor's hypothesis

In this opening chapter, some reference to Taylor's hypothesis is necessary since the study of ABL turbulence is greatly facilitated from an experimental viewpoint by measurements made in time at a fixed point in space. Turbulence theory involves correlations and spectra of quantities formally described in terms of wavenumber or spatial variables. In practice, the measurements of spatial correlations and spectra in wavenumber space are difficult to make. Likewise, measurements of the covariances as spatial averages are equally difficult. Taylor's hypothesis, otherwise known as the frozen turbulence hypothesis, states that if the turbulent intensity is low and the turbulence is approximately stationary and homogeneous, then the turbulent field is unchanged over the ABL time scales of interest and advected with the mean wind. That is, we can use the Eulerian transformations

$$x = \bar{u}t \quad \text{and} \quad 2\pi/\kappa = \bar{u}/n \qquad (1.15)$$

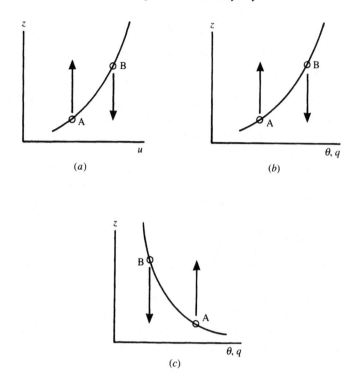

Fig. 1.4 Schematic representation of typical near-surface profiles of (*a*) wind speed, (*b*) potential temperature and specific humidity in stable conditions with surface condensation and (*c*) potential temperature and specific humidity in unstable conditions with surface evaporation. The eddies or parcels designated A (B) have positive (negative) vertical velocity fluctuation w' and carry negative (positive) property fluctuations u', θ' and q' in (*a*) and (*b*), and positive (negative) fluctuations in (*c*).

By convention, the covariances $u'w'$, $w'\theta'$ and $w'q'$ in (*a*) and (*b*) are negative (fluxes or transfer downwards), and in (*c*) are positive (fluxes upwards).

where x and t are space and time coordinates, x being in the direction of the mean wind speed $\bar{u}$, κ is wavenumber and n is natural frequency. The validity of Taylor's hypothesis allows, for example, a frequency spectrum measured at a fixed point in space to be interpreted as the wavenumber spectrum measured at a point in time, and for aircraft data to be compatible with data obtained on a tower.

Notes and bibliography

Section 1.1

Detailed texts on turbulence and the aerodynamic boundary layer from the theoretical and laboratory perspective, relevant to ABL problems, include

Hinze, J. O. (1975), *Turbulence: An Introduction to Its Mechanism and Theory*, 2nd edition. McGraw-Hill, New York, 790 pp.

Schlichting, H. (1979), *Boundary-Layer Theory* (trans. J. Kestin), 7th edition. McGraw-Hill, Hamburg, 817 pp.

Tennekes, H. and J. L. Lumley (1972), *A First Course in Turbulence*. MIT Press,

Cambridge, MA, 300 pp.

The classical specialized texts relevant to the ABL are

Sutton, O. G. (1953), *Micrometeorology*. McGraw-Hill, London, 333 pp.

Priestley, C. H. B. (1959), *Turbulent Transport in the Lower Atmosphere*. University of Chicago Press, Chicago, 130 pp.

Lumley, J. L. and H. A. Panofsky (1964), *The Structure of Atmospheric Turbulence*. Wiley Interscience, New York, 239 pp.

Haugen, D. A., ed. (1973), *Workshop on Micrometeorology*. American Meteorological Society, Boston, MA, 392 pp.

Advanced texts dealing with specialized aspects of the ABL include

Wyngaard, J. C., ed. (1980), *Workshop on the Planetary Boundary Layer*. American Meteorological Society, Boston, MA, 322 pp.

Nieuwstadt, F. T. M. and H. van Dop, eds (1982), *Atmospheric Turbulence and Air Pollution Modelling*. Reidel, Dordrecht, 358 pp.

More recent texts that are both informative and comprehensive include

Stull, R. B. (1988), *An Introduction to Boundary Layer Meteorology*. Kluwer Academic Publishers, Dordrecht, 666 pp.

Sorbjan, Z. (1989), *Structure of the Atmospheric Boundary Layer*. Prentice Hall, New Jersey, 317 pp.

Section 1.2

A major historical review of boundary-layer theory can be found in

Tani, I. (1977), History of boundary-layer theory. *Ann. Rev. Fluid Mech.* **9**, 87–111.

An interesting recent article on the history of the Reynolds number is

Rott, N. (1990), Note on the history of the Reynolds number. *Ann. Rev. Fluid Mech.* **22**, 1–11.

For a history of turbulence research in the atmosphere, the reader is directed to the Introduction in

Monin, A. S. and A. M. Yaglom, (1971), *Statistical Fluid Mechanics: Mechanics of Turbulence*, Vol. 1, ed. J. L. Lumley. MIT Press, 769 pp.

A brief summary of important field experiments may be of interest to the reader. In the early 1950s, the Scilly Isles field experiment of Sheppard *et al.* (1952) focussed on vertical momentum transfer processes over the ocean in the northern hemisphere westerlies. Major land-based ABL experiments commenced with the US Great Plains observations in 1953 (Lettau and Davidson, 1957), followed by the Australian Wangara observations in 1967 (Clarke *et al.*, 1971) and the Koorin observations in 1974 (Clarke and Brook, 1979) and by the US Minnesota experiment in 1973 (Izumi and Caughey, 1976). Interspersed with the above were the surface-layer experiments of Swinbank and co-workers in the 1960s at Hay and Kerang in southern Australia (Swinbank, 1968), and the Kansas experiment in 1968 (Izumi, 1971). Major international efforts with a strong ABL focus include BOMEX in 1969 (Kuettner and Holland, 1969), GATE in 1974 (Kuettner and Parker, 1976), AMTEX in 1974 and 1975 (Lenschow and Agee, 1976) and JASIN in 1978 (Charnock and Pollard, 1983) – all relating to work carried out over the ocean – HAPEX in 1986 (André *et al.*, 1986) and FIFE in 1987 (American Meteorological Society, 1990). In Stull's book, a list of major ABL field experiments can be found with associated references.

Section 1.3

Schlichting's book gives a very comprehensive description of the aerodynamic boundary layer based on wind-tunnel observations. In addition, a number of texts present a synthesis of atmospheric observations covering stable and convective conditions, over the sea and over the land, including the Lumley and Panofsky, Haugen, Nieuwstadt and Van Dop texts and

McBean, G. A., Bernhardt, K., Bodin, S., Litynska, Z., Van Ulden, A. P. and J. C. Wyngaard (1979), *The Planetary Boundary Layer*, WMO Tech. Note No. 165. World Meteorological Organization, Geneva, 201 pp.

A comprehensive introduction to instruments and techniques for making measurements in the ABL is

Lenschow, D. H., ed. (1986), *Probing the Atmospheric Boundary Layer*. American Meteorological Society, Boston, MA, 269 pp.

Section 1.7

For discussions on the validity of Taylor's hypothesis in the ABL, the reader should consult Lumley and Panofsky's book and

Wyngaard, J. C. and S. F. Clifford (1977), Taylor's hypothesis and high-frequency turbulence spectra, *J. Atmos, Sci.* **34**, 922–9.

The length scale L, the Obukhov length, is often called the Monin–Obukhov stability length. The original concept was first discussed in the literature by A. M. Obukhov in 1946; a translation of his paper can be found in

Obukhov, A. M.(1971), Turbulence in an atmosphere with a non-uniform temperature, *Bound. Layer Meteor.*, **2**, 7–29.

Businger, J. A. and A. M. Yaglom (1971), Introduction to Obukhov's paper on 'Turbulence in an atmosphere with a non-uniform temperature', *Bound. Layer Meteor.*, **2**, 3–6.

2

Basic equations for mean and fluctuating quantities

Because the ABL is generally turbulent, a statistical approach rather than a deterministic approach is necessary. This applies both to the measurement of boundary-layer quantities and to the equations that describe boundary-layer behaviour. However, averaging the equations leads to a situation where there are more unknowns than equations – the so-called closure problem. In this chapter, we consider the basic equations that describe the flow properties and the evolution of mean and turbulent quantities. The set of equations forms the basis of any numerical model of the atmosphere. After an introduction to the topic of turbulence, we describe the governing equations for the mean and fluctuating quantities u_i, T and q, and the simplified form of the mean equations in common use in boundary-layer work. The problem of turbulence closure leads on to consideration of the second-moment equations with particular attention to the turbulent kinetic energy (TKE) equation and the introduction of thermal stability (buoyancy) parameters.

2.1 Turbulence and flow description

Most turbulent flows of interest, including those in the ABL have a number of common characteristics:

(i) the flows are rotational and three dimensional (vorticity fluctuations are therefore important);
(ii) the flows are dissipative, so that energy must be supplied to maintain the turbulence;
(iii) the fluid motions are unpredictable in detail;
(iv) the rates of transfer and mixing are several orders of magnitude greater than the rate of molecular diffusion.

Non-turbulent flows are called laminar. In laminar flow, a perfect frictionless fluid would experience no tangential force at a boundary – the condition of *free slip*. In contrast, a real fluid experiences such tangential forces, with a condition of *no-slip* at a boundary. Such tangential forces are also referred to as *shearing stresses*, themselves related to a property of the fluid called viscosity. Viscosity is formally defined through Newton's law of friction:

$$\tau = \mu \partial u / \partial z \qquad (2.1)$$

where τ is the tangential frictional force per unit area (frictional shearing stress) in $N\,m^{-2}$; $\partial u/\partial z$ is the shear normal to the surface (s^{-1}) and μ is the dynamic viscosity ($kg\,m^{-1}\,s^{-1}$). In all fluid motions in which frictional and inertia forces interact it is important to consider the ratio of μ to the density ρ, known as the kinematic viscosity (ν, in $m^2\,s^{-1}$), where

$$\nu = \mu / \rho. \qquad (2.2)$$

The above applies to a simple one-dimensional sheared system (e.g. in parallel flow between two walls, known as Couette flow). Equation 2.1 can be generalized for three-dimensional flow to give Stokes' law of friction which, in Cartesian coordinates, is (Bachelor, 1967, Chapter 3)

$$\sigma_{ij} = - p\delta_{ij} + \mu(e_{ij} - (2/3)\partial u_k/\partial x_k) \qquad (2.3)$$

where $e_{ij} = \partial u_i/\partial x_j + \partial u_j/\partial x_i$ and p is the pressure related directly to the normal stresses in the fluid. Here σ_{ij} is a stress tensor (with six independent component stresses), and δ_{ij} is the Kronecker delta tensor ($\delta_{ij} = 0$ for $i \neq j$ and $\delta_{ij} = 1$ for $i = j$).

In the atmospheric boundary layer, and in the absence of strong thermal stratification effects, the flow is generally turbulent. Early work in the laboratory, by Reynolds (1883) and later by others (e.g. see Schlichting, 1979, Chapter 16), identified the conditions and criteria under which the transition from laminar to turbulent flow occurs. In fact, this work showed that the transition in the boundary layer at a solid body (or in a pipe or channel) depends on several factors, including the pressure distribution in the external flow, the nature of disturbances in the external flow, the nature (roughness) of the boundary or wall, and the value of the Reynolds number, *Re*. This nondimensional number is defined as the ratio of inertia to friction (or viscous) forces:

$$Re = v_c l_c / \nu \qquad (2.4)$$

where v_c is a characteristic velocity scale of the flow and l_c is a characteristic length scale of the flow. Two steady flows with the same Reynolds number are said to be in dynamic similarity, which means that streamlines around bodies of similar geometry are also geometrically similar. Typical *Re* values for flow near the surface and throughout the ABL are usually well above the critical values defining the transition to turbulence.

There are several concepts that are of value when we deal with turbulence in the ABL.

(i) Turbulence is stationary if the statistical properties are independent of time; stationarity implies statistical invariance with respect to translation of the time axis. Thus, for example, the mean wind or wind-speed variance are the same for a given averaging period irrespective of where in a long time series the averaging is made.

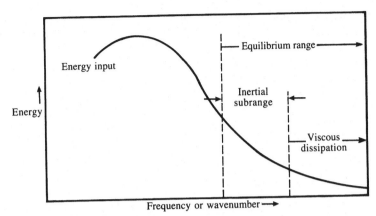

Fig. 2.1 Schematic representation of the energy spectrum of turbulence.

(ii) Turbulence is homogeneous if the field is statistically invariant to translation of the spatial axes. The ABL is far from being so in the vertical, but may be approximately homogeneous in the horizontal.

(iii) Turbulence is isotropic if the field is statistically independent of translation, rotation and reflection of the spatial axes. ABL turbulence is in fact quite anisotropic, because of the effects of the earth's surface and buoyancy, for example, but at sufficiently small scales these effects are negligible, and the flow is locally isotropic.

Within the ABL, where high Reynolds numbers are typical, the spectrum of turbulent eddies extends over a wide range of sizes. There is a strong interaction between these eddies due to the nonlinear and three-dimensional character of turbulence. Turbulent energy is gained at the expense of instabilities in a mean flow having characteristic length and velocity scales l_s and v_s respectively. The smallest scales are responsible for viscous dissipation, and have a length scale (called the Kolmogorov scale) $\eta_k = (v^3/\varepsilon)^{1/4}$, where ε is the rate of dissipation of turbulent kinetic energy by viscosity (see later). The spectrum itself is produced by a cascade process (Tennekes and Lumley, 1972, Chapter 8) in which smaller and smaller eddies result from the instability of larger ones, continuing down to molecular scales where viscous dissipation converts the kinetic energy of motion into heat.

In 1941, Kolmogorov developed a similarity theory of turbulence based on the concept of the cascade process (see Friedlander and Topper, 1961). In this, the three-dimensional energy spectrum (the distribution of kinetic energy with frequency or wavenumber) at large Reynolds number is sufficiently wide ($l_s \gg \eta_k$) that a progressive decoupling of the turbulent structure from the original anisotropic form occurs, with a consequent tendency to *local isotropy* in an *equilibrium range* of wavenumbers (Fig. 2.1). In short, the direct influence of the largest eddies is lost and the small eddies tend to have independent properties and are isotropic.

Kolmogorov's two main hypotheses (e.g. Batchelor, 1953, Chapter 6 and Tennekes and Lumley, 1972, Chapter 8) state the following.

(i) There exists an equilibrium range in which the average properties of the small-scale components of any turbulent motion at large Re are determined uniquely by v and ε.

Appropriate reference length and velocity scales are $\eta_k = (v^3/\varepsilon)^{1/4}$ and $v_\eta = (v\varepsilon)^{1/4}$ (noting that the Reynolds number with reference to this length and velocity is $\eta_k v_\eta / v = 1$). With $v = 1.5 \times 10^{-5}\,\mathrm{m^2\,s^{-1}}$ and taking $\varepsilon \approx 0.01\,\mathrm{m^2\,s^{-3}}$ near the surface, $\eta_k \approx 1\,\mathrm{mm}$ and $v_\eta \approx 0.02\,\mathrm{m\,s^{-1}}$.

(ii) At large enough Reynolds number, there exists an *inertial subrange*, within the equilibrium range but removed from the viscous region, in which eddy structure is independent of energy input or viscous dissipation and where only the inertial transfer of energy is important. In this subrange of wavenumbers average properties are independent of v and determined solely by ε. In this inertial subrange the wavenumber κ satisfies $1/l_s \ll \kappa \ll 1/\eta_k$.

The consequences of the second hypothesis on the form of the spectrum will be described in Chapter 3 when we are dealing with observed spectra in the ABL.

The complexity of a turbulent flow is so formidable that a description of the flow at all points in space and time is not feasible. Consequently, any study of turbulent flows (either in the form of observations or solution of the conservation equations) is directed towards describing their statistical characteristics, usually in terms of *moments* and *spectra*. Following Reynolds (1895), we suppose that any process $s(x, t)$ can be decomposed into a mean flow component, or average, $\langle s \rangle$ and a "rapidly varying" turbulent component, or fluctuation, s'. This Reynolds convention is written as

$$s = \langle s \rangle + s' \tag{2.5}$$

with the following properties,

$$\langle s' \rangle = 0 \tag{2.6a}$$

$$\langle ws \rangle = \langle w \rangle \langle s \rangle + \langle w's' \rangle \tag{2.6b}$$

$$\langle u + v \rangle = \langle u \rangle + \langle v \rangle \tag{2.6c}$$

$$\langle as \rangle = a\langle s \rangle \tag{2.6d}$$

where a is a constant. In the above example, we have introduced the vertical velocity w because the covariance $\langle w's' \rangle$ is an important quantity appearing in the averaged equations. The mean value $\langle s \rangle$ is called the first-order moment of s, the variance $\langle s'^2 \rangle$ is called the second-order moment and the covariance $\langle w's' \rangle$ is a second-order moment of the joint process w and s. Equations of the first- and second-order moments of velocity, temperature and humidity are of crucial importance to many aspects of the ABL, and are referred to extensively throughout the book.

There are several types of average that may be relevant to both laboratory and atmospheric application.

(i) The *ensemble average*: this displays all the properties defined by Eq. 2.6. Let β_e be the ensemble parameter, or experiment index, and consider a random function in space and time $s(\mathbf{r}, t; \beta_e)$, which might be the velocity vector, for example. The ensemble average is then defined by

$$\langle s \rangle = \langle s(\mathbf{r}, t; \beta_e) \rangle = \langle s(\mathbf{r}, t) \rangle = \lim_{N \to \infty} N^{-1} \sum s(\mathbf{r}, t; \beta_e) \quad (2.7)$$

where N is the number of realizations of the random function with $\beta_e = 1$, 2, 3, ..., N. In principle this sort of averaging could be done in laboratory flows by repeating an experiment N times or by generating synthetic data in a large-eddy simulation model, for example. In practice, ensemble averages are not easily made, and averages over space or time are made. Note that whereas s is random the ensemble average is completely non-random, and all the turbulence is contained in the fluctuation field.

(ii) The *time average*: defining T as an averaging period, the time average can be written

$$\bar{s}^t = T^{-1} \int_0^T s(\mathbf{r}, t) \, dt. \quad (2.8)$$

The time average is often applied to fixed-point measurements of s on a tower. It only satisfies Eq. 2.6 if we let $T \to \infty$, but if $T \gg \tau^i$, where τ^i is the correlation or integral time scale of the turbulence, then the time average can closely approximate Eq. 2.6. This is the ergodic hypothesis.

(iii) The *one-dimensional space (line) average*: defining L as a suitable averaging length, the line average can be written

$$\bar{s}^x = L^{-1} \int_0^L s(\mathbf{r}, t) \, dx. \quad (2.9)$$

The line average is often associated with aircraft measurements along a horizontal flight path. The same restrictions apply to this as to the time average.

(iv) The *volume average*: defining suitable grid lengths Δx, Δy and Δz, the volume average can be written

$$\tilde{s} = (\Delta x \Delta y \Delta z)^{-1} \int \int \int s(\mathbf{r}, t) \, d\mathbf{r} \quad (2.10)$$

which does not strictly obey Eq. 2.6. Volume averages can be obtained with ground-based remote sensing systems such as acoustic or Doppler radars.

Throughout the remainder of the book, *averages will be denoted by a simple overbar* with the implication that these may represent time or other averages for observations or ensemble averages for equations. The averaging process in a turbulent flow carries with it the notion of the scale of turbulence. One such scale – the integral scale – tells us something about the length and time scale of the energy-containing eddies, and is defined by

$$\tau^i = \int_0^\infty \rho_s(\tau) \, d\tau \quad (2.11)$$

$$l^i = \int_0^\infty \rho_s(\xi) \, d\xi. \quad (2.12)$$

Here, τ is a time lag and ξ is the streamwise separation, with the autocorrelation function ρ_s defined as

$$\rho_s(\tau) = \overline{s'(t)s'(t + \tau)}/\overline{s'^2} \tag{2.13}$$

$$\rho_s(\xi) = \overline{s'(x)s'(x + \xi)}/\overline{s'^2}. \tag{2.14}$$

The variance $\overline{s'^2}$ is shorthand for $\overline{s'(t)s'(t)}$ or $\overline{s'(x)s'(x)}$ and is independent of either t or x only for stationary and homogeneous turbulence. The spectrum of s is then written as

$$\overline{s'^2} = \int_0^\infty \phi_{ss}(n)\,dn \tag{2.15}$$

so that $\phi_{ss}(n)\,dn$ gives the energy contribution to the variance in the frequency range n to $n + dn$. In analogy to the above, the covariance between w and s, $\overline{w's'}$, is the same as $\overline{w'(t)s'(t)}$, with the cospectrum written as

$$\overline{w's'} = \int_0^\infty \phi_{ws}(n)\,dn. \tag{2.16}$$

Much more detailed considerations can be found in Lumley and Panofsky (1964, Chapter 1) and Tennekes and Lumley (1972, Chapters 6 and 8).

2.2 Governing equations for mean and fluctuating quantities

2.2.1 Approximations

The set of equations governing the flow consists of three equations for the *conservation of momentum* (i.e. the Navier-Stokes equations), an equation for the *conservation of mass* (the continuity equation), an equation for the *conservation of thermal energy* (the thermodynamic or enthalpy equation), an equation for the *conservation of water vapour* (the humidity equation) and the *equation of state* (the gas law). Consideration of these equations will be crucial to many aspects of ABL description considered throughout the book. This system of seven basic equations (an equation for liquid water is omitted at this stage) describes the x, y, z, t dependence of the ABL variables u (longitudinal velocity component), v (transverse velocity component), w (vertical velocity component), ρ (air density), T (absolute temperature), q (specific humidity) and p (pressure).

For most purposes in ABL flow, the full set of equations for the mean and fluctuating quantities applied to a rotating Earth can be simplified to the so-called Boussinesq set by a series of approximations (Businger, 1982). In summary, these are as follows.

(i) The dynamic viscosity ($\mu = \rho\nu$) and molecular thermal conductivity (k_T) are constant throughout the fluid. That is, the small dependence on temperature and pressure (see Appendix 2) that these molecular properties do have can be neglected. This is an excellent assumption for the ABL.

(ii) The heat generated by viscous stresses is neglected in the thermodynamic equation; in the ABL this source of heat is negligible.

(iii) The flow is treated as incompressible; in the ABL this is an excellent assumption.

(iv) Fluctuations in fluid properties are much less than the reference or mean quantities, i.e. p'/p_0, T'/T_0, ρ'/ρ_0 and θ'/θ_0 are all $\ll 1$. In the ABL this is an excellent assumption.

(v) p'/p_0 can be neglected in relation to T'/T_0 and ρ'/ρ_0. The neglect of pressure fluctuations is normally valid except in extremely strong winds and represents the first part of the Boussinesq approximation (*density changes resulting from pressure changes are negligible*).

(vi) Fluctuations in density become significant only when occurring in combination (as a product) with the acceleration due to gravity g. This represents the second part of the Boussinesq approximation (*density changes resulting from temperature changes are important only as they directly affect buoyancy*).

2.2.2 The continuity equation: conservation of mass

The full compressible form of the equation for instantaneous density is written

$$\partial\rho/\partial t + \partial/\partial x_j(\rho u_j) = 0 \tag{2.17a}$$

or in the alternative form, introducing the substantive derivative for the instantaneous advective velocity u_j, $\mathrm{d}/\mathrm{d}t = \partial/\partial t + u_j\partial/\partial x_j$,

$$\partial u_j/\partial x_j = -\rho^{-1}\mathrm{d}\rho/\mathrm{d}t. \tag{2.17b}$$

Compressibility effects embodied in $\mathrm{d}\rho/\mathrm{d}t$ can be neglected in the ABL, so that the "shallow convection", incompressible form of the continuity equation is obtained:

$$\partial u_j/\partial x_j = 0 \tag{2.18}$$

i.e. the instantaneous velocity divergence can be taken as zero for ABL flow.

We now introduce the Reynolds convention for decomposition of the velocity (u_j) into a mean component $\overline{u_j}$ and a fluctuating component u_j'. For this and all future applications, we would for preference choose the ensemble average but in practice may well have to settle for a time or space average. The instantaneous velocity component can be written

$$u_j = \overline{u_j} + u_j' \tag{2.19}$$

and substitution into Eq. 2.18 gives

$$\partial\overline{u_j}/\partial x_j + \partial u_j'/\partial x_j = 0. \tag{2.20}$$

By averaging this equation, and noting that since $\overline{u_j'} = 0$, then $\overline{\partial u_j'/\partial x_j} = 0$, we have

$$\partial\overline{u_j}/\partial x_j = 0 \tag{2.21}$$

i.e. the mean velocity divergence is zero for ABL flow. From combining Eqs. 2.20 and 2.21 it follows that

$$\partial u_j'/\partial x_j = 0 \tag{2.22}$$

i.e. the fluctuating velocity divergence is zero for ABL flow.

2.2.3 The equation of state: the gas law

The ideal gas law for moist air of instantaneous density ρ, can be written

$$p = \rho R_w T \tag{2.23}$$

where T is the absolute temperature of the moist air sample. Here, R_w is the gas constant for moist air, very closely approximated by $R_d(1 + 0.61q)$ where R_d is the gas constant for dry air, of value 287 J kg^{-1} K^{-1} and is related to the universal gas constant R by $R_d = R/M_d$, where M_d is the mean molecular weight of dry air. It thus follows that, by defining a *virtual temperature* (Iribarne and Godson, 1981)

$$T_v = T(1 + 0.61q) \tag{2.24}$$

with q the specific humidity (the mass of water vapour per unit mass of moist air), an alternative form of Eq. 2.23 for moist air is

$$p = \rho R_d T_v. \tag{2.25}$$

In addition to the above relations, it is convenient at this point to introduce the *potential* and *virtual potential temperatures*, which can be derived from the gas law and first law of thermodynamics. The potential temperature θ is the temperature that would result if a parcel of air were brought adiabatically to a standard or reference pressure, taken as $p_R = 1000$ hPa:

$$\theta = T(p/p_R)^{-R_d/c_p} \tag{2.26}$$

and the virtual potential temperature θ_v, is defined in a similar way:

$$\theta_v = T_v(p/p_R)^{-R_d/c_p}. \tag{2.27}$$

Combining Eqs. 2.24, 2.26 and 2.27 gives

$$\theta_v \approx \theta(1 + 0.61q) \approx \theta + 0.61\bar{\theta}q, \tag{2.28}$$

where the mean temperature $\bar{\theta}$ serves as a reference temperature in this linearized form of θ_v. By applying the Reynolds decomposition to each of the state variables in Eq. 2.25, the gas law for the mean variables can be written

$$\bar{p} = \bar{\rho} R_d \overline{T_v} \tag{2.29}$$

where a term in $\overline{\rho' T_v'}$ has been neglected. Note that Eqs. 2.24, 2.26–2.28 can be written in identical form for the mean temperatures, with instantaneous quantities replaced by their mean values.

For the fluctuations, with $p'/\bar{p} \approx 0$,

$$T_v'/\overline{T_v} \approx \theta_v'/\overline{\theta_v} \approx - \rho'/\bar{\rho} \tag{2.30a}$$

$$T_v' \approx T' + 0.61\bar{T}q' \tag{2.30b}$$

$$\theta_v' \approx \theta' + 0.61\bar{\theta}q'. \tag{2.30c}$$

In the ABL, both static and dynamic pressure fluctuations are generally no more than 0.1 hPa, so that $p'/\bar{p} < 10^{-4}$; in comparison, $T'/\bar{T} \approx 3 \times 10^{-3}$. Equation 2.30a is a most important approximation so far as the effects of buoyancy on the dynamics of the ABL are concerned; density fluctuations (difficult to measure) can be replaced by temperature fluctuations (easier to measure). Equations 2.30b and 2.30c will be found to be useful approximations for later purposes (e.g. a conservation equation for θ_v can be readily deduced from separate equations for θ and q). Note also that we expect that $T' \approx \theta'$ and $T_v' \approx \theta_v'$.

By differentiating Eq. 2.26, and using the hydrostatic equation (see later this chapter), the vertical gradients can be related; thus for $\bar{\theta}$ we obtain

$$(T/\theta)\partial\bar{\theta}/\partial z = \partial\bar{T}/\partial z + g/c_p \qquad (2.31)$$

and for $\overline{\theta_v}$ we obtain

$$(T_v/\theta_v)\partial\overline{\theta_v}/\partial z = \partial\overline{T_v}/\partial z + g/c_p. \qquad (2.32)$$

For $q = 0$, $\theta_v = \theta$ and $T_v = T$, whilst for $\partial q/\partial z = 0$, $\partial T_v/\partial z = \partial T/\partial z$. The importance of the vertical gradients of θ and θ_v in the ABL lies in the relation with thermal stability criteria. For an ABL where $\bar{q}$ is zero or constant with height, we have

(i) a neutral ABL if $\partial\bar{\theta}/\partial z = 0$ ($\bar{\theta}$ is constant with height from the surface to the top of the ABL);
(ii) an unstable, or convective, ABL if $\partial\bar{\theta}/\partial z < 0$;
(iii) a stable ABL if $\partial\bar{\theta}/\partial z > 0$.

For a moist ABL in which $\partial\bar{q}/\partial z$ is not zero, the same applies, with $\partial\overline{\theta_v}/\partial z$ replacing $\partial\theta/\partial z$.

2.2.4 The thermodynamic equation: conservation of enthalpy

The conservation of enthalpy (or sensible heat) per unit mass, $c_p\theta$, can be deduced from the first law of thermodynamics. The resulting equation for θ can be written (Businger, 1982)

$$d(c_p\theta)/dt = \kappa_T\partial^2(c_p\theta)/\partial x_j^2 + \rho^{-1}\partial R_j/\partial x_j. \qquad (2.33)$$

Here R_j is the radiative heat flux, and the molecular term contains the molecular thermal diffusivity κ_T ($\kappa_T = k_T/\rho c_p$). In Eq. 2.33, we have omitted the effects of phase changes (but see Chapter 7 on ABL clouds). This equation suggests that in the absence of radiative transfer and of any phase change $c_p\theta$, or simply θ since c_p can be taken as constant, can be considered a conservative property of the air.

A Reynolds decomposition is now applied, with $\theta = \bar{\theta} + \theta'$, $u_j = \overline{u_j} + u_j'$ and $R_j = \overline{R_j} + R_j'$. In the absence of any phase change, the equation for mean potential temperature may be written as

$$\partial\bar{\theta}/\partial t + \overline{u_j}\partial\bar{\theta}/\partial x_j = -\partial(\overline{u_j'\theta'})/\partial x_j + \kappa_T\partial^2\bar{\theta}/\partial x_j^2 + (\rho c_p)^{-1}\partial\overline{R_j}/\partial x_j \quad (2.34)$$

with terms like $\overline{u_j}\ \overline{\partial\theta'/\partial x_j}$ and $\overline{\theta'\partial u_j'/\partial x_j}$ vanishing. The first term on the right-hand side is a new term representing heat transport by the turbulence. For the fluctuating temperature, we have

$$\partial\theta'/\partial t + \partial(u_j'\theta' - \overline{u_j'\theta'} + \overline{u_j}\theta' + u_j'\bar{\theta})/\partial x_j = \kappa_T\partial^2\theta'/\partial x_j^2 + (\rho c_p)^{-1}\partial R_j'/\partial x_j.$$

$$(2.35)$$

2.2.5 The humidity equation: conservation of water vapour

The equation representing conservation of water vapour can be written in terms of the specific humidity q and, if phase changes are omitted, reads

$$dq/dt = \kappa_V \partial^2 q/\partial x_j^2. \tag{2.36}$$

In the absence of phase transitions in the ABL, water vapour is a conservative scalar quantity. For this case, and in analogy with the θ equation, we set $q = \bar{q} + q'$ and have, for the conservation of mean q,

$$\partial \bar{q}/\partial t + \overline{u_j}\partial \bar{q}/\partial x_j = -\partial(\overline{u_j'q'})/\partial x_j + \kappa_V \partial^2 \bar{q}/\partial x_j^2, \tag{2.37}$$

and for the conservation of q'

$$\partial q'/\partial t + \partial(u_j'q' - \overline{u_j'q'} + \overline{u_j}q' + u_j'\bar{q})/\partial x_j = \kappa_V \partial^2 q'/\partial x_j^2. \tag{2.38}$$

The reader should note that separate rate equations for θ_v, $\overline{\theta_v}$ and θ_v' are not necessary. The linearized forms of Eqs. 2.28 and 2.30c allow these quantities to be evaluated from the solutions to Eqs. 2.33 and 2.36 (θ and q respectively), Eqs. 2.34 and 2.37 ($\bar{\theta}$ and $\bar{q}$) and Eqs. 2.35 and 2.38 (θ' and q'). Nevertheless, since the molecular terms in Eqs. 2.34 and 2.37 are negligible in comparison with the turbulence terms, the equation for $\overline{\theta_v}$ is readily deduced by replacing $\bar{\theta}$ and θ' in Eq. 2.34 by $\overline{\theta_v}$ and θ_v' respectively.

2.2.6 The Navier–Stokes equations: conservation of momentum

For a non-rotating fluid system Newton's second law of motion may be written (e.g. Batchelor, 1967, Chapter 3):

$$du_i/dt = F_i + \rho^{-1}\partial\sigma_{ij}/\partial x_j \tag{2.39}$$

where F_i represents body forces (e.g. gravity) and σ_{ij} is the stress tensor given by Eq. 2.3. For the ABL, Eq. 2.39 can be written in incompressible form with the viscosity constant. With the effects of the Earth's rotation included, the Navier–Stokes equation for the instantaneous velocity component can then be written:

$$du_i/dt = -\rho^{-1}\partial p/\partial x_i - g\delta_{i3} - 2\Omega\varepsilon_{ijk}\eta_j u_k + \nu\partial^2 u_i/\partial x_j^2. \tag{2.40}$$

This equation of motion gives the acceleration of the air in terms of the sum of several forces. The first term on the right-hand side represents the pressure gradient force; the second term represents the effect of gravity; the third term describes the effect of rotation in the form of Coriolis forces. Here, Ω is the angular velocity of the Earth's rotation (having a value of $7.29 \times 10^{-5}\,\text{rad s}^{-1}$), and η_j is the jth component of a unit vector parallel to the axis of rotation, i.e. $\boldsymbol{\eta} = (0, \cos\phi, \sin\phi)$ where ϕ is the latitude. It can readily be shown that only terms involving η_3 are significant, with the quantity $2\Omega\eta_3 = 2\Omega\sin\phi$ being defined as f, the Coriolis parameter (positive in the northern hemisphere and negative in the southern). The fourth term describes the influence of viscous stresses.

We now set $u_i = \overline{u_i} + u_i'$, and write Eq. 2.40 for the mean velocity component (representing the conservation of mean momentum) as

$$\partial \overline{u_i}/\partial t + \overline{u_j}\partial \overline{u_i}/\partial x_j$$
$$= -\partial(\overline{u_i'u_j'})/\partial x_j - \rho^{-1}\partial \overline{p}/\partial x_i - g\delta_{i3} - 2\Omega\varepsilon_{ijk}\eta_j\overline{u_k} + \nu\partial^2 \overline{u_i}/\partial x_j^2. \quad (2.41)$$

Because of the Boussinesq approximation used here, we can make ρ and $\overline{\rho}$ interchangeable so that, *henceforth, ρ will designate the mean density*. If Eq. 2.41 is written for the vertical component w of the velocity the dominant terms, in the absence of strong vertical accelerations, will be ρg and $\partial \overline{p}/\partial z$ (Pielke, 1984, Chapter 3). If we assume that the mean state is hydrostatic equilibrium then

$$\partial \overline{p}/\partial z = -\rho g. \quad (2.42)$$

Equation (2.42) is also referred to as the *hydrostatic equation*. It gives a very good approximation to the vertical distribution of mean pressure, even under the strongly turbulent conditions likely to be met in convection, for example.

The equation for the fluctuating velocity u_i' can be written

$$\partial u_i'/\partial t + \partial(\overline{u_j}u_i' + \overline{u_i}u_j' + u_i'u_j' - \overline{u_i'u_j'})/\partial x_j$$
$$= -\rho^{-1}\partial p'/\partial x_i + (\rho'/\rho^2)\partial \overline{p}/\partial x_i - 2\Omega\varepsilon_{ijk}\eta_j u_k' + \nu\partial^2 u_i'/\partial x_j^2 \quad (2.43)$$

where the term $(\rho'/\rho^2)\partial \overline{p}/\partial x_i$ can also be written as $(T_v'/\overline{T_v})g\delta_{i3}$. We will return to Eq. 2.43 shortly.

2.3 The simplified mean equations

Equations 2.34 for the mean temperature $\overline{\theta}$, Eq. 2.37 for the mean humidity $\overline{q}$ and Eq. 2.41 for the mean velocity component $\overline{u_i}$ contain new terms, $\partial(\overline{u_j'\theta'})/\partial x_j$, $\partial(\overline{u_j'q'})/\partial x_j$ and $\partial(\overline{u_i'u_j'})/\partial x_j$. These flux divergences act as source terms to change the mean concentration. They are a direct result of the nonlinearity in the terms $u_j\partial u_i/\partial x_j$, $u_j\partial\theta/\partial x_j$ and $u_j\partial q/\partial x_j$ appearing in the equations for the instantaneous quantities. The covariances are called *fluxes* by analogy with molecular transport, so that $\overline{u_j'\theta'}$ represents the turbulent heat flux, $\overline{u_j'q'}$ represents the turbulent moisture flux and $\overline{u_i'u_j'}$ represents the turbulent momentum flux or *Reynolds stress*. They indicate that the velocity, temperature and humidity fluctuations cause transport of momentum, heat and water vapour across a surface in a fluid. The velocity covariances represent tangential stresses $\overline{\rho u'w'}$, $\overline{\rho v'w'}$, $\overline{\rho u'v'}$ and are of fundamental importance in problems of air flow within the ABL.

In the ABL where the Reynolds number is very large (typically $Re \sim 10^7$) we find that the turbulent terms in the conservation equations for mean quantities are orders of magnitude greater than the molecular terms. For example, in Eq. 2.41 for mean momentum, the viscous term can be written in scaled form as $Re^{-1}(v_s l_s\partial^2 \overline{u_i}/\partial x_j^2)$, where $Re = v_s l_s/\nu$. The term in brackets is the same order of magnitude as all the other terms in Eq. 2.41 so that the viscous term itself (with the factor Re^{-1}) is negligible.

In component form, the mean equations for the horizontally homogeneous ABL can be simplified to

$$\partial\bar{u}/\partial t = -\rho^{-1}\partial\bar{p}/\partial x + f\bar{v} - \partial(\overline{u'w'})/\partial z \qquad (2.44)$$

$$\partial\bar{v}/\partial t = -\rho^{-1}\partial\bar{p}/\partial y - f\bar{u} - \partial(\overline{v'w'})/\partial z \qquad (2.45)$$

$$\partial\bar{\theta}_{\mathrm{v}}/\partial t = (\rho c_{\mathrm{p}})^{-1}\partial\bar{R}_{\mathrm{N}}/\partial z - \partial(\overline{w'\theta_{\mathrm{v}}'})/\partial z \qquad (2.46)$$

$$\partial\bar{q}/\partial t = -\partial(\overline{w'q'})/\partial z. \qquad (2.47)$$

Equations 2.44–2.47 will serve as a focus for much of our later discussion on the mean properties of the one-dimensional ABL. For steady-state conditions ($\partial\bar{u}/\partial t = \partial\bar{v}/\partial t = 0$) the momentum equation can be written as

$$0 = -\rho^{-1}\partial\bar{p}/\partial x + f\bar{v} - \partial(\overline{u'w'})/\partial z \qquad (2.48)$$

$$0 = -\rho^{-1}\partial\bar{p}/\partial y - f\bar{u} - \partial(\overline{v'w'})/\partial z. \qquad (2.49)$$

Above the ABL, the turbulence terms are zero so that Eqs. 2.48 and 2.49 reduce to a simple two-force balance between the Coriolis and pressure-gradient terms. This condition serves as a definition of the *geostrophic wind*, whose components are

$$fv_{\mathrm{g}} = \rho^{-1}\partial\bar{p}/\partial x \qquad (2.50)$$

$$fu_{\mathrm{g}} = -\rho^{-1}\partial\bar{p}/\partial y. \qquad (2.51)$$

Consequently, Eqs. 2.48 and 2.49 may be written as

$$0 = f(\bar{v} - v_{\mathrm{g}}) - \partial(\overline{u'w'})/\partial z \qquad (2.52)$$

$$0 = -f(\bar{u} - u_{\mathrm{g}}) - \partial(\overline{v'w'})/\partial z \qquad (2.53)$$

with the mean *ageostrophic wind* having components $\bar{u} - u_{\mathrm{g}}$ and $\bar{v} - v_{\mathrm{g}}$. Thus, Eqs. 2.52 and 2.53 show that mean ageostrophic winds in the steady-state, one-dimensional ABL are the result of the vertical gradients of eddy covariances or stresses. Note that for this special case $\bar{w} = 0$. The balance of forces represented by Eqs. 2.50–2.53 is shown in Fig. 2.2, where $\mathbf{F}_1$, $\mathbf{F}_2$ and $\mathbf{F}_3$ are the pressure gradient, Coriolis and stress divergence terms respectively. The cross-isobar flow angle, α, is also indicated and the ABL flow is shown for both hemispheres (remember that $f < 0$ in the southern hemisphere and $f > 0$ in the northern hemisphere).

2.4 The turbulence closure problem

Analytical solutions of Eqs. 2.33, 2.36 and 2.40 are generally not possible, and numerical solutions for high Reynolds number flow are impracticable. To see this, consider that a given turbulent flow has a range of eddy sizes or scales of motion all coupled through Eq. 2.40. The smallest scales are responsible for viscous dissipation and are dynamically significant so far as the turbulent kinetic energy budget is concerned and so must be resolved by the numerical scheme. Their length scale is η_{k}. Measurements show that $\varepsilon \sim v_{\mathrm{s}}^3/l_{\mathrm{s}}$, so that the ratio of the length scales of the largest and smallest eddies to be resolved is given by

$$l_{\mathrm{s}}/\eta_{\mathrm{k}} \sim (v_{\mathrm{s}}l_{\mathrm{s}}/v)^{3/4} = Re^{3/4} \qquad (2.54)$$

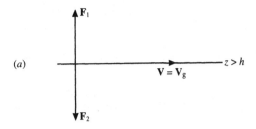

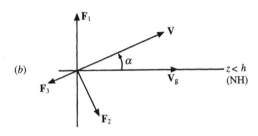

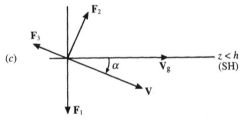

Fig. 2.2 Representation of the simplified steady-state balance of forces (a) above and (b), (c) within the ABL; the x-axis is chosen along the geostrophic wind direction, with flow in both the northern (b) and southern (c) hemisphere boundary layers illustrated. Refer to Eqs. 2.48 and 2.49, where F_1 is the pressure-gradient force, F_2 is the Coriolis force and F_3 is the turbulence term (zero above the ABL). The wind V and the cross-isobar flow angle are also indicated.

Here, Re represents a turbulent Reynolds number and, in the ABL with $v_s \sim 1\,\mathrm{m\,s}^{-1}$ and $l_s \sim 300\,\mathrm{m}$, $Re \approx 10^7$ and $l_s/\eta_k \approx 10^5$. For a three-dimensional numerical model designed to resolve explicitly eddies of scale from 1 mm to 300 m, 10^{15} grid points or more would be required!

Therefore we turn to statistical considerations, and to averaged flow fields represented by Eqs. 2.34, 2.37 and 2.41, and the simplified ABL forms represented by Eqs. 2.44–2.47. However, the very application of the averaging operator has resulted in the appearance of new (unknown) terms in $\overline{u_j'\theta'}$, $\overline{u_j'q'}$ and $\overline{u_i'u_j'}$. We could construct equations for these second-order moments (see later) only to find that further unknown terms in the form of third-order moments would arise. In general, the equation for the *n*th order moment contains terms of the $(n + 1)$th order; thus, it is fundamentally impossible to close the set and there will always be more unknowns than equations. This is the *closure problem* – the higher-order moment terms must be *parameterized* in terms of known quantities. The closure assumption or approximation is named after the highest-order moments retained in the problem. For example, if only mean (first-order) equations are to be solved, and the covariance (second-order) terms parameterized, the closure is termed first order.

In ABL problems, two approaches have been followed, in the form of local and non-local closure schemes. Local closure involves relating the unknown turbulence quantities at a given point in space (e.g. $\overline{u_i'u_j'}$) to known quantities of the flow at the same point. This usually involves some kind of "flux-gradient" relation and so assumes an analogy between turbulent transfer and molecular diffusion. In practice, local closure schemes have been extended to third order, but for most ABL applications first- and second-order schemes have been found to be sufficient. In contrast, non-local closure relates unknown quantities over a region of space. In practice, non-local schemes are not widely used, except in vertically integrated form, and have been limited to first order and no higher. At this early stage in the book, a brief discussion on local closure schemes would seem appropriate.

A first-order scheme uses eddy transfer coefficients or diffusivities K (following Boussinesq), in analogy with Eqs. 2.1 and 2.3 for laminar flow, to relate the mean turbulent stress $\overline{u_i'u_j'}$ to the mean strain, and in general to relate any turbulent flux to the local mean gradient of the quantity being transferred. Scalar representation of the eddy diffusivities is quite suitable for many ABL applications. Thus for any quantity s, the turbulent fluxes or covariances can be written in terms of the following flux-gradient relation:

$$\overline{u_j's'} = - K_s \partial\bar{s}/\partial x_j \tag{2.55}$$

which, for K_s positive, implies downgradient flow. In the main we are concerned with vertical transfer in the ABL, as represented by the turbulent terms in Eqs. 2.44–2.47, so that for the quantities $\bar{u}$, $\bar{v}$, $\overline{\theta_v}$ and $\bar{q}$ the vertical fluxes are written as

$$\tau_x = - \rho\overline{u'w'} = \rho K_M \partial\bar{u}/\partial z \tag{2.56}$$

$$\tau_y = - \rho\overline{v'w'} = \rho K_M \partial\bar{v}/\partial z \tag{2.57}$$

$$H_v = \rho c_p \overline{w'\theta_v'} = - \rho c_p K_H \partial\overline{\theta_v}/\partial z \tag{2.58}$$

$$E = \rho\overline{w'q'} = - \rho K_W \partial\bar{q}/\partial z \tag{2.59}$$

where K_M is the eddy viscosity, K_H is the eddy thermal diffusivity and K_W is the eddy diffusivity for water vapour. In contrast with the molecular case, the eddy diffusivity is not a property of the fluid but may be a function of many quantities, including position and flow velocity. Since K has dimensions of $m^2 s^{-1}$ we expect it to scale with the product of a turbulent velocity scale and a suitable length scale characterizing the dominant eddy size. Typically, with $u_* \approx 0.3 \, m\,s^{-1}$ and $\partial\bar{u}/\partial z \approx 1 \, m\,s^{-1} \, (10 \, m)^{-1}$ near the surface, $K_M \approx 1 \, m^2\,s^{-1}$ and is therefore much greater than v (note that K/v is comparable to a

Reynolds number). The dependence of K on the flow structure is a major disadvantage and, as we will see later, the K-closure (first-order closure) assumption is not always physically acceptable. First-order closure works quite well when the transfer is dominated by "small eddies" (e.g. in the neutral and stably stratified ABL) of a length scale smaller than that of the mean gradient, but fails in turbulent flows dominated by "large eddies" (e.g. in the highly convective ABL). An improvement on the first-order scheme, whilst not requiring the full set of second-order equations, is referred to as one-and-a-half-order closure, in which equations for variances, usually in the form of the turbulent kinetic energy, are carried in addition to the equations for mean quantities. This allows the eddy diffusivities to be described in terms of the TKE (and other parameters), supposedly in a more realistic way than in the first-order scheme. Formulations for K will be discussed further in Chapter 8.

Higher(nth)-order closure schemes include the full set of equations up to nth order, with unknown terms involving pressure–velocity, pressure–temperature and pressure–humidity correlations, molecular dissipation and $(n + 1)$th-order quantities needing to be parameterized. For second-order closure, the second-order prognostic equations will be developed in the next section, and aspects of these relevant to ABL modelling will be discussed further in Chapter 8. Overall, it is generally expected that numerical solutions for lower-order quantities become more accurate and physically realistic as higher-order closure schemes are introduced. However, any parameterization scheme relies on experimental data for validation and such data for higher-order statistics are very difficult to obtain, particularly in the ABL. In part, the problem is one of achieving averaging times (or lengths) that are large enough to achieve a given accuracy in the higher-order statistics. The higher the order, the larger the required averaging time or length (Lumley and Panofsky, 1964, Chapter 1); this in turn may be severely limited by experimental and other constraints (aircraft time, day length, finite data acquisition capability, stable instrument calibration) under real atmospheric conditions.

2.5 The second-moment equations

Equations for the second-order moments $\overline{u_i'u_j'}$, $\overline{u_i'\theta'}$ and $\overline{u_i'q'}$, with suitable treatment (parameterization) of the resulting third-order terms, allow closure of the mean equations discussed previously. In addition, they provide a means of studying the processes affecting the growth, maintenance and decay of turbulence in the ABL.

2.5.1 The $\overline{u_i'u_k'}$ equation

We derived equations for u_i' and θ' earlier in the Chapter (refer to Eqs. 2.43 and 2.35). These are now utilized to construct equations for variances (such as the kinetic energy) and covariances (such as the stress or momentum flux, and heat flux). The procedure for the velocity covariance is as follows. We take the u_i' equation, multiply by u_k' throughout and add this to the corresponding equation for u_k' multiplied by u_i'. It should be noted that, by continuity, terms

$u_i'u_k'\partial\overline{u_j}/\partial x_j = u_i'u_k'\partial u_j'/\partial x_j = \overline{u_i}u_k'\partial u_j'/\partial x_j = \overline{u_k}u_i'\partial u_j'/\partial x_j = 0$.   After   Reynolds averaging and considerable algebraic manipulation we obtain (e.g. Businger, 1982)

$$\partial\overline{u_i'u_k'}/\partial t + \overline{u_j}\,\partial\overline{u_i'u_k'}/\partial x_j = -\overline{u_k'u_j'}\partial\overline{u_i}/\partial x_j - \overline{u_i'u_j'}\partial\overline{u_k}/\partial x_j - \partial\overline{u_i'u_j'u_k'}/\partial x_j$$
$$+ (g/\overline{\theta_v})(\overline{u_i'\theta_v'}\delta_{3k} + \overline{u_k'\theta_v'}\delta_{3i}) - 2\Omega\eta_j(\varepsilon_{ijl}\overline{u_k'u_l'} + \varepsilon_{kjl}\overline{u_l'u_i'})$$
$$- \rho^{-1}(\overline{u_k'\partial p'/\partial x_i} + \overline{u_i'\partial p'/\partial x_k}) + \nu(\overline{u_k'\partial^2 u_i'/\partial x_j^2} + \overline{u_i'\partial^2 u_k'/\partial x_j^2}).$$

$$(2.60)$$

The various terms can be interpreted as follows: terms on the left-hand side represent the local time rate of change and advection of $\overline{u_i'u_k'}$, and together equal the substantive derivative $D(\overline{u_i'u_k'})/Dt$ following the mean motion $(D/Dt = \partial/\partial t + \overline{u_j}\partial/\partial x_j)$. On the right-hand side terms 1 and 2 are *production* terms resulting from the interaction of the turbulence and the mean flow, whilst term 3, a third-moment (triple correlation) term, can be interpreted as a transport of turbulent second moments by turbulent fluctuations, with a local loss or gain due to the divergence of the turbulent flux. Term 4 represents buoyant production or destruction, i.e. a conversion of turbulent kinetic energy to turbulent potential energy. Term 5 is a rotation term and can usually be neglected for averaging times less than one hour; term 6 represents the interaction of the fluctuating pressure and velocity fields; and term 7 is the molecular term.

For many purposes it is of interest to consider the second-moment equations in simplified form, applicable to the *horizontally homogeneous* case (this is achieved by letting $\partial/\partial x = \partial/\partial y = 0$). Equation 2.60 then becomes, for the $\overline{u'w'}$ covariance,

$$\partial\overline{u'w'}/\partial t = -\overline{w'^2}\partial\bar{u}/\partial z + (g/\overline{\theta_v})\overline{u'\theta_v'} - \partial\overline{u'w'^2}/\partial z$$
$$- \rho^{-1}(\overline{u'\partial p'/\partial z} + \overline{w'\partial p'/\partial x}).   (2.61a)$$

A similar equation holds for $\overline{v'w'}$. Note that the molecular term is negligible in the case of the covariance, because viscosity is dominant only at high wavenumbers, where the turbulence is locally isotropic and so the covariance is zero. In steady-state neutral conditions, for example, where $\overline{u'\theta_v'} = 0$ and the transport term is likely to be small (Sorbjan, 1989, Chapter 3), the equation simplifies further to

$$0 = -\overline{w'^2}\partial\bar{u}/\partial z - \rho^{-1}(\overline{u'\partial p'/\partial z} + \overline{w'\partial p'/\partial x}).   (2.61b)$$

Equations 2.61 illustrates the importance of the pressure–velocity correlation terms and the need for their accurate parameterization in higher-order closure schemes. Equation 2.61b shows that the pressure–velocity correlation terms destroy stress at the same rate as it is produced by the gradient–normal-stress interaction.

Equation 2.61b can be used to provide some insight into the nature of the transfer process. Let us assume that the pressure covariance terms can be approximated as (Wyngaard, 1982)

$$\rho^{-1}(\overline{u'\partial p'/\partial z} + \overline{w'\partial p'/\partial x}) = \overline{u'w'}/\tau_1   (2.62a)$$

where τ_1 is a suitable time scale characteristic of the larger eddies (e.g. defined in terms of the turbulent kinetic energy and integral length scale or identified with the integral time scale τ^i). In neutral conditions, Eq. 2.61b becomes

$$- \overline{u'w'} = \tau_1 \overline{w'^2} \, \partial \bar{u}/\partial z, \qquad (2.62b)$$

so, in comparison with Eq. 2.56, $\tau_1 \overline{w'^2}$ behaves as the eddy coefficient K_M. With $K_M \approx 1 \, \mathrm{m^2 \, s^{-1}}$ and $\overline{w'^2} \approx u_*^2 = 0.1 \, \mathrm{m^2 \, s^{-2}}$, $\tau_1 \approx 10 \, \mathrm{s}$, which is typical of eddy time scales near the surface.

2.5.2 The $\overline{u_i' \theta'}$ equation

The equation for $\overline{u_i' \theta'}$ is derived in a similar manner to Eq. 2.60 for the velocity covariance (the derivation of an equation for $\overline{u_i' q'}$ is identical, and is not given here). The result is

$$\partial \overline{u_i'\theta'}/\partial t + \bar{u}_j \partial \overline{u_i'\theta'}/\partial x_j = - \overline{u_i'u_j'} \, \partial \bar{\theta}/\partial x_j - \overline{u_j'\theta'} \, \partial \bar{u}_i/\partial x_j - \partial \overline{u_i'u_j'\theta'}/\partial x_j$$
$$+ (g/\overline{\theta_v})\overline{\theta'\theta_v'}\delta_{3i} - \rho^{-1}\overline{\theta'\partial p'/\partial x_i} + \nu\overline{\theta'\partial^2 u_i'/\partial x_j^2} + \kappa_T\overline{u_i'\partial^2\theta'/\partial x_j^2}.$$
$$(2.63)$$

The terms in this equation are analogous to those in Eq. 2.60: on the right-hand side we have, respectively, two production terms, a transport term, a buoyancy production term, a pressure–temperature interaction term and two molecular terms.

With the molecular terms negligible, Eq. 2.63 for the horizontally homogeneous case simplifies to

$$\partial \overline{w'\theta'}/\partial t = - \overline{w'^2} \, \partial\bar{\theta}/\partial z - \partial \overline{w'^2\theta'}/\partial z + (g/\overline{\theta_v})\overline{\theta'\theta_v'} - \rho^{-1}\overline{\theta'\partial p'/\partial z}. \quad (2.64a)$$

In *steady-state* conditions, and not too close to the boundaries, the transport term can be neglected (Sorbjan, 1989, Chapter 3), so that

$$0 = - \overline{w'^2} \, \partial\bar{\theta}/\partial z + (g/\overline{\theta_v})\overline{\theta'\theta_v'} - \rho^{-1}\overline{\theta'\partial p'/\partial z}. \qquad (2.64b)$$

This shows a balance between production and destruction of heat flux terms. The pressure covariance term acts to destroy heat flux, as can be seen in the near-neutral limit when $\overline{\theta'\theta_v'} \to 0$ and this term then balances the only production term.

As with Eqs. 2.61, 2.62 for the velocity covariance, Eq. 2.64b can be rewritten to illustrate the nature of the eddy coefficient for heat. Taking

$$\rho^{-1}\overline{\theta'\partial p'/\partial z} = \overline{w'\theta'}/\tau_2 \qquad (2.65a)$$

with τ_2 a time scale ($\sim \tau_1$), Eq. 2.64b becomes

$$\overline{w'\theta'} \approx - \tau_2 \overline{w'^2}[\partial\bar{\theta}/\partial z - (g/\overline{\theta_v})\overline{\theta'\theta_v'}/\overline{w'^2}]. \qquad (2.65b)$$

Again, by comparison with Eq. 2.58 we expect $\tau_2 \overline{w'^2}$ to serve as an eddy coefficient K_H, although Eq. 2.65b does not reveal a straightforward flux-gradient relation. The significance of the second term on the right-hand side of the equation is apparent in the context of the K approach applied to the middle and upper regions of the convective boundary layer, where it acts to preserve an upwards (positive) heat flux in the presence of near-zero or even slightly positive $\partial\bar{\theta}/\partial z$ (see Sections 6.1.6 and 8.7.1).

Solutions of the rate equations for $\overline{u_i'\theta'}$ (Eq. 2.63) and $\overline{u_i'q'}$ (not given) allow the covariance $\overline{u_i'\theta_v'}$ to be evaluated without recourse to a separate rate equation for $\overline{u_i'\theta_v'}$. This follows from the linear form of Eq. 2.30c which, when multiplied through by u_i' and averaged gives

$$\overline{u_i'\theta_v'} = \overline{u_i'\theta'} + 0.61\bar{\theta}\,\overline{u_i'q'}. \tag{2.66}$$

2.5.3 The $\overline{\theta'^2}$ equation

The temperature variance equation is obtained from Eq. 2.35 to give

$$\partial(\overline{\theta'^2}/2)/\partial t + \overline{u}_j\partial(\overline{\theta'^2}/2)/\partial x_j$$
$$= -\overline{u_j'\theta'}\,\partial\bar{\theta}/\partial x_j - (1/2)\partial\overline{\theta'^2 u_j'}/\partial x_j + (\rho c_p)^{-1}\overline{\theta'\partial R_j'/\partial x_j} + \kappa_T\overline{\theta'\partial^2\theta'/\partial x_j^2}$$
$$\approx -\overline{u_j'\theta'}\,\partial\bar{\theta}/\partial x_j - (1/2)\partial\overline{\theta'^2 u_j'}/\partial x_j - \kappa_T\overline{(\partial\theta'/\partial x_j)^2} \tag{2.67}$$

where the rate of molecular destruction of temperature fluctuations is given by

$$\chi = \kappa_T\overline{(\partial\theta'/\partial x_j)^2}. \tag{2.68}$$

For the horizontally homogeneous case, Eq. 2.67 simplifies to

$$\partial(\overline{\theta'^2}/2)/\partial t = -\overline{w'\theta'}\,\partial\bar{\theta}/\partial z - (1/2)\partial\overline{w'\theta'^2}/\partial z - \chi \tag{2.69a}$$

which, for the *steady state*, becomes

$$0 = -\overline{w'\theta'}\,\partial\bar{\theta}/\partial z - (1/2)\partial\overline{w'\theta'^2}/\partial z - \chi. \tag{2.69b}$$

There is observational and modelling evidence that near the surface (Hogstrom, 1990) and throughout the ABL (Sorbjan, 1989, Chapter 3) the transport term is small, so that a close approximation to Eq. 2.69b is

$$-\overline{w'\theta'}\,\partial\bar{\theta}/\partial z = \chi. \tag{2.69c}$$

This shows a simple balance between production due to the mean θ gradient and destruction due to molecular conduction. Taking $\overline{w'\theta'} = 0.4\,\mathrm{m\,s^{-1}\,K}$ and $\partial\bar{\theta}/\partial z = -1\,\mathrm{K}\,(10\,\mathrm{m})^{-1}$ near the surface in unstable conditions, we obtain $\chi \approx 0.04\,\mathrm{K^2\,s^{-1}}$.

An analogous equation for the humidity variance can readily be derived starting with Eq. 2.38. The details of this are left to the reader.

As with the quantities $\bar{\theta}_v$ and $\overline{u_i'\theta_v'}$, the variance $\overline{\theta_v'^2}$ does not need a separate rate equation. It is given by (cf. Eq. 2.30c)

$$\overline{\theta_v'^2}/2 = \overline{\theta'^2}/2 + (0.61\bar{\theta})^2\,\overline{q'^2}/2 + 0.61\bar{\theta}\,\overline{\theta'q'} \tag{2.70}$$

where the covariance $\overline{\theta'q'}$ can be evaluated through a rate equation based on Eqs. 2.35 and 2.38.

2.6 Turbulent kinetic energy and stability parameters

2.6.1 The $\overline{u_i'^2}$ equation and turbulent kinetic energy (TKE)

We define TKE as $\bar{e}$, where

$$\bar{e} = \overline{u_i'^2}/2 = (\overline{u'^2} + \overline{v'^2} + \overline{w'^2})/2. \tag{2.71}$$

Then with $k = i$ in Eq. 2.60 the equation for $\bar{e}$ can be written (by summing for $i = 1, 2$ and 3),

$$\partial\bar{e}/\partial t + \overline{u_j}\partial\bar{e}/\partial x_j$$
$$= -\overline{u_i'u_j'}\,\partial\overline{u_i}/\partial x_j + (g/\overline{\theta_v})\overline{u_i'\theta_v'}\delta_{3i} - \partial\overline{eu_j'}/\partial x_j - \rho^{-1}\partial\overline{p'u_i'}/\partial x_i - \varepsilon \quad (2.72)$$

where we have used $\overline{u_i'\partial p'/\partial x_i} = \overline{\partial u_i'p'/\partial x_i}$. Note that the rotation terms sum identically to zero since the Coriolis force does no work on a fluid parcel.

In Eq. 2.72 we have defined the rate of molecular (viscous) dissipation of TKE as ε, with

$$\varepsilon = -\overline{vu_i'\partial^2 u_i'/\partial x_j^2} \quad (2.73a)$$

$$\approx \overline{v(\partial u_i'/\partial x_j)^2} \quad (2.73b)$$

in high Reynolds number turbulence.

The quantity ε is a significant atmospheric parameter since it is related ultimately to the dissipation of the kinetic energy of all atmospheric motions, with maximum values occurring in the ABL near the surface. It has been measured indirectly in the atmosphere using known properties of the velocity spectrum at high wavenumber and indirectly using approximate forms of the TKE equation (Panofsky and Dutton, 1984, Chapter 8). Figure 2.3 shows the profile of normalized ε for both day and night conditions. Above the boundary layer, dissipation rates decrease rapidly to near zero, and attain maximum values near the surface where wind shear tends to be greatest.

For the horizontally homogeneous case, Eq. 2.72 simplifies to

$$\partial\bar{e}/\partial t = -\overline{u'w'}\,\partial\bar{u}/\partial z - \overline{v'w'}\,\partial\bar{v}/\partial z + (g/\overline{\theta_v})\overline{w'\theta_v'} - \partial(\overline{w'e} + \overline{w'p'}/\rho)/\partial z - \varepsilon$$
$$(2.74a)$$

where on the right-hand side, terms 1 and 2 represent shear production of TKE (in the horizontal components), term 3 represents buoyant production or destruction (by conversion to turbulent potential energy) of the vertical component, term 4 represents vertical transport, or a redistribution of TKE, within the ABL (integrated over the whole depth of the ABL it is zero), and term 5 is a pressure–velocity correlation term.

For the *steady-state surface layer*, Eq. 2.74a becomes (with $\bar{v} = 0$),

$$0 = -\overline{u'w'}\,\partial\bar{u}/\partial z + (g/\overline{\theta_v})\overline{w'\theta_v'} - \partial(\overline{w'e} + \overline{w'p'}/\rho)/\partial z - \varepsilon. \quad (2.74b)$$

The budgets of TKE in the clear-sky unstable and stable boundary layers are summarized in Fig. 2.4. In the stable case, shear production approximately balances dissipation at all levels. In the convective case, in the middle to upper regions of the boundary layer, vertical gradients of u and v are near zero so that TKE production is mostly due to buoyancy and transport from other levels. Just below the inversion, the heat flux is directed downwards due to entrainment and the only source of TKE is the transport term. Close to the surface, the turbulent transport appears to be significant and negative, but higher up it changes sign. This implies a net export of TKE from the lower half to the upper half of the boundary layer. Its role is to redistribute TKE with height. In the strongly

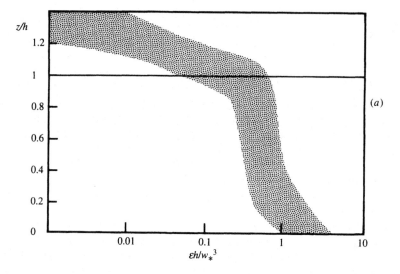

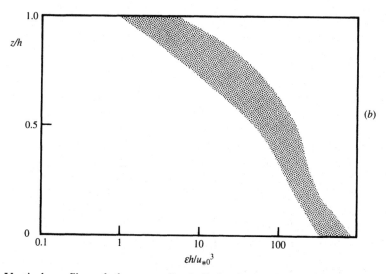

Fig. 2.3 Vertical profiles of the normalized dissipation rate for (a) daytime and (b) night-time conditions over land, with the shaded regions indicating the likely spread of individual observations. Data are from Caughey *et al.* (1979) and Kitchen *et al.* (1983), with h the ABL depth, u_{*0} the surface friction velocity and w_* the convective velocity scale. After Stull (1988); reprinted by permission of Kluwer Academic Publishers.

unstable surface layer, according to observations (Hogstrom, 1990), the pressure transport term $\partial \overline{p'w'}/\partial z$ is a significant source of TKE with $\overline{p'w'}$ being zero at a rigid surface and becoming increasingly negative with increasing height.

In neutral conditions near the surface, the observational evidence suggests that

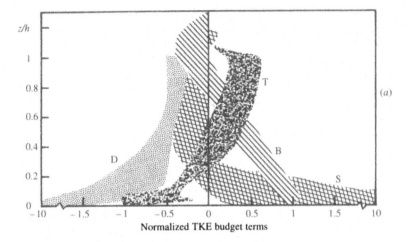

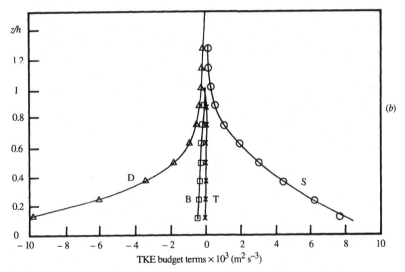

Fig. 2.4 Terms in the TKE equation (2.74b) as a function of height, normalized in the case of the clear daytime ABL (*a*) through division by w_*^3/h; actual terms are shown in (*b*) for the clear night-time ABL. Profiles in (*a*) are based on observations and model simulations as described in Stull (1988; Figure 5.4), and in (*b*) are from Lenschow *et al.* (1988) based on one aircraft flight. In both, B is the buoyancy term, D is dissipation, S is shear generation and T is the transport term. Reprinted by permission of Kluwer Academic Publishers.

$$\varepsilon \approx - \overline{u'w'} \, \partial \bar{u} / \partial z, \qquad (2.74c)$$

giving a near balance between shear production and viscous dissipation. Equation 2.74c allows a ready estimate of ε: taking $u_* = 0.3 \, \mathrm{m\,s^{-1}}$ and $\partial \bar{u}/\partial z = 0.1 \, \mathrm{s^{-1}}$ gives $\varepsilon \approx 0.01 \, \mathrm{m^2\,s^{-3}}$.

Before proceeding it is worthwhile to compare the simplest form of the TKE equation (2.74c) with the analogous equations for the kinetic energy of the three velocity components individually. These equations can be obtained in the same way as the equation for $\overline{u_i'^2}$. If the Reynolds number is so large that the dissipative structure can be assumed to be isotropic, the individual equations can be written as

$$0 = -\overline{u'w'}\,\partial\bar{u}/\partial z + \rho^{-1}\overline{p'\partial u'/\partial x} - \varepsilon/3 \qquad (2.75)$$

$$0 = \qquad\qquad\qquad \rho^{-1}\overline{p'\partial v'/\partial y} - \varepsilon/3 \qquad (2.76)$$

$$0 = \qquad\qquad\qquad \rho^{-1}\overline{p'\partial w'/\partial z} - \varepsilon/3 \qquad (2.77)$$

The sum of these three equations equals (2.74c), as it should. This is so since, because of incompressibility,

$$\overline{p'\partial u'/\partial x} + \overline{p'\partial v'/\partial y} + \overline{p'\partial w'/\partial z} = \overline{p'\partial u_i'/\partial x_i} = 0. \qquad (2.78)$$

In the above it follows that the individual pressure–velocity correlation components are not necessarily zero. Comparing Eq. 2.74c with Eqs. 2.75–2.77, we see that the entire production of TKE occurs in the $\overline{u'^2}$ equation. The v and w components must thus receive their energy from the pressure interaction terms listed in Eq. 2.78. Because the sum of the pressure terms is equal to zero, the pressure terms exchange energy between components, without changing the total amount of energy. Also, if $\overline{v'^2}$ and $\overline{w'^2}$ are to be maintained against dissipation, $\overline{p'\partial v'/\partial y}$ and $\overline{p'\partial w'/\partial z}$ must be positive, so that $\overline{p'\partial u'/\partial x}$ must be negative. Of course, in unstable conditions, for example, the w component (Eq. 2.77) receives all of the buoyancy production $(g/\overline{\theta_v})\overline{w'\theta_v'}$.

For moist air of density ρ, the buoyancy term in the TKE equation is $(g/\overline{\theta_v})\overline{w'\theta_v'} = -(g/\rho)\overline{w'\rho'}$. The buoyancy flux is an important quantity, though its measurement is usually indirect using known relations with the sensible heat $(\overline{w'\theta'})$ and moisture $(\overline{w'q'})$ fluxes. From Eq. 2.66, the vertical component is

$$\overline{w'\theta_v'} = \overline{w'\theta'} + 0.61\bar{\theta}\,\overline{w'q'}$$

$$= \overline{w'\theta'}(1 + 0.61\bar{\theta}\,\overline{w'q'}/\overline{w'\theta'}). \qquad (2.79)$$

For the surface fluxes,

$$(\overline{w'\theta_v'})_0 = (\overline{w'\theta'})_0(1 + 0.61\bar{\theta}\gamma/B) \qquad (2.80)$$

where $B = H_0/\lambda E_0 = c_p(\overline{w'\theta'})_0/\lambda(\overline{w'q'})_0$ is known as the Bowen ratio, and $\gamma = c_p/\lambda$ is called the psychrometric constant. Values of the parenthetic term δ are shown in Table 2.1 for a range of values of B and absolute temperature. In the table, the expressions and values for B are based on simple considerations for a saturated surface (see Chapter 5, Eq. 5.27), and those for B_0 are arbitrary values covering moist (0.75) to very dry (10) surfaces. In the latter, associated δ values show an expected decrease towards a value of unity as the surface dries.

2.6.2 Stability parameters

In Eq. 2.74a for moist air, the terms describing shear production and buoyant production (if $\partial\overline{\theta_v}/\partial z < 0$) or destruction (if $\partial\overline{\theta_v}/\partial z > 0$) are rather important in

Table 2.1. *Values of the Bowen ratio B* $(= B_1) = \gamma/s$ *(Eq. 5.27) and the associated humidity correction term* $\delta = 1 + 0.61\bar{\theta}\gamma/B$ *appearing in Eq. 2.80, as functions of temperature T. Values of* δ *for arbitrary values of* B $(= B_0)$ *are also shown for* $T = 20\,°C$. *See also Appendix 2, Table A2 for further related information.*

T (°C)	B_1	δ	$T = 20\,°C$	
			B_0	δ
0	1.45	1.05		
5	1.06	1.06		
10	0.78	1.09		
15	0.59	1.12		
20	0.45	1.16	0.75	1.10
25	0.34	1.22	1.0	1.07
30	0.26	1.29	2.0	1.04
35	0.21	1.37	5.0	1.01
40	0.16	1.50	10.0	1.01

determining the intensity of turbulence. Their ratio, called the flux Richardson number, Rf, can be used to define the local structure and evolution of the turbulence. Thus, the dimensionless number

$$Rf = (g/\overline{\theta_v})\overline{w'\theta_v'}/(\overline{u'w'}\,\partial\bar{u}/\partial z + \overline{v'w'}\,\partial\bar{v}/\partial z) \tag{2.81}$$

characterizes the thermal stability of the flow. Alternatively, we can define a gradient Richardson number, Ri, by use of the flux-gradient relations (Eqs. 2.56 to 2.59); it follows that

$$Rf = (K_H/K_M)Ri \tag{2.82}$$

where

$$Ri = (g/\overline{\theta_v})\partial\overline{\theta_v}/\partial z/[(\partial\bar{u}/\partial z)^2 + (\partial\bar{v}/\partial z)^2] \tag{2.83}$$

The gradient Richardson number has been widely used as a thermal stability parameter in the ABL; in unstable conditions both shear and buoyancy terms are positive (production of TKE) and Rf, Ri are negative. In stable conditions buoyancy acts to destroy TKE, and Rf, Ri are positive. In practice, the gradient Richardson number is often approximated in finite difference form and the resulting parameter is sometimes referred to as the bulk Richardson number. In the present text, however, the formal bulk Richardson number (Ri_B) will be defined in terms of differences in properties at height z in the air and at the surface (see Eq. 3.45).

It is convenient to introduce an additional stability parameter, deduced historically by Obukhov using the TKE equation. In Eq. 2.74a, as height z increases from the surface the shear term will tend to decrease more rapidly than the buoyancy term. The height at which the buoyancy term equals the shear term is a relevant length scale in non-neutral conditions; in the unstable case the terms are equal when $Rf = -1$. If we take axes along the mean wind direction ($\bar{v} = 0$), then for $Rf = -1$ we have, near the surface

$$- \overline{(u'w')}_0\, \partial\bar{u}/\partial z = (g/\overline{\theta_v})\overline{w'\theta_v'}$$

so that (see Section 3.3.1) with $\partial\bar{u}/\partial z = (u_{*0}/kz)\Phi$, the terms are equal at the height $z = -\Phi L$. The length scale L is the Obukhov stability length defined as

$$L = -\,u_{*0}{}^3/[k(g/\overline{\theta_v})\overline{w'\theta_v'}] = u_{*0}{}^2/[k(g/\overline{\theta_v})\theta_{v*0}] \qquad (2.84)$$

and the nondimensional height $\zeta = z/L$ is used extensively as a thermal stability parameter. It is a primary parameter of the Monin–Obukhov similarity theory of the atmospheric surface layer, and through flux-gradient relations can be readily related to Rf and Ri.

We can now summarize Eq. 2.74 in terms of the sign of Ri (and ζ).

  (i) When buoyancy is zero, $Ri = \zeta = 0$ and we have a condition of neutral stratification or forced convection.

 (ii) When the shear production term is small or zero and Ri and ζ are negative, then as Ri and ζ tend to $-\infty$, the condition of free convection is approached.

(iii) If Ri and ζ are positive, turbulence will cease if the buoyancy term is sufficiently negative. When Ri reaches a critical positive value, Ri_c, the transition from turbulent to laminar flow occurs. Throughout the ABL, this critical gradient Richardson number $Ri_c \approx 0.2$–0.25; the value $Ri_c = 1/4$ was deduced formally for inviscid stability by Miles (1961).

(iv) With stratified laminar flow, the transition to turbulent flow (at large enough Reynolds number) occurs at $Ri \approx 0.25$ and above this value the flow is stable to infinitesimal perturbations.

Notes and bibliography

Section 2.1

The practical application of the ergodic hypothesis to measuring turbulence moments to a required degree of accuracy is discussed in

Wyngaard, J. C. (1973), On surface-layer turbulence, in *Workshop on Micrometeorology*, ed. D.A. Haugen, pp. 101–49. American Meteorological Society, Boston, MA, and in

Wyngaard, J. C. (1983), Lectures on the planetary boundary layer, in *Mesoscale Meteorology – Theories, Observations, and Models*, eds T. Gal-Chen and D. K. Lilly, pp. 603–50. Reidel, Dordrecht.

For a theoretical treatment of turbulence spectra, the reader should consult Hinze's book, Chapter 1. For additional detailed discussions on the properties of spectral and correlation functions, see Lumley and Panofsky's book (pp. 14–35) and

Busch, N. E. (1973), On the mechanics of atmospheric turbulence, in *Workshop on Micrometeorology*, ed. D. A. Haugen, pp. 1–28. American Meteorological Society, Boston, MA, and also Chapter 3 in

Panofsky, H. A. and J. A. Dutton (1984), *Atmospheric Turbulence – Models and Methods for Engineering Applications*. John Wiley and Sons, New York, 397 pp.

Section 2.2

The Boussinesq approximation to the full equations, and conditions required for its validity in ABL flow, are discussed by

Businger, J. A. (1982), Equations and concepts, in *Atmospheric Turbulence and Air Pollution Modelling*, eds F. T. M. Nieuwstadt and H. van Dop, pp. 1–36. Reidel, Dordrecht, and by

Mahrt, L. (1986), On the shallow motion approximations, *J. Atmos. Sci.* **43**, 1036–44.

Important references will be found in both papers.

Discussion on the Boussinesq approximation in free convection can be found in the appendix to Chapter 14 of

Tritton, D. J. (1988), *Physical Fluid Dynamics*, 2nd edition. Clarendon Press, Oxford, 519 pp.

The development of the Navier–Stokes equation of motion with reference to both non-rotating and rotating fluid systems can be found in

Batchelor, G. K. (1967), *An Introduction to Fluid Mechanics*. Cambridge University Press, 615 pp.

3

Scaling laws for mean and turbulent quantities

The present chapter deals primarily with similarity theories and scaling laws. A similarity theory uses dimensional analysis as the basis for expressing relationships between different quantities in nondimensional form, so as to reveal underlying scaling laws. This approach involves the choice of suitable scaling variables and the organization of these into nondimensional groups. The Reynolds similarity law is a classic example, involving the use of the nondimensional quantity called the Reynolds number (Monin and Yaglom, 1971, Chapter 1).

One of the major goals of any similarity theory applied to the atmosphere is the correct scaling of characteristic features of the ABL (the wind profile, turbulence variances) through the choice of appropriate length, velocity and temperature scales (called similarity scaling). Several theories will be described (Rossby-number similarity, Monin–Obukhov (surface-layer) similarity, mixed-layer similarity, local similarity) with the express purpose of formulating profile relations (in particular, the wind profile) and associated drag, heat and mass transfer relations. These relations are of primary importance in numerical modelling of the atmosphere where it is required to relate unknown surface fluxes (second-order moments) to known model variables (first-order moments).

One housekeeping matter should be stated at this point – **henceforth, the overbar will be omitted when identifying averaged quantities other than the products of fluctuating quantities.**

3.1 The wind profile: simple considerations

Before considering in detail the various scaling laws for the wind profile in the ABL, we can deduce some simple properties of the wind variation throughout the ABL under somewhat restrictive conditions. We take the ABL to be steady state and horizontally homogeneous.

3.1.1 Surface layer
The eddy viscosity K_M has dimensions of velocity × length, so that we expect $K_M \sim u_{*0}l$, where l is a mixing length equal to kz (Prandtl's hypothesis; see

Schlichting, 1979, Chapter 19). Here, k is the von Karman constant. The relevant turbulent velocity scale is the surface *friction velocity*, u_{*0}, related to the magnitude of the surface stress τ_0 by

$$u_{*0}^2 = \tau_0/\rho = [(\overline{u'w'})_0^2 + (\overline{v'w'})_0^2]^{1/2} \qquad (3.1)$$

so that if the x-axis is chosen such that $v = \overline{u'v'} = 0$, then Eq. 3.1 reduces to $u_{*0}^2 = -(\overline{u'w'})_0$.

Substitution of the above expression for K_M into Eq. 2.56 gives $\partial u/\partial z = u_{*0}/kz$. The assumptions implicit in this relation are that the stress is constant throughout the surface layer (a "constant-flux" layer) and that z is the sole length scale determining the shear. Integration gives

$$ku/u_{*0} = \ln z + \text{constant} \qquad (3.2)$$

yielding the classical logarithmic velocity law of the "constant-flux" layer. Digressing briefly, there are two issues that need stating here. Firstly, it is important to distinguish between "smooth" and "rough" surfaces, particularly in the aerodynamic sense. Immediately adjacent to a smooth surface there exists a very thin *viscous sublayer*, of depth δ_1, given by (Schlichting, 1979, Chapter 20)

$$\delta_1 \approx 5v/u_*. \qquad (3.3)$$

The smallness of δ_1 can be seen by taking $u_* = 0.15 \, \text{m s}^{-1}$ which, with $v = 1.5 \times 10^{-5} \, \text{m}^2 \, \text{s}^{-1}$, gives $\delta_1 \approx 0.5 \, \text{mm}$. For $z < \delta_1$ the shearing stress is caused predominantly by viscosity ($|\overline{u'w'}| \ll v\partial u/\partial z$). Above the viscous sublayer exists the turbulent boundary layer within which the stress is mainly related to turbulent velocity fluctuations and where the viscous stress is negligible.

In specifying the degree of roughness of a particular surface at least two lengths may be involved: the height of the elements and their mean distance apart. The notions of smooth flow and rough flow can be illustrated by reference to the work of Nikuradse, who investigated turbulent flow in pipes roughened uniformly by grains of sand of different diameter h_r. The definitions of smooth and rough surfaces *in the aerodynamic sense* are based on the relative size of h_r (the height of the roughness elements) and δ_1. An aerodynamically smooth surface is one for which $h_r \ll \delta_1$ so that a viscous sublayer exists. In contrast, an aerodynamically rough surface is one where $h_r \gg \delta_1$ so that the flow is turbulent right down to the surface and is Reynolds-number independent, and the frictional drag (stress) is due predominantly to "form drag" i.e. pressure differences across the individual roughness elements. A surface may change from being aerodynamically smooth to aerodynamically rough by an increase in the flow velocity. Experiments over flat plates and in pipes suggest that a surface is aerodynamically smooth for $u_*h_r/v < 5$, and aerodynamically rough for $u_*h_r/v > 75$ (Sutton, 1953, Chapter 3). For most natural surfaces, and under typical environmental conditions, flow is generally aerodynamically rough. Note that even at large Reynolds number heat and mass transfer in the immediate vicinity of a surface (within the interfacial sublayer) is controlled by molecular diffusion.

The second issue concerns the definition of the origin of z for a rough surface.

This needs considering since the protrusion of roughness elements above the substrate surface displaces the entire flow upwards. To account for this, we define the displaced height $z = Z - d$. Here, d is the fluid-dynamic height origin or zero-plane displacement, discussed in more detail in Chapter 4, and Z is the height above the substrate surface.

3.1.2 The Ekman spiral and solutions of the simplified momentum equation

We combine Eqs. 2.52 and 2.53 for the u and v components with Eqs. 2.56 and 2.57 and assume constant K_M throughout the boundary layer. Setting $\mathbf{V} = (u, v) = u + iv$, where $i = (-1)^{1/2}$, and taking the x-axis along the geostrophic wind direction so that $\mathbf{V_g} = (u_g, 0)$, we can write

$$\partial^2 \mathbf{V}/\partial z^2 = (if/K_M)(\mathbf{V} - u_g). \tag{3.4}$$

Because we must deal with flow in both the northern and southern hemispheres, it is necessary to distinguish between f and its magnitude, $|f|$. To do this, we write $f = |f| \operatorname{sgn} f$ where $\operatorname{sgn} f$ is $+1$ and -1 for the northern and southern hemispheres respectively. The boundary conditions are that $\mathbf{V} \to u_g$ as $z \to \infty$ and $\mathbf{V} = 0$ at $z = 0$. Then the solution to Eq. 3.4 is the Ekman spiral

$$u = u_g[1 - \exp(-a_0 z)\cos(a_0 z)] \tag{3.5a}$$

$$v = \operatorname{sgn} f u_g \exp(-a_0 z)\sin(a_0 z) \tag{3.5b}$$

where

$$a_0^2 = |f|/2K_M. \tag{3.6}$$

Features of the Ekman spiral are shown in Fig. 3.1, for both hemispheres. By expanding the exponential and trigonometrical functions in a series for small $a_0 z$, the cross-isobar flow at the surface, α_0, is found to be

$$\tan \alpha_0 = v/u \to \operatorname{sgn} f \text{ as } z \to 0, \text{ i.e. } |\alpha_0| = \pi/4 \text{ at } z = 0.$$

In the above, a_0 is the reciprocal of a scale height; when $a_0 z = \pi/2$, $u = u_g$, $v = \operatorname{sgn} f u_g \exp(-\pi/2)$ and $\tan \alpha = \operatorname{sgn} f \exp(-\pi/2)$; when $a_0 z = \pi$, $u = u_g[1 + \exp(-\pi)]$, $v = 0$ and $\tan \alpha = 0$.

The solution produces a surface wind parallel to the surface-stress vector and at 45° to the geostrophic wind, a flow angle that is somewhat larger than that observed in a neutral, barotropic atmosphere (see Fig. 3.12b). In addition, there is no unique level that can be identified as the ABL top. For $K_M = 12.5 \text{ m}^2 \text{ s}^{-1}$ and $|f| = 10^{-4} \text{ s}^{-1}$, $a_0^{-1} = 500$ m, so that at $a_0 z = \pi/2$ where $u = u_g$, $z \approx 750$ m, which is comparable with a neutral ABL depth. The solution Eq. 3.5 was first described by Ekman (1905) for laminar ocean currents, with K replaced by the viscosity for water (so that the depth scale implied in Eq. 3.6 is much smaller than that for the ABL estimated above). Even with its limitations, the Ekman solution does give a useful first approximation to the neutral wind profile above the surface layer.

3.2 Wind profile laws: the neutral case

In order to derive the form of the wind profile in the inner and outer layers of the ABL, we consider a steady-state situation with horizontal homogeneity and

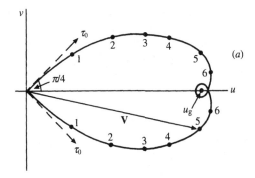

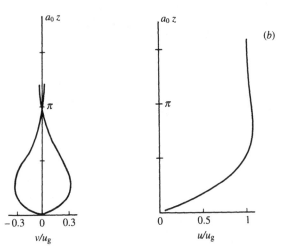

Fig. 3.1 Features of the Ekman spiral based on Eq. 3.5, with a_0 defined by Eq. 3.6. In (a), wind hodographs are shown for both hemispheres, together with the surface-stress vector and the wind vector at height z; points 1 to 6 correspond to values of $a_0 z$ of (1) $\pi/12$, (2) $\pi/6$, (3) $\pi/4$, (4) $\pi/3$, (5) $\pi/2$ and (6) $2\pi/3$. The x-axis lies in the direction of the geostrophic wind vector. Vertical profiles of the wind components are shown in (b).

the geostrophic wind constant with height (i.e. barotropic). In addition, the surface heat flux and evaporation are taken as zero so that conditions are neutral. The inner layer, in which fluxes do not deviate too far from their surface values and where the shear is large and the generation of TKE greatest, is about 10 per cent of the ABL depth. The initial problem is to determine the most appropriate velocity and length scales with which to scale the wind profile. Our task is not only to derive the wind profile law, but also to use this to formulate a suitable drag law, i.e. a relation between the surface stress (or friction velocity u_{*0}) and external flow variables (e.g. the geostrophic wind).

3.2.1 Rossby-number similarity

We start with some deductions on the form of the drag law. There are three external parameters that are potentially important, G, f and z_0 (here we anticipate the use of the aerodynamic roughness length, which is a representation of the surface geometry; for smooth flow, z_0 is replaced by v/u_{*0}). This gives two external length scales; a large one given by $G/|f|$ ($\approx 10^5$ m for $G = 10$ m s^{-1} and $|f| = 10^{-4}$ s^{-1}) and a smaller one z_0 (≈ 0.1 m for general land surfaces), and an external velocity scale G, the geostrophic wind speed (we let $|\mathbf{V}_g| = G$). The relation

$$u_{*0} = GF_g(G/|f|z_0) \tag{3.7}$$

is in the form of a drag law, giving the surface stress, or internal ABL velocity scale u_{*0}, as a function of the geostrophic wind and a quantity $Ro = G/|f|z_0$ known as the *surface Rossby number*. This is a central nondimensional parameter in ABL flows, combining the three important external parameters. The quantity $(u_{*0}/G)^2$ is known as the *geostrophic drag coefficient*. There is, in fact, another way of writing Eq. 3.7, utilizing the alternative length scale $u_{*0}/|f|$. Thus, the relation $u_{*0}/G = F(u_{*0}/|f|z_0)$ is also a drag law giving the geostrophic drag coefficient as an *implicit* function of surface Rossby number. The parameter $u_{*0}/|f|z_0$ is called the friction Rossby number (Tennekes and Lumley, 1972).

In order to explore further the above drag law, we turn our attention to the wind profile and the simplified mean momentum equations (Eqs. 2.52 and 2.53). The x-axis is chosen to give $v = 0$ (i.e. to lie along the surface wind or surface stress direction), with $-(\overline{u'w'})_0 = u_{*0}{}^2$, $(\overline{v'w'})_0 = 0$. This lower boundary condition, and the fact that the velocity defects $(u - u_{g0}, v - v_{g0})$ are the result of turbulent friction, suggests choosing u_{*0} as the characteristic velocity scale (for more detailed considerations see Blackadar and Tennekes, 1968; Tennekes, 1973, 1982). Equations 2.52 and 2.53 are thus written in nondimensional form as

$$(u - u_{g0})/u_{*0} = -\partial(\overline{v'w'}/u_{*0}{}^2)/\partial(zf/u_{*0}) \tag{3.8}$$

$$(v - v_{g0})/u_{*0} = \partial(\overline{u'w'}/u_{*0}{}^2)/\partial(zf/u_{*0}). \tag{3.9}$$

Note that both the stress and ageostrophic wind components (the velocity defects) are zero above the boundary layer, so that boundary conditions at the ABL top are homogeneous and independent of external parameters. However, at the surface the stress divergence has components v_{g0}/u_{*0} and u_{g0}/u_{*0}, which are likely to depend on the surface Rossby number (through Eq. 3.7). Since G/u_{*0} increases with Ro (the surface stress decreases with decreasing surface roughness), Eqs. 3.8 and 3.9 cannot apply at the surface in the formal limit process $Ro \to \infty$. Their use is thus restricted to levels away from the surface; this is consistent with the idea that close to the surface, $u_{*0}/|f|$ (used to scale z in Eqs. 3.8 and 3.9) is not the relevant length scale. Scaled as above, the equations of motion become independent of any parameters, in particular the surface Rossby number, so their form suggests writing

$$(u - u_{g0})/u_{*0} = F_x(zf/u_{*0}) \tag{3.10}$$

$$(v - v_{g0})/u_{*0} = F_y(zf/u_{*0}) \tag{3.11}$$

This is called the *Rossby-number similarity* of the ageostrophic wind in the neutral, barotropic ABL and represents a *velocity defect* law. It suggests that the depth of the neutral ABL is directly proportional to the quantity $u_{*0}/|f|$. Because of the restriction close to the surface, this Ro similarity does not extend to the lower regions of the surface layer.

Near the surface, we write Eqs. 2.52 and 2.53 in the form

$$(fz_0/u_{*0})(u - u_{g0})/u_{*0} = - \partial(\overline{v'w'}/u_{*0}{}^2)/\partial(z/z_0) \tag{3.12}$$

$$- (fz_0/u_{*0})(v - v_{g0})/u_{*0} = - \partial(\overline{u'w'}/u_{*0}{}^2)/\partial(z/z_0). \tag{3.13}$$

The left-hand sides of Eqs 3.12 and 3.13 can be at most of order $|f|z_0 G/u_{*0}{}^2$ or $Ro^{-1}(G/u_{*0})^2$. Thus, in the limit as $Ro \to \infty$, the left-hand sides vanish and the stress components are constant with height (Tennekes, 1982). The surface layer in the large Ro limit is thus a constant-flux layer that does not feel the turning effects of the Coriolis force. Because the stress at the surface has been assumed to have no y-component, the wind in the surface layer also has no y-component. The wind profile must then have components

$$u/u_{*0} = f_x(z/z_0) \tag{3.14}$$

$$v/u_{*0} = 0 \tag{3.15}$$

which are also independent of the surface Rossby number. As in the outer-layer scaling, this Rossby-number similarity has been achieved at the expense of having to use the internal velocity scale u_{*0} instead of the external scale G. The above equations represent the only set with the correct nondimensionalization, for which the properties are independent of external parameters (e.g. Ro) and for which a set of universal profiles are generated.

Now Eqs. 3.10 and 3.11 do not apply as $z|f|/u_{*0} \to 0$ and Eqs. 3.14 and 3.15 do not apply as $z/z_0 \to \infty$. Asymptotic matching (Van Dyke, 1975) requires that both laws are identical in the double limit process (consistent with $Ro \to \infty$)

$$z/z_0 \to \infty \quad \text{and} \quad z|f|/u_{*0} \to 0.$$

That is, there exists an overlap region in which the descriptions of the outer- and inner-layer scaling are valid simultaneously: this is called the *matched layer* or *inertial sublayer* (Tennekes, 1982); see Fig. 1.1. The procedure yields, for the u-component (Blackadar and Tennekes, 1968; Tennekes, 1973),

$$(z/u_{*0})\partial u/\partial z = \text{constant} = 1/k. \tag{3.16}$$

For historical reasons, the constant is written as $1/k$, where k is the von Karman constant. Equation 3.16 is consistent with the results in Section 3.1.1 (leading to Eq. 3.2) that z is the only relevant length scale for the surface-layer shear in neutral conditions. Mathematically, it is the result of finding a layer free of explicit dependence on external length scales as a consequence of the double limit process. Integration of Eq. 3.16 yields two solutions: the first must be consistent with Eq. 3.14, hence

$$ku/u_{*0} = \ln(z/z_0) \quad (z/z_0 \gg 1) \tag{3.17}$$

and the second must be consistent with Eq. 3.10, hence

$$k(u - u_{g0})/u_{*0} = \ln(z|f|/u_{*0}) + A. \quad (z|f|/u_{*0} \ll 1) \tag{3.18}$$

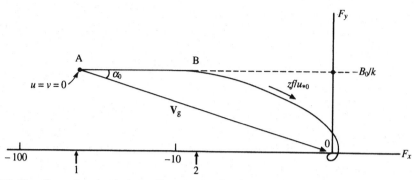

Fig. 3.2 Nondimensional wind profile plotted schematically in hodograph form for the northern hemisphere with the nondimensional height $z|f|/u_{*0}$ increasing towards the right along the curve. Here, $F_x = (u - u_{g0})/u_{*0}$ and $F_y = (v - v_{g0})/u_{*0}$. Point A corresponds to the surface (where $u = v = 0$), and the geostrophic wind vector $\mathbf{V_g}$ is drawn as a straight line joining point A and the origin 0. Point B corresponds to the approximate top of the surface layer. The wind profile between points A and B depends upon Ro, and in this region the wind vector does not rotate with height (F_y is constant). To the right of point B the profile is Ro-independent and given by Eqs. 3.10 and 3.11. The z-dependence in this outer layer could be approximated as an Ekman spiral (Fig. 3.1(a)) or as a parabolic curve, as described by Hinze (1975, p. 631, Fig. 7–11). Thus, changing Ro affects the end point A but does not affect the wind "spiral" – this expresses the existence of Ro similarity.

Along the abscissa, a logarithmic scale is used; point 1 corresponds to $Ro = 10^{10}$ (open sea), whence $F_x(0) \approx -43$ based on the pecked curve in Fig. 3.12(a). Point 2 corresponds to the approximate top of the surface layer (where $z \approx 0.02(u_{*0}/|f|)$), with $F_x(0.02) \approx -6$. Vertical arrows identify the corresponding locations of points A and B on the plotted wind profile.

For the v-component, asymptotic matching gives $v = 0$ (Eq. 3.15) with

$$kv_{g0}/u_{*0} = \text{constant} = -B\,\text{sgn}\,f. \qquad (3.19)$$

In schematic form, the nondimensional wind profile (Eqs. 3.10, 3.11) has the structure shown in Fig. 3.2: the outer-layer spiral (to the right of point B) is matched to the wind profile in the surface layer (represented by the logarithmic relation of Eq. 3.18). Above the surface layer, the profile resembles the Ekman spiral shown in Fig. 3.1 for the northern hemisphere. The important point is that the wind profile can be constructed for any value of Ro by sliding the point A ($u = v = 0$) horizontally along the fixed line $F_y = -B_0/k$, without affecting the profile to the right of point B. All neutral wind profiles are the same when plotted in this form, except for the length of the straight-line extension along their left asymptote (at point A, the value of F_x is $-u_{g0}/u_{*0}$ which is a function of Ro; see Eq. 3.20). The derivation of Eq. 3.17 reveals that assumptions on constant flux and mixing length hypotheses are not needed. Further, the log law is a property both of the neutral inner-(surface)layer scaling and of the outer-(Ekman)layer scaling. In practice, the inertial sublayer (and hence the log law) extends to about one-tenth the ABL depth (≈ 100 m), with stress decreasing and turning with height, but with wind turning very little with height.

Equation 3.17 also serves to define z_0, the aerodynamic roughness length; thus, extrapolation of the wind profile, taking into account the zero-plane displacement, gives $z = z_0$ as the level where $u = 0$.

The logarithmic velocity law has been verified extensively in large Reynolds number, wind-tunnel flow, for both aerodynamically smooth and aerodynamically rough cases (see e.g. Hinze, 1975, Chapter 7 and Schlichting, 1979, Chapter 20) and also in the atmosphere when true neutral conditions have occurred (e.g. Plate, 1971, Chapter 1). The reader should note that the wind-tunnel boundary layer is unaffected by Coriolis effects, and flow in the outer layer scales with the boundary-layer depth itself rather than with $G/|f|$ or $u_{*0}/|f|$ as in the ABL. The von Karman constant k is assumed to be a universal constant, with extensive wind-tunnel data suggesting $k \approx 0.40$. Atmospheric determinations have been less precise, with the majority lying in the range 0.38 to 0.42 (see Appendix 4, Table A5). In the atmosphere, even when neutral conditions occur, evaluation of k requires accurate measurements of $u(z)$, $\partial u/\partial z$ and u_{*0}, though often stability effects must be taken into account and extrapolation to neutrality made.

3.2.2 Variation of stress with height

Equations 2.52 and 2.53 imply a variation of stress with height that is dependent upon the velocity defect profile. Written in nondimensional form (Eqs. 3.8, 3.9), the normalized stress profiles well away from the surface, and in the limit of $Ro \to \infty$, are predicted to be universal functions of zf/u_{*0}. Figure 3.3(a) shows scaled momentum flux profiles measured from aircraft over the ocean for a slightly unstable ABL beneath a capping inversion (in this diagram, the x-axis is taken along the surface stress ($v = 0$) direction). They agree qualitatively with profiles in a neutral ABL (Fig. 3.3(b)), computed in a LES model (where the x-axis is taken in the geostrophic wind direction). According to the $\overline{u'w'}$ profiles, the ABL depth in the model is about $0.3u_{*0}/|f|$ but is $0.18u_{*0}/|f|$ as observed. This difference could be related to the effects of the capping inversion, which tends to limit the ABL depth, and to unknown model constraints. The monotonic decrease of $\overline{u'w'}$ with height, and the $\overline{v'w'}$ maximum, are notable features of a neutral, barotropic ABL. In addition, the data in Fig. 3.3(a) tend to collapse towards a single profile, supporting the scaling described by Eqs. 3.8 and 3.9.

3.2.3 Geostrophic wind relations

Since Eqs. 3.17 and 3.18 are simultaneously satisfied, the following result must apply:

$$ku_{g0}/u_{*0} = \ln(u_{*0}/|f|z_0) - A_0 \tag{3.20}$$

which, together with Eq. 3.19, is the drag law sought (see Eq. 3.7). Here, A_0 and B_0 are to be interpreted as universal constants for the neutral, barotropic steady-state ABL. Combining Eqs. 3.19 and 3.20 gives the neutral geostrophic drag coefficient C_{gn} as an implicit function of Ro:

$$
\begin{aligned}
C_{gn} &= u_{*0}^2/G^2 = k^2/\{[\ln(u_{*0}/|f|z_0) - A_0]^2 + B_0^2\} \\
&= k^2/\{[\ln(RoC_{gn}^{1/2}) - A_0]^2 + B_0^2\}
\end{aligned} \tag{3.21}
$$

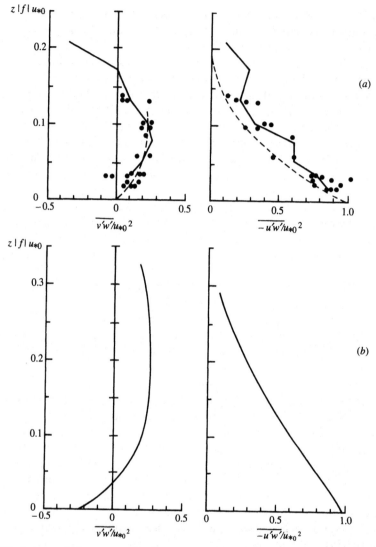

Fig. 3.3 Variation of the normalized stress components with nondimensional height $z|f|/u_{*0}$. (*a*) Aircraft data for seven separate flights over the north-east Atlantic during JASIN in slightly unstable conditions. The continuous curve connects mean values for several dimensionless height classes, whilst the pecked curve represents Eqs. 2.52 and 2.53 based on the observed velocity profiles. The *x*-axis is along the surface stress. (Modified from Nicholls, 1985). (*b*) LES results for a neutral, barotropic ABL, with the *x*-axis along the geostrophic wind direction. Modified from Mason and Thomson (1987).

Equation 3.21 allows the stress to be determined in terms of external parameters G, f and z_0 alone. There is an extensive literature on the determination of the constants A_0 and B_0 (e.g. Zilitinkevich, 1989), suggesting that $A_0 \approx 2$ and $B_0 \approx 4.5$. The angle between the surface and geostrophic winds (the cross-isobar flow) is given by

$$\tan \alpha_0 = v_{g0}/u_{g0} = \operatorname{sgn} f \, B_0 u_{*0}/k u_{g0} \qquad (3.22)$$

showing that $|\alpha_0|$ increases with increased roughness. The variations of C_{gn} and α_0 with Ro, using the above values of A_0 and B_0, are shown as pecked curves in Figs. 3.12(*a*), (*b*). The neutral drag coefficient varies by about a factor of five between the relatively small values of Ro (10^5) characteristic of flow over forests and the large values (10^9) found over the sea. The cross-isobar flow curves are consistent with observations of relatively large values ($\approx 30°$) over very rough surfaces, decreasing towards zero over the sea.

The idealized Rossby-number similarity theory allows specific deductions to be made about the wind relations in a neutral, barotropic boundary layer. The results form the basis for extensions of the theory to the non-neutral, baroclinic ABL.

3.3 Monin–Obukhov similarity theory: the non-neutral surface layer

3.3.1 Flux–gradient relations

General
The influence of buoyancy on the TKE budget (Section 2.7) can be represented through the flux Richardson number Rf (Eq. 2.81), or through the scaling length L (Eq. 2.84) characterizing the buoyancy in a stratified surface layer. In order to include the effect of stratification in the description of turbulent transport and mean profiles Monin and Obukhov (1954) hypothesized that any dimensionless characteristic of the turbulence can depend only upon u_{*0}, z, g/θ_v and $(\overline{w'\theta_v'})_0$, i.e. upon $\zeta = z/L$.

The reader is reminded that the x-axis has been chosen to lie along the mean surface wind direction, so that $v = 0$. The gradient form of the wind profile, suitably nondimensionalized, can be written for the non-neutral case as a function of ζ:

$$(kz/u_{*0})\partial u/\partial z = \Phi_M(\zeta) \qquad (3.23)$$

Likewise for the nondimensional gradients of θ_v and q, using the scaling parameters θ_{v*0} and q_{*0} we expect

$$(kz/\theta_{v*0})\partial \theta_v/\partial z = \Phi_H(\zeta) \qquad (3.24)$$

$$(kz/q_{*0})\partial q/\partial z = \Phi_W(\zeta) \qquad (3.25)$$

From these relations it follows that

$$Ri = \zeta \Phi_H/\Phi_M{}^2 = f_1(\zeta) \qquad (3.26)$$

$$K_H/K_M = \Phi_M/\Phi_H = f_2(\zeta) \qquad (3.27)$$

where the ratio of the eddy diffusivities K_M/K_H is called the turbulent Prandtl number, P_t. In addition, we have

$$K_M = ku_{*0}z/\Phi_M(\zeta) \tag{3.28a}$$

$$K_H = ku_{*0}z/\Phi_H(\zeta). \tag{3.28b}$$

Note that, in order that Eq. 3.23 reduces to the neutral form, we must have $\Phi_M(0) = 1$. In addition, we expect enhanced mixing in unstable conditions where buoyancy aids the generation of TKE, and hence $\Phi(\zeta) < 0$. With reduced mixing in stable conditions, we expect $\Phi(\zeta) > 0$. The nature of the Φ functions can be predicted only in an asymptotic sense; otherwise, experimental data are required to evaluate them. Useful information on the Φ functions can be obtained by considering their asymptotic behaviour.

Asymptotic limits
There are three asymptotic limits of interest.

(i) The neutral limit: in this limit the parameter ζ is a measure of the influence of buoyancy. For $|\zeta| \ll 1$, we require $\Phi \to 1$ so that the neutral form of the wind profile (Eqs. 3.16, 3.17) is recovered. For small values of ζ, Φ can be expanded as a power series in ζ, assuming it is analytic:

$$\Phi(\zeta) = 1 + \beta_1\zeta + \beta_2\zeta^2 + \ldots$$

$$\approx 1 + \beta_1\zeta. \tag{3.29}$$

For $\zeta \to 0$, the correct limit of $\Phi = 1$ is achieved. The linear form given by Eq. 3.29 has been verified observationally for very small ζ on both sides of neutrality, and for a larger ζ range in stable conditions (e.g. Webb, 1970, 1982).

(ii) The stable limit: in highly stable conditions ($\zeta \to \infty$), eddy vertical motion is strongly limited by the positive stratification, so we expect eddy sizes to be limited entirely by the stability and not by distance from the surface. Eddies will scale with L rather than z as in the neutral case. This is referred to as local-height-independent (z-less) scaling, and will occur for $L \lesssim z \ll h$, where h is the boundary-layer depth (Wyngaard, 1973). This postulate is supported by observation, in that Eq. 3.29 is verified for ζ as large as unity, or more. In this case, the profiles tend towards $\partial u/\partial z \propto u_{*0}/L$ and $\partial\theta_v/\partial z \propto \theta_{v*0}/L$ behaviour, which is confined to a fairly narrow height range.

At larger values of ζ and z/h, outer-layer scaling of profiles and turbulence variances with L fails (z is large enough for surface fluxes to be irrelevant locally), and local-similarity concepts are needed (e.g. Nieuwstadt, 1984, and Section 3.5).

(iii) The unstable limit: in highly unstable conditions ($\zeta \to -\infty$) where winds become light and a state of local free convection (Tennekes, 1970) is approached, the stress becomes insignificant and u_{*0} ceases to be a scaling parameter. Monin–Obukhov scaling then fails, but can be replaced by free convective scaling, using scaling parameters for velocity (u_f) and temperature (θ_f) defined by (Wyngaard *et al.*, 1971)

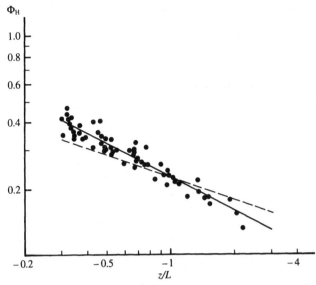

Fig. 3.4 Values of Φ_H based on the Kansas experimental data, plotted as a function of z/L. Curves are shown with slopes of $-1/2$ and $-1/3$ (see Eq. (3.32b)). After Businger *et al.* (1971), *Journal of Atmospheric Sciences*, American Meteorological Society.

$$u_f = [z(g/\theta_v)(\overline{w'\theta_v'})_0]^{1/3} \qquad (3.30a)$$

$$\theta_f = [(\overline{w'\theta_v'})_0^2/z(g/\theta_v)]^{1/3}. \qquad (3.30b)$$

We recall that the Monin–Obukhov four-variable (u_{*0}, z, (g/θ_v), $(\overline{w'\theta_v'})_0$) similarity theory gives the prediction of flow properties (e.g. the mean gradients or variances) to within an unknown *function* of ζ. In contrast, the free convection three-variable (z, (g/θ_v), $(\overline{w'\theta_v'})_0$) scaling theory gives the gradients to within an unknown *constant*, as required by Buckingham's theory (see Stull, 1988, Chapter 9). In the limit $\zeta \rightarrow -\infty$, the wind shear becomes negligible and the main interest lies in the θ_v profile, given by

$$(z/\theta_f)\partial\theta_v/\partial z = -\alpha_1 \qquad (3.31)$$

where α_1 is a positive constant. Use of Eq. 3.30b gives a minus four-thirds power law,

$$\partial\theta_v/\partial z = -\alpha_1(\overline{w'\theta_v'})_0^{2/3}(g/\theta_v)^{-1/3}z^{-4/3} \qquad (3.32a)$$

as predicted from simple dimensional arguments by Priestley (1954) for example. Unfortunately, measurements in free convective conditions are notoriously difficult, so that observations tend to be analysed at finite values of ζ or Ri and any asymptotic approach represented by Eq. 3.32a tends to be identified by inspection. Under such conditions u_{*0} is still a relevant scaling parameter. In Fig. 3.4, the data are presented in terms of

$\Phi_H(\zeta)$; thus, we note that rearrangement of Eq. 3.32a, utilizing Eq. 3.24, gives

$$\Phi_H(\zeta) = \alpha_1 k^{4/3} (-\zeta)^{-1/3}. \qquad (3.32b)$$

However, this ζ dependence is not confirmed unambiguously in the plot, although the data do suggest a value of α_1 close to 0.7. In summary the observations probably indicate that free convection has not been achieved at $\zeta \approx -1$ to -2 (see also Kader and Yaglom, 1990).

Experiments

The analytic forms of the Φ functions have been extensively studied in the past using surface-layer observations from many experiments. Most observations have been obtained over bare soil or grassy surfaces, and many semi-empirical relations exist, including interpolation formulae for different ranges of ζ (e.g. Yaglom, 1977). Determinations of Φ_H have usually been based on θ profile observations, and the reader should note that assigning this Φ function to both θ and θ_v profiles formally requires that $\Phi_H = \Phi_W$. Overall, and for moderate ranges of ζ (positive and negative), observations suggest that

$$\Phi_M(\zeta) = (1 - \gamma_1 \zeta)^{-1/4} \qquad (3.33a)$$

$$\Phi_H(\zeta) = \Phi_W(\zeta) = (1 - \gamma_2 \zeta)^{-1/2} \qquad (3.33b)$$

for $-5 < \zeta < 0$, and (Eq. 3.29)

$$\Phi_M = \Phi_H = \Phi_W = 1 + \beta_1 \zeta$$

for $0 < \zeta < 1$, with $\gamma_1 \approx \gamma_2 \approx 16$ and $\beta_1 \approx 5$ (see Table A5 in Appendix 4 for a summary of experimental values). With the above forms, $\Phi_M = \Phi_H = \Phi_W = 1$ when $\zeta = 0$, so that $K_H/K_M = 1$ at $\zeta = 0$ giving the turbulent Prandtl number $P_t = 1$.

For unstable conditions ($\zeta < 0$), Eq. 3.33 implies $\Phi_H = \Phi_W = \Phi_M{}^2$ so that $f_1(\zeta) = \zeta$ in Eq. 3.26, and $f_2(\zeta) = (1 - \gamma_1 \zeta)^{1/4}$ in Eq. 3.27, i.e. K_H increases relative to K_M as ζ becomes more negative, due to the increasing dominance of buoyancy on the vertical heat transfer. Note should be made that Eq. 3.33b for $\zeta \ll -1$ tends to a $\zeta^{-1/2}$ form in contrast to the $\zeta^{-1/3}$ behaviour expressed in Eq. 3.32b.

For stable conditions ($\zeta > 0$), combination of Eqs. 3.26 and 3.29 yields $Ri = \zeta(1 + \beta_1 \zeta)^{-1}$ so that in the limit $\zeta \to \infty$, $Ri = \beta_1{}^{-1} \approx 0.2$. This limit occurs as both the heat flux and the surface stress approach zero, i.e. as the turbulence diminishes and the flow becomes laminar. The limiting value of Ri can thus be interpreted as the critical Richardson number (Section 2.6.2).

3.3.2 Integral forms of the flux-gradient relations

Profile relations

Equation 3.17 (the logarithmic wind law) represents an integral profile relation for the neutral case. For the general, non-neutral case, the integration of Eq. 3.23 yields

$$ku/u_{*0} = \int (\zeta')^{-1}\Phi_M(\zeta')\mathrm{d}\zeta'$$

$$= \ln(z/z_0) - \int [1 - \Phi_M(\zeta')]\mathrm{d}(\ln \zeta').$$

$$\approx \ln(z/z_0) - \Psi_M(\zeta). \tag{3.34}$$

This defines the function $\Psi_M(\zeta)$, and with the wind speed written in this form the effects of buoyancy can be interpreted as a deviation of the wind speed from the neutral value. The approximation in Eq. 3.34 involves replacing the lower limit of integration $\zeta_0(= z_0/L)$ with zero, and assumes that $z/z_0 \gg 1$. Since $\Psi_M(0)$ is zero, Eq. 3.34 reverts to the logarithmic wind law in neutral conditions. In unstable conditions, $0 < \Phi < 1$ and $\Psi > 0$, whilst in the stable case, $\Phi > 1$ and $\Psi < 0$. This influence of buoyancy on the wind profile is demonstrated in Fig. 3.5 using observed wind profiles above a low roughness surface. With $\Phi_M(\zeta)$ given by Eq. 3.33a for $\zeta < 0$, $\Psi_M(\zeta)$ can be evaluated by letting $x = (1 - \gamma_1\zeta)^{1/4} = \Phi_M(\zeta)^{-1}$. Simple integration then yields

$$\Psi_M(\zeta) = 2\ln[(1 + x)/2] + \ln[(1 + x^2)/2] - 2\tan^{-1}x + \pi/2. \tag{3.35}$$

With $\Phi_M(\zeta)$ given by Eq. 3.29 for $\zeta > 0$, substitution into Eq. 3.34 gives

$$\Psi_M(\zeta) = -\beta_1\zeta \tag{3.36}$$

so that

$$ku/u_{*0} = \ln(z/z_0) + \beta_1\zeta. \tag{3.37}$$

This is the well-known log-linear wind law (Webb, 1970; Monin and Yaglom, 1971, Chapter 4).

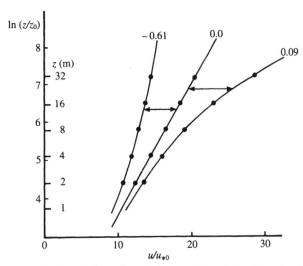

Fig. 3.5 Three wind profiles from the Kansas field data (Izumi, 1971) plotted in normalized form at three values of the gradient Ri ($z = 5.66$ m). Both normalized and absolute heights are shown, whilst the magnitude of the horizontal arrows indicates the effect of buoyancy on the wind relative to the neutral profile (see Eq. 3.34).

Integration of Eq. 3.24 for the θ profile gives, in analogy to Eq. 3.34 for wind speed,

$$k(\theta_v - \theta_0)/\theta_{v*0} = \ln(z/z_T) - \int[1 - \Phi_H(\zeta')]d(\ln\zeta')$$

$$\approx \ln(z/z_T) - \Psi_H(\zeta) \qquad (3.38)$$

where θ_0 is the surface potential temperature (T_0 is the absolute surface temperature) and z_T is a surface scaling length for temperature. Formally $T = T_0$ at $z = z_T$ since there is no *a priori* reason for taking $T_0 = T(z_0)$. Over many natural land surfaces, z_0 is almost an order of magnitude greater than z_T and these two temperatures may differ by several degrees kelvin. This statement is based on the measurement of T_0 radiatively, though the actual (unknown) surface temperature and the radiative (measured) surface temperature need not be the same. Thus, for rough surfaces such as open canopies there is a wide spread of T_0 values on different parts of the surface, though even here a spatially integrated T_0 value can be made to match the temperature at z_T.

For $\zeta < 0$, $\Psi_H(\zeta)$ can be evaluated using $y = (1 - \gamma_2\zeta)^{1/2} = \Phi_H(\zeta)^{-1}$ giving

$$\Psi_H(\zeta) = 2\ln[(1 + y)/2] \qquad (3.39)$$

whilst with $\zeta > 0$

$$\Psi_H(\zeta) = -\beta_1\zeta. \qquad (3.40)$$

As with θ_v, the q gradient given by Eq. 3.25 can be integrated to yield

$$k(q - q_0)/q_{*0} = \ln(z/z_q) - \Psi_W(\zeta) \qquad (3.41)$$

where q_0 is the surface specific humidity (defined as the humidity at $z = z_q$). Observations and theory suggest $\Phi_W = \Phi_H$, $\Psi_W = \Psi_H$, $z_q \approx z_T$, the latter being valid only if sources and sinks of heat and water vapour are coincident or nearly so.

Bulk transfer relations

For many practical purposes, drag and bulk transfer coefficients are required when relating fluxes to mean properties of the flow. A drag coefficient C_D can be defined as

$$C_D = (u_{*0}/u)^2 = k^2/[\ln(z/z_0) - \Psi_M(\zeta)]^2 \qquad (3.42)$$

using Eq. 3.34. The neutral drag coefficient is given from Eq. 3.42, with $\zeta = 0$:

$$C_{DN} = k^2/[\ln(z/z_0)]^2 \qquad (3.43)$$

so that combining Eqs. 3.42 and 3.43 gives the ratio

$$C_D/C_{DN} = [1 - \Psi_M(\zeta)/\ln(z/z_0)]^{-2}. \qquad (3.44)$$

The stability parameter appearing in Eq. 3.44 is formally ζ, but this may be replaced by Ri, or even a bulk Richardson number Ri_B defined as

$$Ri_B = (g/\theta_v)z(\theta_v - \theta_0)/(u^2 + v^2) = f_3(\zeta, z/z_0, z/z_T). \qquad (3.45)$$

Now, whereas Ri and ζ are uniquely related, at least according to Monin–Obukhov similarity theory (Eq. 3.26), Ri_B and ζ are not. According to Eq. 3.45, Ri_B is a function also of the nondimensional variables z/z_0 and z_0/z_T. Sample curves of $Ri_B(\zeta)$, for two values of z/z_0 and z_0/z_T, are shown in Fig. 3.6, where f_3 is determined from a combination of Eqs. 3.34 and 3.38 using $\Psi(\zeta)$ functions described by Eqs. 3.35, 3.36, 3.39 and 3.40. In Figure 3.7(*a*), curves of C_D/C_{DN} at $z = 10$ m are given, as functions of ζ, for two values of z/z_0, to illustrate the strong effect of stability.

By analogy with the drag coefficient, a heat transfer coefficient C_H can be defined as

$$H_{v0}/\rho c_p = \overline{(w'\theta_v')}_0 = C_H u(\theta_0 - \theta_v) \qquad (3.46a)$$

$$H_0/\rho c_p = \overline{(w'\theta')}_0 = C_H u(\theta_0 - \theta) \qquad (3.46b)$$

where we have distinguished between the surface buoyancy flux H_{v0} and the sensible heat flux H_0, and have assumed that both are associated with the same C_H. The latter is given by combining Eqs. 3.34 and 3.38 to obtain

$$C_H = k^2/[\ln(z/z_0) - \Psi_M(\zeta)][\ln(z/z_T) - \Psi_H(\zeta)] \qquad (3.47)$$

and noting that if Eq. 3.46b is used to determine H_0 in moist air, ζ (i.e. L) in Eq. 3.47 is still defined in terms of H_{v0}. The neutral value, C_{HN}, is given by

$$C_{HN} = k^2/[\ln(z/z_0)\ln(z/z_T)] \qquad (3.48)$$

so that the ratio C_H/C_{HN} can be expressed as

$$C_H/C_{HN} = [1 - \Psi_M(\zeta)/\ln(z/z_0)]^{-1}[1 - \Psi_H(\zeta)/\ln(z/z_T)]^{-1} \qquad (3.49)$$

Curves of C_H/C_{HN} at $z = 10$ m are given, as functions of ζ, for two values of z/z_0 and z_0/z_T, in Fig. 3.7(*b*). Note that combining Eqs. 3.43 and 3.48 gives

$$C_{DN}/C_{HN} = \ln(z/z_T)/\ln(z/z_0). \qquad (3.50)$$

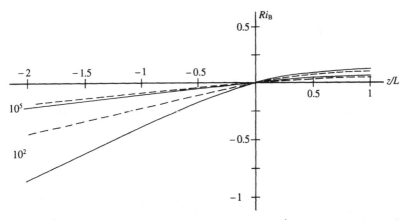

Fig. 3.6 The bulk Richardson number as a function of z/L, for two values of z_0/z_T. Pecked curves, $z_0/z_T = 1$; solid curves, $z_0/z_T = 7.4$. For the upper curves, $z/z_0 = 10^5$; for the lower curves, $z/z_0 = 10^2$.

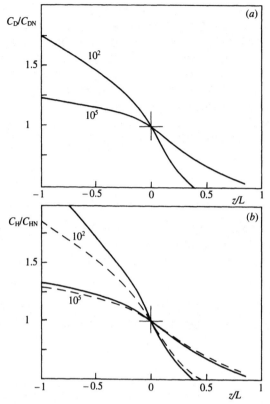

Fig. 3.7 Values of (*a*) C_D/C_{DN} and (*b*) C_H/C_{HN} as functions of z/L for two values of z/z_0 as indicated. In (*b*), the solid curves have $z_0 = z_T$, and the pecked curves have $z_0/z_T = 7.4$ (see Chapter 4).

For a reference height of 10 m, values of this ratio are given in Table 3.1 for $z_0/z_T = 7.4$ and a range of values of z_0. Values greater than unity reflect the more efficient transfer of momentum relative to heat at rough surfaces, to be discussed further in Chapter 4. The bulk transfer coefficient for water vapour is defined by

$$E_0/\rho = \overline{(w'q')}_0 = C_E u(q_0 - q) \tag{3.51}$$

where C_E can be expressed in the same way as C_H, through equations analogous to 3.47 and 3.48.

In Eq. 3.46, the surface quantity θ_0 is related to the absolute temperature T_0 through Eq. 2.26; when the surface pressure differs from p_R, differences between these "surface temperatures" will exist.

3.3.3 Aerodynamic resistances

The drag, heat and mass transfer coefficients discussed in Section 3.3.2 above take into account both turbulent transfer and molecular transfer of a property s between the surface (property concentration s_0) and reference height z in the

Table 3.1. *Values of C_{DN}/C_{HN} based on Eq. 3.50, for five values of z_0, with $z_0/z_T = 7.4$ and a reference height of 10 m*

z_0 (m)	z/z_0	C_{DN}/C_{HN}
10^{-4}	10^5	1.17
10^{-3}	10^4	1.22
10^{-2}	10^3	1.29
0.1	10^2	1.43
1	10	1.87

surface layer (concentration s). For some applications, e.g. in the fields of micrometeorology and plant physiology, and over vegetated surfaces in particular, it is often more convenient to replace the transfer coefficients by quasi-resistance parameters. In this approach, the linking together of molecular transfer in the vicinity of the surface (within the so-called interfacial sublayer) and turbulent transfer in the surface layer is simplified. This partly relates to the additive property of resistances in series.

By analogy with Ohm's law in electricity (resistance = potential difference ÷ current), we define an aerodynamic resistance r_a in terms of a property flux F_s and concentration difference $\psi_s - \psi_a$, so that

$$r_a = (\psi_s - \psi_a)/F_s, \qquad (3.52)$$

with r_a having dimensions of $s\,m^{-1}$. In the context of mass transfer, the reciprocal, r_a^{-1}, is often referred to as a deposition velocity or a conductance. From the definition of C_D (Eq. 3.42), we define the *bulk aerodynamic resistance* to the transfer of momentum from a level z to the surface, i.e. to $z = z_0$, as

$$r_{aM} = \rho u(z)/\tau_0 = u(z)/u_{*0}^2 = (C_D u)^{-1}, \qquad (3.53)$$

and so as C_D increases (higher z_0 or more unstable conditions) or u increases, the resistance decreases. For a reference height of 10 m, and with $u = 5\,m\,s^{-1}$, it is readily deduced (Eq. 3.43) that r_{aM} typically lies in the range 10–100 $s\,m^{-1}$ for land surfaces. The reader should note that the quantity $C_D u$ is comparable to conductance, and that the concept of conductance rather than resistance is equally valid in the following discussion. The essential simplicity of Eq. 3.53 arises from the basic zero-slip boundary condition requiring $u = 0$ at $z = z_0$.

No such simplicity exists in the case of heat and water vapour transfer where quantities θ_{v0} and q_0, which define the surface boundary condition for θ_v and q, are not easily estimated. With these surface values, expressions for the bulk aerodynamic resistance to sensible and latent heat exchange can be defined as

$$r_{aH} = \rho c_p(\theta_0 - \theta_v)/H_{v0} = (\theta_v - \theta_0)/u_{*0}\theta_{v*0} \qquad (3.54)$$

$$r_{aV} = \rho(q_0 - q)/E_0 = (q - q_0)/u_{*0}q_{*0} \qquad (3.55)$$

Comparison with Eqs. 3.46 and 3.51 gives

$$r_{aH} = (C_H u)^{-1} \tag{3.56}$$

$$r_{aV} = (C_E u)^{-1} \tag{3.57}$$

so that for land surfaces in particular under near-neutral conditions, where C_H, $C_E < C_D$, it follows that $r_{aH} \approx r_{aV} > r_{aM}$. For canopy surfaces, it is useful to define an excess resistance to heat and water vapour transfer, r_b, such that

$$r_{bH} = r_{aH} - r_{aM} \tag{3.58}$$

$$r_{bV} = r_{aV} - r_{aM}. \tag{3.59}$$

Thus, in neutral conditions, $r_{bH} \approx r_{bV} > 0$ is a measure of the additional resistance to the transfers of heat and water vapour arising from molecular diffusion in the vicinity of each leaf comprising the canopy. Under non-neutral conditions, C_H and C_E will differ from C_D (and, thus, so will the respective resistances) for reasons having nothing to do with the canopy. Use of Eqs. 3.43 and 3.48 combined with Eqs. 3.53 and 3.56–3.59 gives

$$r_{bH} = (ku_{*0})^{-1} \ln (z_0/z_T) \tag{3.60}$$

$$r_{bV} = (ku_{*0})^{-1} \ln (z_0/z_q) \tag{3.61}$$

showing that the excess resistances *in neutral conditions only* are exclusively the result of differences between the aerodynamic roughness length and the analogous scaling lengths for heat or mass transfer (excess, that is, for surfaces with $z_0 > z_T$, $z_0 > z_q$). The aerodynamic resistances, and their relation to surface concentration values, are schematically represented in Fig. 3.8.

Alternative forms of the bulk transfer relations, Eqs. 3.46 and 3.51, can be written:

$$H_{v0}/\rho c_p = (\overline{w'\theta_v'})_0 = (\theta_0 - \theta_v)/r_{aH} \tag{3.62a}$$

$$H_0/\rho c_p = (\overline{w'\theta'})_0 = (\theta_0 - \theta)/r_{aH} \tag{3.62b}$$

$$E_0/\rho = (\overline{w'q'})_0 = (q_0 - q)/r_{aV}. \tag{3.63}$$

The use of resistances or transfer coefficients is based on the validity of K-theory, and ceases to be useful when this fails, e.g. near and within the roughness elements.

3.3.4 The roughness sublayer or transition layer

Even when allowance is made for the zero-plane displacement (i.e. neutral velocity profiles are constrained to be logarithmic, at sufficiently large values of z/z_0, by an optimum choice of d, where $z = Z - d$), deviations from the log law close to the surface are observed in wind-tunnel flow and also in the atmosphere. Wind-tunnel or pipe data for aerodynamically smooth flow (Hinze, 1975, Chapter 7 and Schlichting, 1979, Chapter 20) show that the log law tends to fail when zu_{*0}/ν is less than about 20 (see Fig. 4.3), whilst for flow over rough surfaces (e.g. Plate, 1971, Chapter 1 and Raupach *et al.*, 1980) deviations from the log law are also observed when z/z_0 becomes small. Some recent experimental data are illustrated in Fig. 3.9.

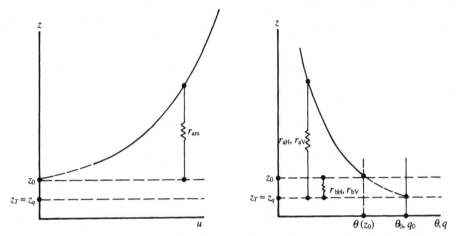

Fig. 3.8 Schematic representation of aerodynamic resistances to the transfer of momentum and to the transfer of scalar properties, showing the excess resistance r_b due to molecular effects and the relation between the surface temperature θ_0 and the temperature $\theta(z_0)$.

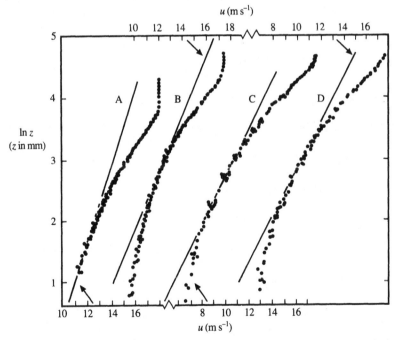

Fig. 3.9 Sample velocity profiles over smooth and rough surfaces in a wind tunnel showing a logarithmic region (where data coincide with the lines of slope u_{*0}/k) and deviations from this at lower levels. Physical and aerodynamic characteristics of the experimental surfaces are as follows. A, smooth; B, rough, with $z_0 = 0.054$ mm; C, rough, with $z_0 = 0.346$ mm; D, rough, with $z_0 = 0.395$ mm. From Raupach *et al.* (1980): reprinted by permission of Kluwer Academic Publishers.

In the atmosphere, such deviations from the logarithmic wind and temperature law have been inferred from measurements of the Monin–Obukhov stability functions (Eqs. 3.29 and 3.33) in non-neutral conditions. The established functional forms shown in Table A5 (which are consistent with the log law) have been determined mainly from observations over low-roughness surfaces (short grass, bare soil, water), where minimum values of z/z_0 in most experiments lie in the range 100–1000. Deviations from these stability functions (e.g. Garratt, 1980, 1983) are apparent in measurements made over large-roughness surfaces (e.g. tall crops and forests) where, typically, $z_0 \approx 1$ m and minimum values of z/z_0 are between 5 and 10. The breakdown of the inertial sublayer profile forms at low z/z_0 occurs even when allowance is made for the effects of the zero-plane displacement. It is consistent with the constraints of the asymptotic analysis described in Section 3.2.1.

Atmospheric observations (wind and temperature profiles) tend to support the following modifications to Eqs. 3.23 and 3.24:

$$(kz/u_{*0})\partial u/\partial z = \Phi(\zeta, z/z_*) \approx \Phi_M(\zeta)\phi(z/z_*) \tag{3.64}$$

$$(kz/\theta_{*0})\partial\theta/\partial z = \Phi(\zeta, z/z_*) \approx \Phi_H(\zeta)\phi(z/z_*) \tag{3.65}$$

where z_* is the depth of the roughness sublayer within which inertial sublayer laws break down. The Φ functions are given by Eqs. 3.29 and 3.33, whilst the ϕ function seems to be independent of stability, as is to be expected close to the surface, and to have the same form for both momentum and heat transfer. Observations over tall crops and trees, in both unstable and stable conditions, can be represented by the purely empirical relation

$$\phi(z/z_*) = \exp[-0.7(1 - z/z_*)] \qquad z < z_* \tag{3.66}$$

whose main failing is the implied discontinuity in $\partial\phi/\partial z$ at $z = z_*$. The observations also suggest that the quantity z_*/z_0 is a function of roughness density. Values of z_*/z_0 tend to vary between 10 and about 150, so that taking $z_*/z_0 = 100$ implies that for short grass ($z_0 \approx 0.005$ m), $z_* \approx 0.5$ m and for an open forest ($z_0 = 0.5$ m), $z_* \approx 50$ m.

In neutral conditions, Eq. 3.64 can be written

$$(kz/u_{*0})\partial u/\partial z = \phi(z/z_*) \tag{3.67}$$

where we expect ϕ to be less than unity because of enhanced turbulent mixing within the roughness sublayer (note that the neutral eddy diffusivity is given by $K_M = ku_{*0}z/\phi(z/z_*)$). Equation 3.67 can be integrated to give the neutral wind relation, in analogy with Eq. 3.34,

$$ku/u_{*0} = \ln(z/z_0) + \Psi_M^*(z/z_*) \qquad z < z_* \tag{3.68}$$

where Ψ_M^* is positive, and gives the excess wind or deviation from the extrapolated log profile below $z = z_*$; for $z \geqslant z_*$, $\Psi_M^* = 0$.

3.4 Generalized ABL similarity theory

In Section 3.3, we looked at the extension of Rossby-number similarity theory to the non-neutral surface layer, where the effects of buoyancy were represented through the nondimensional stability parameter z/L. We now consider its extension to the whole ABL.

3.4.1 Extension to non-neutral conditions

The velocity defect law (Eqs. 3.10, 3.11) and geostrophic drag relations (Eqs. 3.19, 3.20) are of value in providing a theoretical framework for boundary-layer analysis, but are not necessarily suitable for application to the real world. In particular, some consideration needs to be given to the effects of buoyancy and baroclinity (i.e. the geostrophic wind shear) in particular, and to the appropriateness of the geostrophic wind components as wind scales. This is particularly important when the drag or heat transfer laws are to be utilized for evaluating surface fluxes from external or large-scale parameters.

As with the surface-layer analysis, it seems appropriate to represent the effect of buoyancy (i.e. due to a non-zero surface heat flux) by L, the Obukhov stability length. The height scale is designated as h (as yet undefined), so that the scaling problem contains an additional stability parameter $\mu = h/L$ and a scale-height ratio (Arya, 1977) $R = |f|h/u_{*0}$. In the neutral limit ($\mu \to 0$), R equals a constant. For flow well away from the surface, the velocity defects are written (e.g. Kazanski and Monin, 1961; Yamada, 1976)

$$(u - u_{g0})/u_{*0} = F_x(z/h, \mu, R) \tag{3.69}$$

$$(v - v_{g0})/u_{*0} = F_y(z/h, \mu, R) \tag{3.70}$$

where the possible dependence upon R is included. For $\mu = 0$, these two equations reduce to Eqs. 3.10 and 3.11.

For flow in the surface layer, u and v satisfy Eqs. 3.14 and 3.15. It is commonly assumed that there exists, as in the neutral case, an overlap layer where Eqs. 3.14 and 3.69, and Eqs. 3.15 and 3.70, are simultaneously satisfied. An analogous matching procedure to that described in Section 3.2.1 yields the result (Hess, 1973; Sorbjan, 1989, Chapter 4)

$$ku_{g0}/u_{*0} = \ln(h/z_0) - A_1(\mu, R) \tag{3.71}$$

$$kv_{g0}/u_{*0} = -B_1(\mu, R)\,\text{sgn}\,f \tag{3.72}$$

where A_1 and B_1 are unknown functions to be determined from experiment.

The main problem with the scaling in Eqs. 3.69–3.72 is that both observations and numerical simulations suggest that the rotational height scale $u_{*0}/|f|$ does not seem to be appropriate in non-neutral conditions (Deardorff, 1972). A possible similarity relation for h is

$$h = (u_{*0}/|f|)G(u_{*0}/|f|L) \tag{3.73}$$

but this seems to be confined to the stable boundary-layer depth in steady state conditions over land (Zilitinkevich, 1972). In unstable conditions, h is affected by factors external to the boundary layer and, if interpreted as the actual ABL depth, is close to the subsidence inversion depth (z_i) during the quasi-steady state conditions usually existing on a sunny afternoon overland, but is somewhat less than z_i in morning hours. The approach is to identify h as the ABL depth, and to make it an independent variable to be determined separately through Eq. 3.73 or otherwise (Chapter 6).

3.4.2 Influence of baroclinity and consideration of wind scales

Baroclinity

The atmosphere is rarely barotropic (in fact, it is rarely steady state or horizontally homogeneous), so it is necessary to consider the likely influence of baroclinity on the universal ABL profiles (Eqs. 3.69, 3.70) and on the drag laws (Eqs. 3.71, 3.72). Intuitively, one would expect the presence of geostrophic wind shear ($\partial u_g/\partial z$, $\partial v_g/\partial z$) to affect the actual boundary-layer wind shear directly, and for this effect to be dependent upon the intensity of turbulent mixing.

In the baroclinic ABL, the modified velocity-defect and drag laws are expected to contain at least one nondimensional baroclinity parameter. One such parameter utilizes the actual ABL depth, and is written in component form (Arya & Wyngaard, 1975)

$$(M_x, M_y) = (h/u_{*0})(\partial u_g/\partial z, \partial v_g/\partial z). \tag{3.74}$$

Thus, the similarity functions F_x, F_y, A_1 and B_1 will depend additionally upon M_x and M_y.

Wind scales

When u_g and v_g vary with height within the ABL, it is no longer clear that the surface geostrophic wind (components u_{g0}, v_{g0}) is the most appropriate wind scale. This follows from the scaling procedures used on the equations of motion for the barotropic case (Section 3.2.1). An alternative scale is the height-averaged geostrophic wind within the ABL, with components ($\hat{u}_g$, $\hat{v}_g$). This scale should be nearly independent of baroclinity (Arya & Wyngaard, 1975), at least in the highly unstable limit, and be more appropriate when the similarity laws represent large-scale areal averages (Arya, 1977). Even then, any one of these geostrophic wind scales, used in suitably modified drag laws, has one major drawback: geostrophic winds cannot often be measured reliably. Therefore, additional wind scales, including the wind components at the top of the ABL (u_h, v_h) and the ABL vertically averaged wind ($\hat{u}$, $\hat{v}$), need to be considered (Yamada, 1976). In fact, we can show that the vertically averaged real and geostrophic winds are related to each other directly.

We take the simplified mean momentum equations (2.52 and 2.53), and integrate vertically between the surface and h. For zero turbulent fluxes at $z = h$ (and using surface-layer coordinates) the result is

$$\hat{u} = \hat{u}_g \tag{3.75}$$

$$\hat{v} - u_{*0}^2/fh = \hat{v}_g \tag{3.76}$$

where, for any variable s, we define

$$\hat{s} = h^{-1} \int_0^h s\,dz. \tag{3.77}$$

From a practical viewpoint, measurements of $\hat{u}$ and $\hat{v}$ are much more reliable than measurements of the geostrophic components. In the barotropic case, for example, Eqs. 3.75 and 3.76 become $\hat{u} = u_{g0}$ and $\hat{v} - u_{*0}^2/fh = v_{g0}$, so that $\hat{u}$ and $\hat{v}$ can be directly incorporated into Eqs. 3.71 and 3.72. In the baroclinic case, we expect that

$$\hat{u}_g = u_{g0} + u_{*0}f_u(M_x) \tag{3.78}$$

$$\hat{v}_g = v_{g0} + u_{*0}f_v(M_y), \tag{3.79}$$

where f_u and f_v are functions of the baroclinity parameters. For example, for a linear geostrophic shear, $f_u(M_x) = M_x/2$ and $f_v(M_y) = M_y/2$. Based on Eqs. 3.78 and 3.79, and using Eqs. 3.75 and 3.76, the baroclinic drag laws can be rewritten in terms of the real winds:

$$k\hat{u}_g/u_{*0} = k\hat{u}/u_{*0} = \ln(h/z_0) - A_2(\mu, M_x, R) \tag{3.80}$$

$$k\hat{v}_g/u_{*0} = k\hat{v}/u_{*0} - ku_{*0}/fh = -B_2(\mu, M_y, R)\,\text{sgn}\,f. \tag{3.81}$$

Likewise, the velocity-defect profiles given in Eqs. 3.69 and 3.70 for the ideal, barotropic case can be generalized to

$$k(u - \hat{u})/u_{*0} = F_x(z/h, \mu, M_x, R) \tag{3.82}$$

$$k(v - \hat{v})/u_{*0} + ku_{*0}/fh = F_y(z/h, \mu, M_y, R). \tag{3.83}$$

The functional dependences shown in Eqs. 3.80–3.83 appear quite complex, and some simplification seems advisable if the relations are to be useful in practice.

Observations

Approximate values of the functions A_2 and B_2 can be deduced for the special case of a well-mixed unstable ABL, of arbitrary baroclinity. The idealized profiles are shown in Fig. 3.10 with the x-axis along the surface stress, so that $\hat{u} \approx u_s$ and $\hat{v} \approx 0$ (see Chapter 6 for further details of the convective boundary layer). Here u_s is the wind at $z = h_s$, the top of the surface layer, and is given by Eq. 3.34, so that

$$k\hat{u}/u_{*0} \approx \ln(h_s/z_0) - \Psi_M(h_s/L)$$

$$\approx \ln(h/z_0) - \Psi_M(\mu/10) - 2.3 \tag{3.84}$$

$$k\hat{v}/u_{*0} = 0. \tag{3.85}$$

For our present purposes, it is sufficient to take h_s, the depth of the surface layer, equal to $0.1h$. Comparison with Eqs. 3.80 and 3.81 suggests that

$$A_2 = \Psi_M(\mu/10) + 2.3 \tag{3.86}$$

$$B_2 \approx ku_{*0}/|f|h \tag{3.87}$$

for the well-mixed convective boundary layer. That is, A_2 is a function of stability and B_2 is a function of the scale–height ratio. In general, such specific dependencies are not readily deduced and experimental determinations must be made.

Many observations have been made in order to evaluate the functions A and B for a range of stability and baroclinic conditions (see e.g. Yamada, 1976). Using vertically averaged winds, or ABL-top winds, baroclinic effects are found

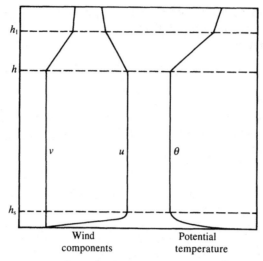

Fig. 3.10 Schematic representation of wind and temperature profiles in a three-layer model of the CBL. The top of the surface layer is at height h_s, the top of the mixed layer is at h, and the top of the interface is at h_1.

to be small, with A_2 and B_2 mostly affected by stability (μ). The functions $A_2(\mu)$ and $B_2(\mu)$ can be represented empirically as follows, based on relations appearing in the literature (e.g. see Arya, 1977; Brutsaert, 1982, Chapter 4):

unstable–neutral ($\mu \leqslant 0$),

$$A_2 = 5 - 4(1 - 0.0084\mu)^{-1/3} \tag{3.88a}$$

$$B_2 = 4.5(1 - 3.3\mu)^{-1/3}; \tag{3.88b}$$

slightly stable ($0 < \mu \leqslant 35$),

$$A_2 = 1 - 0.38\mu \tag{3.89a}$$

$$B_2 = 4.5 + 0.3\mu; \tag{3.89b}$$

strongly stable ($\mu > 35$),

$$A_2 = -3.17(\mu - 20)^{1/2} \tag{3.90a}$$

$$B_2 = 3.17(\mu - 12.5)^{1/2}. \tag{3.90b}$$

Note that, for $\mu = 0$, $B_2 = B_0 = 4.5$, and $A_0 - A_2 = 1$. Based on Eqs. 3.20 and 3.80, this difference should equal $\ln(1/c)$, $= 1.6$ for $c = 0.2$, where $h = cu_{*0}/|f|$ (see Appendix 3, Eq. A23). For $\mu < 0$, Eqs. 3.87 and 3.88b give quite different dependences, but show fair agreement in absolute values of B_2, using typical values of R, of between 0.15 and 0.3. Values of A_2 are overestimated by Eq. 3.86, although, with Ψ_M given by Eq. 3.35 and thus restricted to $h_s/L > -5$, application of this equation for $\mu < -50$ is not recommended.

The lack of any significant baroclinic dependence overall, or dependence of A_2 on the scale–height ratio, at least in unstable conditions, is revealed in Fig. 3.11. This shows several sets of observations made under baroclinic, unstable

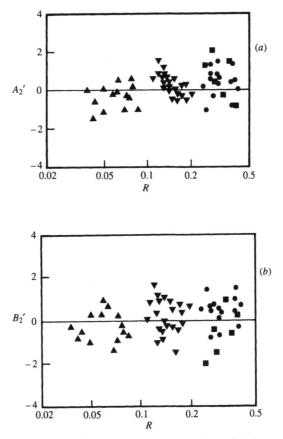

Fig. 3.11 (a) The residual normalized u-component A_2' and (b) the residual normalized v-component B_2', each as a function of the scale–height ratio R, for three data sets where each datum is the average of five individual runs. Squares, AMTEX 1975; circles, WANGARA (afternoon); triangles with apex at top, Koorin (morning); triangles with apex at bottom, Koorin (afternoon).

The residual quantities are given by the difference between the predicted values of ku/u_{*0} and kv/u_{*0} and their observed values, with the predictions based on Eqs. 3.80, 3.81 and 3.86, 3.87, i.e. $A_2' = \ln(h/z_0) - \Psi_M(\mu/10) - 2.3 - (ku/u``3_0)^{obs}$, $B_2' = -(kv/u_{*0})^{obs}$. After Garratt *et al.* (1982), *Journal of Atmospheric Sciences*, American Meteorological Society.

conditions where local accelerations, advection and entrainment were present. The observations are scaled according to Eqs. 3.80 and 3.81, with A_2 and B_2 given by Eqs. 3.86 and 3.87. The ordinates represent the residual dependences (call these A_2' and B_2') upon factors other than μ (for A) and R (for B). Here A_2' is the difference between the predicted (Eq. 3.84) and observed values of $k\hat{u}/u_{*0}$, and B_2' is the difference between the predicted (Eq. 3.85) and observed values of $k\hat{v}/u_{*0}$. The data reveal no significant dependence of A_2 on the scale–height ratio, R, and no systematic effects of baroclinity. The observed data scatter is in fact equivalent to standard deviations in A and B of $\sigma_A \approx 1.1$ and $\sigma_B \approx 1.4$.

Equations 3.80 and 3.81 can be combined into a drag law analogous to Eq. 3.21 for the neutral case, and the above A_2 and B_2 functions (Eqs. 3.88 and 3.90) can be used to generate $C_g(h/z_0, \mu)$ and $\alpha(h/z_0, \mu)$. Sample curves are shown in Fig. 3.12, where they illustrate the strong variation in cross-isobar flow between highly convective (angles less than 5°) and highly stable conditions. In fact, in the limit $\mu \to \infty$, $\alpha_0 \to 45°$ independently of h/z_0 or Ro. This follows from Eqs. 3.22 and 3.80, since $\tan \alpha_0 \approx |B_2/A_2|$ with $A_2 \gg \ln(h/z_0)$, and $A_2 = B_2 = 3.17\mu$ in the limit of very large μ (Eq. 3.90).

Finally, we come to the velocity defect profiles (Eqs. 3.82, 3.83) and the functions F_x and F_y. There profile forms are valid only in the outer regions of the ABL, since within the inner layer and overlap region, u, for example, is given by Eqs. 3.17 and 3.18, with $v = 0$. Qualitative features of these functions are illustrated in Figs. 3.13 and 3.14 for three stability regimes, using the Wangara data (Yamada, 1976). The data show little z/h dependence in F_x and F_y for strongly unstable conditions when vertical mixing is vigorous, but do reveal strong z/h dependence in stable conditions when mixing is weak. The effects of stability are significant, but it should be noted that these data do not have a wide range of variation in M or R.

3.4.3 Heat and moisture transfer relations

Having derived general similarity relations for the ABL wind profiles and associated drag laws (Eqs. 3.21, 3.80 and 3.81), the next step is to describe the analogous relations for temperature and humidity profiles, and their associated transfer relations. Similiar arguments as used for wind scaling may be applied in this case to yield the following formulae (e.g. Yamada, 1976): for the profiles

$$k(\theta_v - \hat{\theta}_v)/\theta_{v*0} = F_\theta(z/h, \mu, \mathbf{M}) \tag{3.91}$$

$$k(q - \hat{q})/q_{*0} = F_q(z/h, \mu, \mathbf{M}) \tag{3.92}$$

and for the transfer laws

$$k(\hat{\theta}_v - \theta_0)/\theta_{v*0} = \ln(h/z_T) - C(\mu, \mathbf{M}, R) \tag{3.93}$$

$$k(\hat{q} - q_0)/q_{*0} = \ln(h/z_q) - D(\mu, \mathbf{M}, R). \tag{3.94}$$

Here, $\mathbf{M}$ is a baroclinity parameter whose magnitude is defined as $|\mathbf{M}|^2 = M_x{}^2 + M_y{}^2$.

As with A and B, approximate values of the function C, for example, can be deduced for the special case of a well-mixed unstable ABL, for arbitrary baroclinity. The idealized profile is sketched in Fig. 3.10, where it is apparent that $\hat{\theta}_v \approx \theta_{vs}$; here θ_{vs} is the value of θ_v at the top of the superadiabatic layer, assumed to coincide with the surface layer for present purposes. The temperature θ_{vs} is therefore given by Eq. 3.38, whence

$$k(\hat{\theta}_v - \theta_0)/\theta_{v*0} \approx \ln(h/z_T) - \Psi_H(\mu/10) - 2.3. \tag{3.95}$$

This implies that $C = \Psi_H(\mu/10) + 2.3$, with Ψ_H given by Eq. 3.39.

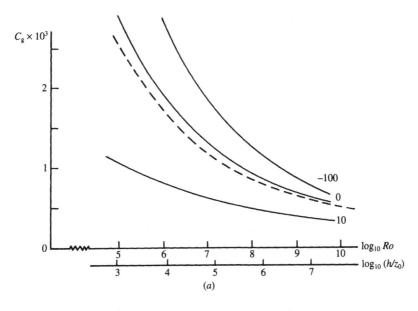

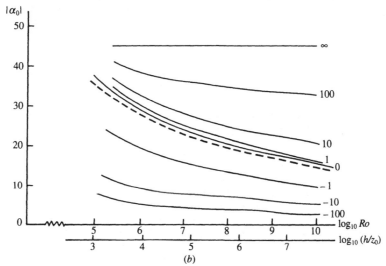

Fig. 3.12 The solid curves show the variations of (*a*) the boundary-layer drag coefficient and (*b*) the cross-isobar flow angle as functions of h/z_0 in unstable and stable conditions. Calculations are based on Eqs. 3.80 and 3.81, with A_2 and B_2 given by Eqs. 3.88 to 3.90. Values of the stability parameter μ are indicated. The pecked curves represent the neutral drag coefficient and cross-isobar flow angle as a function of Ro, based on Eqs. 3.21 and 3.22. The relation between Ro and h/z_0 assumes $h = 0.2u_{*0}/|f|$.

Most observational support for Eqs. 3.91–3.94 involves the potential temperature profile and heat transfer in unstable conditions. As with the wind components, stability effects tend to dominate the functions F_θ and C. Overall, the functional form of C in both unstable and stable conditions can be

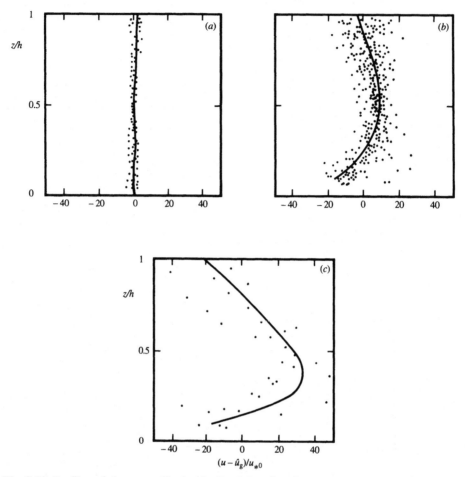

Fig. 3.13 Profiles of the normalized velocity defect for the *u*-component as a function of normalized height z/h, based on Eq. 3.82 and an analysis of Wangara observations. Three stability regimes are presented: (a) $-150 < h/L < -120$; (b) $0 < h < 30$; (c) $180 < h/L < 210$. Curves are drawn by eye. After Yamada (1976), *Journal of Atmospheric Sciences*, American Meteorological Society.

represented empirically as

$$C = 1 + 1.5 \ln (0.5 + y/2) \qquad \mu \leqslant 0 \qquad (3.96a)$$

where $y = (1 - 16\mu)^{1/2}$, and

$$C = 1 - 4 \ln (1 + 0.2\mu^2) \qquad \mu > 0. \qquad (3.96b)$$

In some evaluations appearing in the literature, z_0 has been used in Eq. 3.93 in place of z_T, so differences in experimental C values for land surfaces may be $\approx \ln (z_0/z_T) \approx 2$. Very few determinations of the function D have been made, and it is usual to take $C = D$, and $z_T = z_q$ for most practical purposes. For $-50 < \mu < 0$, Eqs. 3.95 and 3.96a give comparable values of C; more negative

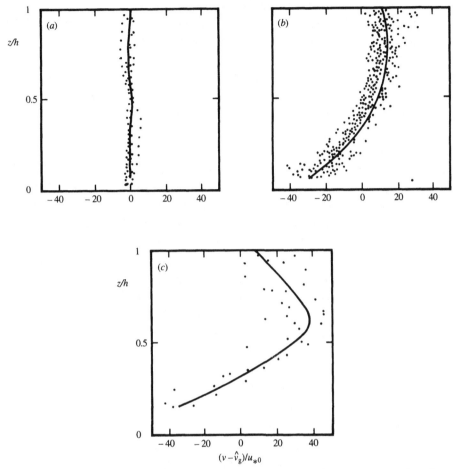

Fig. 3.14 Profiles of the normalized velocity defect for the v-component as a function of normalized height z/h, based on Eq. 3.83 and an analysis of Wangara observations. Three stability regions are presented: (a) $-150 < h/L < -120$; (b) $0 < h < 30$; (c) $180 < h/L < 210$. Curves are drawn by eye. After Yamada (1976), *Journal of Atmospheric Sciences*, American Meteorological Society.

values of μ lie outside the range for which Ψ_H is well known. The reader should note that Eqs. 3.80 and 3.81, when combined with Eqs. 3.93 or 3.94, yield bulk transfer relations for the surface heat flux $u_{*0}\theta_{v*0}$ or moisture flux $u_{*0}q_{*0}$ that are analogous to the geostrophic or boundary-layer drag law. In this case, a boundary-layer heat transfer coefficient C_{gH} is used.

Qualitative features of the profile function F_θ (Eq. 3.91) are illustrated in Fig. 3.15 for observations from the Wangara experiment revealing, as with the wind profiles, a dominance of the stability effect. In addition, strong height variations occur only when the conditions are stably stratified, and vertical mixing is weak producing a well-defined surface-based inversion. Radiative cooling is important under these conditions, as will be discussed in Chapter 6.

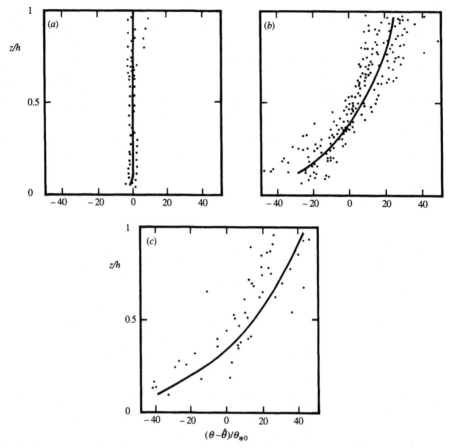

Fig. 3.15 Profiles of the normalized temperature defect, based on Eq. 3.91 and an analysis of Wangara data, for the following stability regimes: (a) $-60 < h/L < -30$; (b) $0 < h < 30$; (c) $60 < h/L < 90$. After Yamada (1976), *Journal of Atmospheric Sciences, American Meteorological Society.*

3.5 Similarity theory and turbulence statistics

Studies of variances and spectra in the ABL are important since they provide, with the relevant observations, a direct test of similarity predictions, and also an indirect means of estimating vertical fluxes. In addition, knowledge of these turbulent properties is required in problems related to atmospheric dispersion (which is very much dependent on the intensity of turbulence and on how these properties depend on stability and height within the ABL).

The standard deviations and variances are related through $\sigma_s^2 = \overline{s'^2}$, where s is any one of u, v, w, θ_v and q. For the one-dimensional spectra and cospectra, evaluations are made from time or space series in s' and w' using fast Fourier transform methods. The variances and convariances can be expressed in terms of the frequency spectra and cospectra through Eqs. 2.15 and 2.16.

In Chapter 2, we discussed briefly Kolmogorov's similarity theory of turbulence and the existence of an inertial subrange of wavenumbers where average properties depend only upon the rate of viscous dissipation ε. The consequences of this require that the one-dimensional velocity spectrum in the inertial subrange has the form (Tennekes and Lumley, 1972, Chapter 8)

$$\phi_{uu}(\kappa) = \beta_u \varepsilon^{2/3} \kappa^{-5/3} \tag{3.97}$$

$$\phi_{ww}(\kappa) = (4\beta_u/3) \varepsilon^{2/3} \kappa^{-5/3}, \tag{3.98}$$

where κ is the wavenumber. The value of the constant $\beta_u \approx 0.6$ has been determined from observations (Deacon, 1988).

Entirely similar arguments apply to the temperature and humidity spectra in their inertial subrange. For temperature, the spectral density will depend on κ, ε and χ (the rate of molecular dissipation of temperature fluctuations; see Eq. 2.68), whence (Corrsin, 1951)

$$\phi_{\theta\theta}(\kappa) = \beta_\theta \chi \varepsilon^{-1/3} \kappa^{-5/3}. \tag{3.99}$$

The constant $\beta_\theta (= \beta_q) \approx 0.8$ based on observations (Dyer & Hicks, 1982; and references contained therein).

Equations 3.97–3.99 can be used as a basis for scaling observed frequency spectra. To do this, we utilize Taylor's hypothesis to transform from wavenumber space to frequency space. Thus, with $2\pi/\kappa = u/n$, where u is the mean wind speed, it follows that $\kappa\phi(\kappa) = n\phi(n)$.

Many of the observations discussed below relate to the ABL over land surfaces. Far fewer data have been obtained over water, but these generally reveal close similarities with features over the land. Much of the information in the present section is derived from Panofsky and Dutton (1984, Chapter 7), Stull (1988, Chapter 9) and Sorbjan (1989, Chapter 4). The reader should consult these for further details.

3.5.1 Surface-layer scaling ($z \ll h$)

Variances

The nondimensional variances formed by using u_{*0}, θ_{*0} and q_{*0} are universal functions of $z/L \, (= \zeta)$ if Monin–Obukhov scaling is correct. Thus, in neutral conditions when $\zeta = 0$ the normalized variances should be constants, independent of height or surface roughness. The quantities of primary concern are σ_u/u_{*0}, σ_v/u_{*0}, σ_w/u_{*0} for velocity and $\sigma_\theta/|\theta_{*0}|$ for temperature.

The neutral values of the velocity ratios for flat terrain are typically 2.4, 1.9 and 1.25 respectively, with the horizontal components having larger ratios in rolling terrain: 3.4 for u and 2.9 for v. Such values illustrate the strong anisotropic nature of the turbulence (see also Raupach *et al.* (1990) for atmospheric and wind-tunnel values). Figure 3.16 shows the variation of σ_w/u_{*0} with ζ that is typical of many sites, with the pecked curve representing the variation in unstable conditions given by (Panofsky *et al.*, 1977)

$$\sigma_w/u_{*0} = 1.25(1 - 3\zeta)^{1/3}. \tag{3.100}$$

The free-convective asymptotic behaviour implied in Eq. 3.100 is consistent with free-convective scaling, i.e. $\sigma_w/u_{*0} \propto |\zeta|^{1/3}$. In this case, turbulent quantities become independent of u_{*0} and θ_{*0}, but scale with u_f and θ_f (Eq. 3.30). With

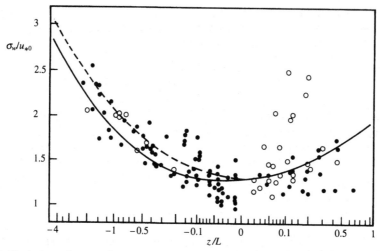

Fig. 3.16 Variation of normalized σ_w with stability parameter z/L, for surface-layer data over flat and rolling terrain; data from Panofsky and Dutton (1984, Fig. 7.1). The pecked curve represents Eq. 3.100 and the solid curve represents the expression

$$\sigma_w/u_{*0} = 1.3(\Phi_M - 2.5\zeta)^{1/3}$$

used by Merry and Panofsky (1976) to describe several sets of data. Reprinted by permission of John Wiley and Sons.

$\sigma_w \propto u_f$ and with $\sigma_\theta \propto \theta_f$, u_{*0} independence requires

$$\sigma_w/u_{*0} = a_1(-\zeta)^{1/3} \tag{3.101}$$

$$\sigma_\theta/|\theta_{*0}| = a_2(-\zeta)^{-1/3}. \tag{3.102}$$

These relations are compared with observations in Fig. 3.17, with $a_1 = 1.8$ and $a_2 = 0.95$.

For the horizontal wind components, observations reveal no dependence on height throughout the surface layer (and most of the ABL), even in highly unstable conditions, so Monin–Obukhov scaling does not apply. This seems to be related to the influence of low-frequency eddies, either through boundary-layer instabilities or convection, and the ABL depth, h, becomes the relevant scaling height. Stability is a strong influence, though, and a good representation of behaviour in unstable conditions is (Panofsky *et al.*, 1977)

$$\sigma_u/u_{*0} \approx \sigma_v/u_{*0} = (12 - 0.5h/L)^{1/3}. \tag{3.103}$$

This is outer-layer scaling where u_{*0}, h and L are relevant. In fact, in the limit of large negative h/L, Eq. 3.103 reduces to $\sigma_u = \sigma_v = 0.6w_*$, which is consistent with mixed-layer scaling.

In stable conditions, observations for all quantities are inconclusive, mainly because of the problem of making reliable measurements of u_{*0} and σ_w, which become small as stability increases. Typically, values of the normalized variances are equal to, or slightly greater than, those in neutral conditions. The lack of any significant dependence upon ζ is strongly suggested in the data of Takeuchi (1961).

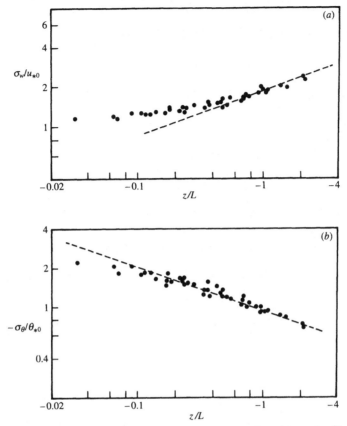

Fig. 3.17 (a) Normalized σ_w and (b) normalized σ_θ as functions of z/L in unstable conditions, using observations from the Kansas experiment. In (a) the pecked line represents Eq. 3.101 with $a_1 = 1.8$; in (b) the pecked line represents Eq. 3.102 with $a_2 = 0.95$. After Wyngaard *et al.* (1971), *Journal of Atmospheric Sciences*, American Meteorological Society.

Spectra and Cospectra

Observations of spectra and cospectra can be represented as a series of curves based on Monin–Obukhov scaling, in analogy with the variances described above. A normalized frequency, $f = nz/u$, is introduced and the spectra are described by

$$f\phi_{\text{vel}}(f)/u_{*0}^2 = F_{\text{vel}}(f, z/L) \tag{3.104}$$

$$f\phi_{\theta\theta}(f)/\theta_{*0}^2 = F_{\theta\theta}(f, z/L). \tag{3.105}$$

The velocity spectra, based on the Kansas field experiment (Kaimal *et al.*, 1972) and covering a wide range in stabilities, are shown in Fig. 3.18. In this diagram, further normalization using a factor $\phi_\varepsilon^{2/3}$ has been used to ensure collapse of the curves in the inertial subrange where $\phi_{\text{vel}} \propto \varepsilon^{2/3} n^{-5/3}$. The quantity $\phi_\varepsilon = \varepsilon kz/u_{*0}^3$ is the nondimensional rate of TKE dissipation and is found by

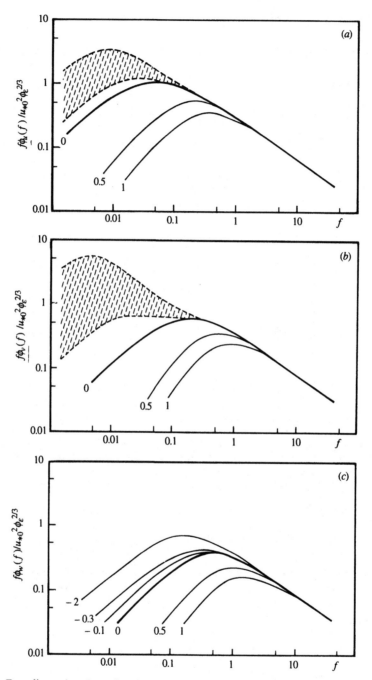

Fig. 3.18 One-dimensional, surface-layer velocity spectra in normalized form for (*a*) the *u*-component, (*b*) the *v*-component and (*c*) the *w*-component; the spectra are based on the Kansas observations and are plotted using the normalized frequency $f = nz/u$ for a range of z/L ($= \zeta$) values as indicated. For the horizontal components, the shaded regions represent the unstable regime for $-2 < \zeta < 0$. After Kaimal *et al.* (1972), *Quarterly Journal of the Royal Meteorological Society*.

setting the transport term in Eq. 2.74b to zero:

$$\phi_\varepsilon(\zeta) = \Phi_M(\zeta) - \zeta \tag{3.106}$$

The normalized spectral density in the inertial subrange then becomes

$$f\phi_{vel}(f)/(u_{*0}^2\phi_\varepsilon^{2/3}) = \beta(2\pi k)^{-2/3}f^{-2/3} \tag{3.107}$$

and thus a single curve. Using the value of β_u given above, the reader can verify that Eq. 3.107 is consistent with the normalized spectral densities in Fig. 3.18. All spectra show a systematic shift to lower frequencies as ζ changes from positive to negative values, reflecting the impact of the larger convective eddies on the turbulent structure. For all three components, the unstable spectra are confined to the shaded area, with no particular regard to ζ. The horizontal velocity spectra illustrate the breakdown of surface-layer scaling in convective conditions, and are excluded from the region between stable and unstable conditions. In Figs. 3.18(a) and 3.18(b), the curve identified by $\zeta = 0$ sets the neutral limit for spectra on the stable side.

The temperature spectra from the Kansas experiment are shown in Fig. 3.19, where normalization using the factor $\Phi_H\phi_\varepsilon^{-1/3}$ has been used to ensure collapse of the curves in the inertial subrange. The relation $\Phi_H = \chi kz/u_{*0}\theta_{*0}^2$ is based on Eq. 2.69c, where $\Phi_H(\zeta)$ is the nondimensional gradient function defined by Eq. 3.24. The normalized spectral density in the inertial subrange then becomes

$$f\phi_{\theta\theta}(f)/(\theta_{*0}^2\Phi_H\phi_\varepsilon^{-1/3}) = \beta_\theta(2\pi k)^{-2/3}f^{-2/3} \tag{3.108}$$

and thus represents a single curve. All spectra show a systematic shift with stability as seen in the u and v spectra, and the breakdown in systematic behaviour in unstable conditions.

Normalized cospectra for the momentum and heat fluxes are shown in Fig.

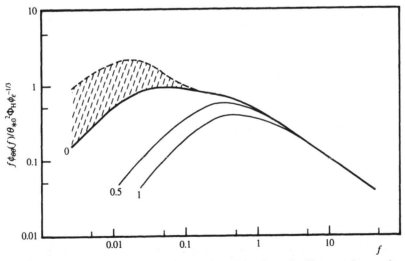

Fig. 3.19 Temperature spectra in normalized form based on the Kansas observations. The curves are plotted using the normalized frequency $f = nz/u$ for a range of z/L values. After Kaimal *et al.* (1972), *Quarterly Journal of the Royal Meteorological Society*.

3.20 revealing similar behaviour with stability to that seen in the spectra. All unstable cospectra belong in the shaded area contained within the pecked curves. There is an apparent overlap between the stable and unstable regions, since the curves for $\zeta = 0$ lie within the shaded area. The cospectra do not vanish across the complete inertial subrange, as required by local isotropy. Rather, the tendency towards isotropy at the higher frequencies forces the cospectrum to decrease more rapidly than the spectrum. Theoretical predictions of an $n^{-7/3}$ dependence (Wyngaard and Coté, 1972) are well supported in the observations.

Knowledge of the distribution of cospectral energy with frequency has an important practical benefit. It allows better specification of instrument response times and the averaging periods needed to measure vertical fluxes (i.e. the

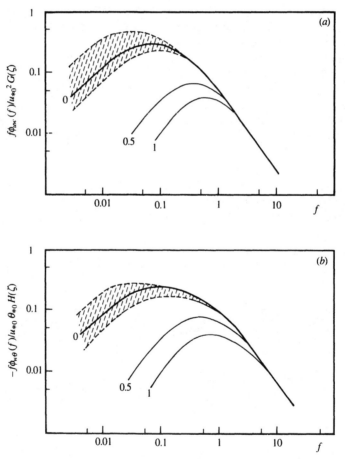

Fig. 3.20 One-dimensional cospectra in normalized form for (*a*) $u'w'$ and (*b*) $w'\theta'$, based on the Kansas observations. Curves are plotted for the range of $\zeta = z/L$ values indicated. In both cases the shaded regions represent the unstable regime for $-2 < \zeta < 0$. The ordinate has been partly normalized with stability functions as follows: for $\zeta < 0$, $-G(\zeta) = H(\zeta) = 1$; for $\zeta > 0$, $G(\zeta) = 1 + 7.9\zeta$; for $\zeta > 0$, $H(\zeta) = 1 + 6.4\zeta$. After Kaimal *et al.* (1972), *Quarterly Journal of the Royal Meteorological Society*.

covariances). If frequency response or averaging time is inadequate, known cospectral shapes can be used to compute corrections.

3.5.2 Mixed-layer scaling ($0.1h < z < h$)

Variances

Above the surface layer in unstable conditions, Monin–Obukhov scaling is inappropriate since the structure of turbulence is insensitive to z and u_{*0}, and h emerges as the dominant length scale. Mixed-layer scaling then incorporates convective scales w_* and T_* (Eqs. 1.12, 1.13), so that nondimensional properties are functions of z/h alone.

The behaviour of the velocity variances in unstable conditions is illustrated in Fig. 3.21. The profiles of the horizontal components are consistent with the surface-layer behaviour described by Eq. 3.103 which, for large negative μ, reduces to σ_u, $\sigma_v \approx 0.6w_*$ (i.e. an outer-layer scaling form). For the vertical component, the free-convection form at small z/h (i.e. Eq. 3.101, which can be rewritten as $\sigma_w/w_* = (a_1 k^{1/3})(z/h)^{1/3}$) gives way to nearly constant values through most of the mixed layer, with $\sigma_w \approx 0.6w_*$. That is, the turbulence is closely isotropic within the mixed layer, in contrast to its anisotropic nature in the surface layer. The continuous curve drawn through the data represents the

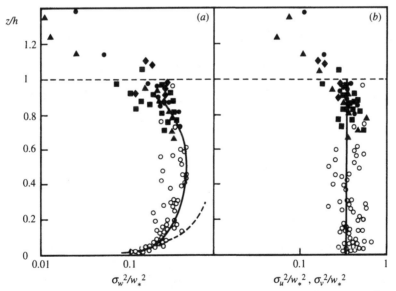

Fig. 3.21 (*a*) Normalized vertical velocity variance in unstable conditions as a function of normalized height throughout the CBL. The pecked curve represents Eq. 3.101 with $a_1 = 1.8$, whilst the solid curve represents Eq. 3.109 according to Sorbjan (1989). The behaviour represented by this expression is similar to that found by Lenschow *et al.* (1980) for the CBL over the sea during cold air outbreaks. (*b*) As in (*a*), but for the horizontal wind components. The vertical straight line represents $\sigma_u^2/w_*^2 = \sigma_v^2/w_*^2 = 0.36$. Atmospheric data from Caughey and Palmer (1979).

relation

$$\sigma_w/w_* = 1.08(z/h)^{1/3}(1 - z/h)^{1/3} \tag{3.109}$$

given in Sorbjan (1989).

The behaviour of the temperature variance in unstable conditions is shown in Fig. 3.22, with the free-convection form at small z/h (i.e. Eq. 3.102 rewritten as $\sigma_\theta/T_* = (a_2 k^{-1/3})(z/h)^{-1/3}$) giving way to values of about 1.5 in mid-regions. The curve drawn through the data represents

$$\sigma_\theta^2/T_*^2 = 2(z/h)^{-2/3}(1 - z/h)^{4/3} + 0.94(z/h)^{4/3}(1 - z/h)^{-2/3} \tag{3.110}$$

as suggested by Sorbjan (1989). The increased values near the inversion are associated with the presence of entrainment, with the entrainment heat flux depending strongly on T_* (see the discussion in Chapter 6 on top-down, bottom-up diffusion).

Overall, for the convective boundary layer, mixed-layer scaling for the variances is broadly supported by both atmospheric and laboratory observations.

Spectra and cospectra
As with the variances just discussed, w_* and h are the relevant scales in convective conditions, so that spectra should be described by

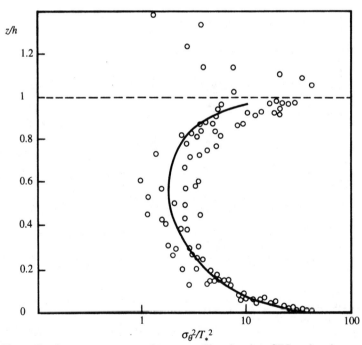

Fig. 3.22 Normalized temperature variance profiles in the CBL, showing atmospheric observations from Caughey and Palmer (1979). The solid line is the prediction given by Eq. 3.110, based on Sorbjan (1989). The expression uses the value -0.2 for the flux ratio β (see Section 6.1.6), based on the work of Moeng and Wyngaard (1984).

$$f_i \phi_{vel}(f_i)/w_*^2 = F_{vel}(f_i, z/h) \qquad (3.111)$$

where $f_i = nh/u$. Observations tend to support this mixed-layer scaling, as illustrated in Fig. 3.23 for the three velocity components. Additional normalization again produces a single curve in the inertial subrange, with ψ_ε given by

$$\varepsilon[(g/\theta_v)(\overline{w'\theta_v'})_0]^{-1} = \psi_\varepsilon = \varepsilon h/w_*^3. \qquad (3.112)$$

The normalized spectral density in the inertial subrange then becomes

$$f_i \phi_{vel}(f_i)/(w_*^2 \psi_\varepsilon^{2/3}) = \beta(2\pi)^{-2/3} f_i^{-2/3} \qquad (3.113)$$

and thus represents a single curve. The spectral peaks show a clear shift to lower frequencies with increases in height through the ABL, particularly for the w-component spectrum.

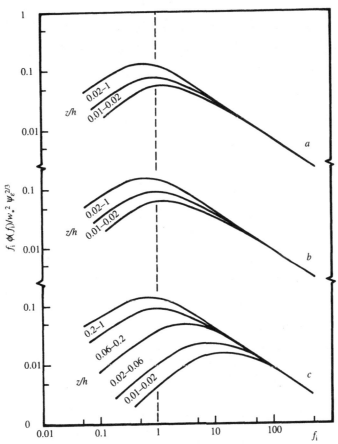

Fig. 3.23 One-dimensional velocity spectra in the mixed layer, suitably nondimensionalized according to mixed-layer scaling arguments, and plotted as functions of the normalized frequency $f_i = nh/u$. The ranges of values of z/h are indicated. (a) u-component spectrum, (b) v-component spectrum, (c) w-component spectrum. After Kaimal *et al.* (1976), *Journal of Atmospheric Sciences*, American Meteorological Society.

3.5.3 Local scaling (stable boundary layer)

The main advantage of local scaling is that it offers a similarity approach for turbulence in a stably stratified region.

Variances

Well above the surface layer, local properties are independent of the surface fluxes. The main prediction, therefore, of local similarity theory (e.g. Nieuwstadt, 1984) is that properties normalized by local scales u_*, θ_* and q_* should be functions of the local stability parameter z/Λ, and tend towards constant values at large z/Λ. Here, Λ is the local Obukhov length. This asymptotic limit of local scaling occurs under conditions of such high stability that the eddy size is limited by stability only and z is no longer important.

In Fig. 3.24 the predictions of local scaling in the nocturnal boundary layer are compared with observations for the quantities σ_w/u_*, σ_θ/θ_* and the gradient Richardson number Ri. In each case, the observations do indeed fall on a single curve, although the scatter for σ_θ is quite considerable. Despite the appreciable scatter, the experimental data agree reasonably well with the theoretical curves. Thus, for z/Λ not too small, $\sigma_w/u_* \approx 1.4$, $\sigma_\theta/\theta_* \approx 3$ and $Ri \to 0.2$ approximately.

In practice, local scaling parameters are rarely available and scaling with surface fluxes produces a strong dependence upon z/h. The following relations were suggested by Nieuwstadt (1984) from observations made in the slowly evolving nocturnal boundary layer (see also later discussion in Chapter 6):

$$\sigma_w/u_{*0} = 1.4(1 - z/h)^{3/4} \tag{3.114}$$

$$\sigma_\theta/\theta_{*0} = 3(1 - z/h)^{1/4}. \tag{3.115}$$

Spectra and cospectra

As with variances, suitably normalized spectra and cospectra should obey local scaling in the outer layer such that

$$f\phi_{vel}(f)/u_*^2 \propto f\phi_{vel}(f)/\sigma_{vel}^2 = F_{vel}(z/\Lambda, f) \tag{3.116}$$

and

$$f\phi_{\theta\theta}(f)/\theta_*^2 \propto f\phi_{\theta\theta}(f)/\sigma_\theta^2 = F_{\theta\theta}(z/\Lambda, f). \tag{3.117}$$

Normalized logarithmic spectra and cospectra (Caughey *et al.*, 1979) are shown in Figs. 3.25 and 3.26 as functions of f/f_m, where f_m is the frequency of the spectral maximum. This procedure forces the different spectra and cospectra to collapse into single curves, not only in the inertial subrange but over much of the frequency range. The data are shown for three z/h ranges, and imply general support for local scaling, with spectra and cospectra varying little with height (z/h) and stability (z/Λ).

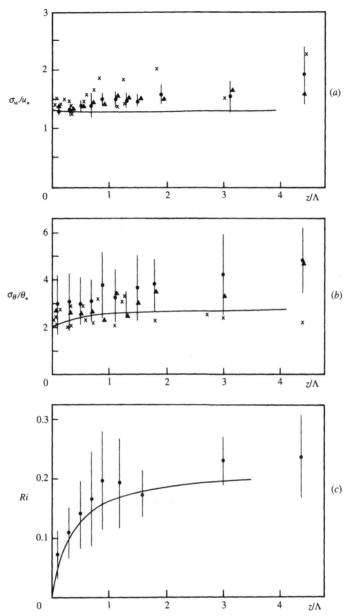

Fig. 3.24 Normalized quantities (a) σ_w/u_*, (b) σ_θ/θ_*, (c) Ri as functions of the local stability parameter z/Λ in the stable boundary layer. The solid curves represent solutions of the variance and covariance equations described by Nieuwstadt (1984), *Journal of Atmospheric Sciences*, American Meteorological Society.

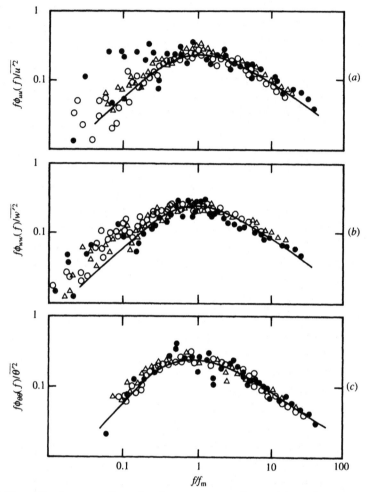

Fig. 3.25 Normalized one-dimensional spectra for (*a*) the *u*-component, (*b*) the *w*-component and (*c*) θ in the stable boundary layer; f_m is the frequency of the spectral maximum. Open circles, $0 < z/h < 0.1$; triangles $0.1 < z/h < 0.5$; solid circles, $0.5 < z/h < 1$. Data and curves (representing theoretical expressions) are from Caughey *et al.* (1979), *Journal of Atmospheric Sciences*, American Meteorological Society.

Notes and bibliography

Section 3.2

For additional information on the determination of the similarity constants *A* and *B*, the reader should consult

Wipperman, F. (1973), *The Planetary Boundary Layer of the Atmosphere*. Deutschen Wetterdienst, Offenbach, 346 pp., and

McBean, G. A., Bernhardt, K., Bodin, S., Litynska, Z., Van Ulden, A. P. and J. C. Wyngaard (1979), *The Planetary Boundary Layer*. WMO Technical Note No. 165, World Meteorological Organization, Geneva, 201 pp.

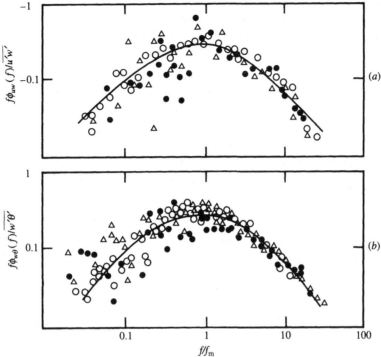

Fig. 3.26 Normalized one-dimensional cospectra for (a) $u'w'$ and (b) $w'\theta'$; f_m is the frequency of the spectral maximum. Open circles, $0 < z/h < 0.1$; triangles, $0.1 < z/h < 0.5$; solid circles, $0.5 < z/h < 1$. After Caughey *et al.* (1979), *Journal of Atmospheric Sciences*, American Meteorological Society.

Section 3.3

For an extensive discussion on the Monin–Obukhov similarity theory, Chapter 4 of the following is recommended:

Monin, A. S. and A. M. Yaglom (1971), *Statistical Fluid Mechanics*, Vol. 1, English trans., ed. J. L. Lumley. MIT Press, Cambridge, MA, 769 pp.

Much new experimental data concerning the unstably stratified surface layer have been obtained in recent years by the Institute of Atmospheric Physics in Moscow. Comparisons of these data with the predictions of similarity theory are described by

Kader, B. A. and A. M. Yaglom (1990), Mean fields and fluctuation moments in unstably stratified turbulent boundary layers, *J. Fluid Mech.* **212**, 637–62.

For general details regarding the form of the stability functions, see Chapter 4 in

Brutsaert, W. (1982), *Evaporation into the Atmosphere*. Reidel, Dordrecht. 299 pp.

Section 3.4

Earlier detailed analyses of Rossby-number similarity theory are described in Wipperman (1973).

For a general summary of parametric relations in the ABL, and references therein, see

Arya. S. P. S. (1984), Parametric relations for the ABL, *Bound. Layer Meteor.* **30**. 57–73.

A summary of the functional relations for the similarity parameters A, B and C can be found in Chapter 4 of Brutsaert (1982).

The reader can readily deduce from Figs. 3.7 and 3.12 (and the formulations upon which the curves are based) the ratio of the 10 m wind speed to the ABL or geostrophic wind speed. This ratio is equal to $(C_g/C_D)^{1/2}$. For example, in neutral to unstable conditions with $h = 1000$ m, the ratio is about 0.72 for $z_0 = 10^{-4}$ m and about 0.42 for $z_0 = 0.1$ m. Some observed values have been discussed in, e.g.,

Deacon, E. L. (1973), Geostrophic drag coefficients, *Bound. Layer Meteor.* **5**, 321–40.

Section 3.5

Integral and spectral statistics for the surface layer and the ABL as a whole are discussed in more detail in Panofsky and Dutton's book, Chapter 7, and in Chapter 4 of

Sorbjan, Z. (1989), *Structure of the Atmospheric Boundary Layer*. Prentice Hall, NJ, 317 pp.

A collection of wind-tunnel and atmospheric data for rough flow is given in

Raupach, M. R., Antonia, R. A. and S. Rajagopalan (1990), Rough-wall turbulent boundary layers, *Appl. Mech. Rev.* **44**, 1–25.

This paper has useful discussions on asymptotic matching and on the roughness sublayer.

4

Surface roughness and local advection

In the previous chapter, the aerodynamic and scalar roughness lengths were introduced as surface parameters in the development of profile relations and drag and bulk transfer formulations. The aerodynamic roughness length z_0 is an important parameter, as is the related quantity C_D, the drag coefficient. They are essential in calculations of the wind profile and the surface stress and also in calculations of the scalar roughness lengths. Much of the present chapter is devoted to simple conceptual models of these roughness lengths, and how they depend on the physical characteristics of the surface. After an initial discussion on z_0 over land and the displacement height, including the dependence of these two aerodynamic quantities upon measurable surface quantities, formulations for the scalar roughness lengths are developed. Emphasis is placed on the vegetation canopy (partly because of its increasing importance in surface schemes in numerical models, issues which are discussed further in Chapters 5 and 8). This then leads on to the topic of flow over the sea, and to the important problems of the wind-speed dependence of roughness and of the behaviour of the bulk transfer coefficients. The main purpose throughout is to give the roughness lengths a firm physical basis.

Internal boundary layers (IBLs) in the atmosphere are associated with the horizontal advection of air across a discontinuity in some property of the surface. The latter part of the chapter introduces the problem of local advection across a sudden change in z_0. This has many applications to the real world, including the micrometeorological problem of adjustment of profiles and eddy fluxes to surface changes, and the fetch–height ratio.

4.1 Aerodynamic characteristics of the land

Wind-profile relations involve two parameters determined, to a large degree, by the physical nature of the surface: the zero-plane displacement and the aerodynamic roughness length.

4.1.1 Zero-plane displacement

Profile relations used in earlier chapters involve the height z measured relative to a reference level termed the zero-plane displacement height, d, rather than the height Z above the actual ground surface. This displacement height can be likened to the level of action of the surface drag on the main roughness elements (Thom, 1971). Historically, d was introduced to retain the logarithmic form of the profile in neutral conditions for measurements made above tall vegetation where the minimum levels of measurement were $\sim h_c$, the height of the canopy (e.g. Paeschke, 1937). That is, d is chosen so that the wind $u(Z)$ satisfies (cf. Eq. 3.17)

$$ku/u_{*0} = \ln[(Z - d)/z_0]$$ (4.1)

with

$$z = Z - d.$$ (4.2)

A measure of caution is required in this approach, however, since such levels of measurement may lie within the roughness sublayer. Even allowing for the presence of this sublayer, the essential concept is that some correction to the wind profile is required within a few element heights of the surface (e.g. Monin and Yaglom, 1971, p. 293).

When measurements are made far above the surface (say, $Z > 10h_c$) then d may be ignored; for example, corrections are often not needed for measurements over short grass. Otherwise, d needs to be incorporated as in Eq. 4.1. In the past, values of d for specific surfaces were determined from observations of neutral wind profiles close to the surface, with d chosen to give a logarithmic profile according to Eq. 4.1. The approach can be extended to non-neutral conditions, and it is generally assumed that d is independent of stability. In addition, the concept of the zero-plane displacement can be extended to all other profile forms, θ and q in particular, so that z in Eqs. 3.38 and 3.41 is given by Eq. 4.2. It is usual to assume that d is the same for all profile forms, although there is no *a priori* reason for this to be so. Conclusive observations are lacking.

The simple relation $d/h_c \approx 2/3$ seems to be fairly representative of many natural vegetated surfaces, and of crops and forests in particular. However, it is clear that this ratio cannot be a constant. For extremely sparsely placed roughness elements, the ground surface is the true reference plane and d should be very close to zero. On the other hand, for very densely placed roughness elements, when the flow skims over the tops, d/h_c should approach unity. Values of d/h_c, for a range of surface types, can be found in Appendix Table A6.

4.1.2 Aerodynamic roughness length

General land surfaces

The aerodynamic roughness length z_0 was introduced in Chapter 3 as a surface length scale, and defined specifically by the logarithmic wind law for neutral conditions (Eq. 3.17). With this definition, the extrapolated wind speed equals zero at the height $z = z_0$. For aerodynamically smooth flow (when the viscous sublayer is deeper than surface roughness protuberances), experiment shows that (Hinze, 1975, Chapter 7)

$$z_0 \approx 0.11 \nu / u_{*0} \tag{4.3}$$

so that in this case z_0 is independent of roughness element geometry. A typical value of z_0 for smooth flow is thus 0.01 mm at a value of $u_{*0} = 0.165 \, \text{m s}^{-1}$. For aerodynamically rough flow over immobile elements, z_0 is a complicated function of surface geometry (the element height being the dominant parameter), and where elements are flexible (crops, grass) z_0 may depend on wind speed (or u_{*0}). Estimates of z_0, in the wind tunnel or in the atmosphere, are usually determined from observations of the wind profile, preferably in neutral conditions.

One crucial requirement in the functional dependence of z_0 upon surface geometry is that z_0/h_c should reach a maximum at some optimum value of the roughness element density. In the absence of major roughness elements, the drag will be due entirely to the underlying surface. As the density of elements increases, the drag will increase and, consequently, so will z_0. At some intermediate density where z_0 is relatively large, flow will cease to enter the inter-element spaces (it becomes "skimming"), and further increases in element density will produce decreases in drag and hence in z_0. The tendency for z_0/h_c to reach a maximum value at an intermediate element density is evident in wind-tunnel and atmospheric data, though the critical density is dependent on other factors, including the shape of the roughness elements. For atmospheric data (see Fig. 4.1), surface structure is often represented by the parameter λ_1 defined as the element silhouette area normal to the wind per unit surface area occupied by each element. Variations in z_0/h_c are generally consistent with the mean curve drawn in the figure to represent results for artificial surfaces, with the maximum occurring at $\lambda_1 \approx 0.4$. The data in Fig. 4.1 suggest that a linear relation between z_0/h_c and λ_1 is an oversimplification for many surfaces. Further, in addition to the possible dependence of z_0 upon u_{*0} for flexible elements, z_0 is a function of time for seasonally varying vegetation, both due to changes in h_c and λ_1. Appendix Table A6 gives values of z_0 for a wide range of natural surfaces, showing that overall $0.02 < z_0/h_c < 0.2$ (corresponding to values of λ_1 in the approximate range 0.04–0.75) with $z_0 = 0.1$ m being a good overall value for land surfaces. This range in values for z_0/h_c is consistent with the rule-of-thumb, found in many texts, that $z_0/h_c = 0.1$.

As well as its direct dependence upon h_c, z_0 can also be related to d, or more precisely to the projection of the roughness elements above the zero-plane displacement. This relation is usually of the form

$$z_0 = \gamma_1 (h_c - d) \tag{4.4}$$

with $\gamma_1 = \text{constant}, \approx 0.2$ to 0.4. With $d/h_c = 2/3$, this gives z_0/h_c between 0.07 and 0.14.

Sand and snow surfaces

The aerodynamic roughness length of flat deserts or surfaces covered in snow increases as the wind speed increases. This effect is apparently related to the

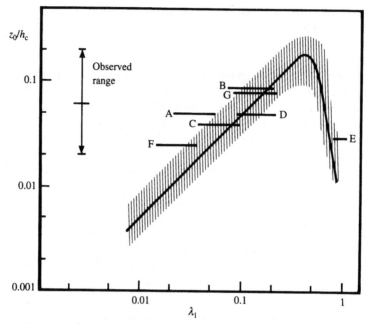

Fig. 4.1 Variation of z_0/h_c with element density, based on the results of Kutzbach (1961), Lettau (1969) and Wooding *et al.* (1973), represented by the shaded area and solid curve. Some specific atmospheric data are also shown as follows: A and B, trees; C and D, wheat; E, pine forest; F, parallel flow in a vineyard; G, normal flow in a vineyard. Analogous wind-tunnel data are described in Seginer (1974). From Garratt (1977b).

increasing movement and raising from the surface of sand or snow particles (Chamberlain, 1983). The variation of z_0 with wind speed is in fact closely described by Charnock's (1955) relation for rough flow over the sea (see Section 4.4), and is given by

$$z_0 = \alpha_c u_{*0}{}^2/g \qquad (4.5)$$

where α_c is referred to as Charnock's constant. The data shown in Fig. 4.2 for blowing sand and blowing snow tend to confirm this with $\alpha_c = 0.016$. Values of z_0 of between 0.05 and 1.5 mm are often quoted in the literature and are seen to be consistent with observations for values of u_{*0} in the range 0.2–1 m s^{-1}. Movement of sand grains is known to become significant only when u_{*0} is greater than about 0.12 m s^{-1} (Segal, 1990), so this value should represent a threshold below which Eq. 4.5 does not apply. At smaller u_{*0}, a constant value of 0.05 mm seems appropriate. For snow-covered rolling terrain (no vegetation) or extensive sand dunes, values of z_0 are likely to be greater than those given by Eq. 4.5. For snow-covered short vegetation, z_0 values are likely to be comparable with those in Eq. 4.2 but for snow-covered tall vegetation values may be comparable with the snow-free case.

The issue of aerodynamic roughness lengths over heterogeneous or complex terrain is discussed in the final chapter.

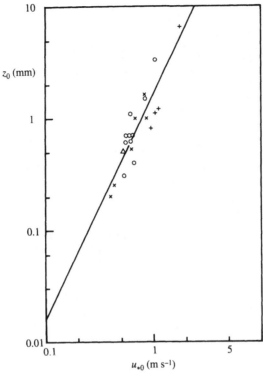

Fig. 4.2 Roughness lengths of blowing sand and of blowing snow as a function of the friction velocity, showing observations for sand (diagonal crosses) and snow (circles, triangles, upright crosses) from several sources. The straight line represents Eq. 4.5 with $\alpha_c = 0.016$. From Chamberlain (1983); reprinted by permission of Kluwer Academic Publishers.

4.2 Scalar roughness lengths

4.2.1 Concepts

The surface-layer temperature and humidity profiles are defined in Eqs. 3.38 and 3.41 respectively with scalar roughness lengths z_T and z_q replacing z_0 in the wind profile relation. Surface temperature and humidity are thus assumed to apply at heights $z = z_T$ and $z = z_q$ respectively. Indeed, with surface temperature defined as the radiative temperature measured using an airborne infrared radiometer, Eq. 3.38 is consistent with observations over a number of different surfaces only if z_T is allowed to differ from z_0.

Conceptually, the distinction between z_T, z_q and z_0 is suggested by consideration of the transport mechanisms for heat, water vapour and momentum in the presence of aerodynamically rough flow close to the surface. The transfer of momentum is effected by pressure fluctuations in the turbulent wakes behind the roughness elements, whilst for heat and water vapour transfer no such dynamical mechanism exists. Rather, heat and water vapour must ultimately be transferred by molecular diffusion across the interfacial sublayer. The identity of transfer mechanisms for heat and momentum is referred to as the *Reynolds analogy* (Monin & Yaglom, 1971, Chapter 3), so that the above argument suggests that the analogy is not valid in flow over rough surfaces in general. In this situation, the *resistance to transfer* of momentum between the surface and some height in the surface layer must be less than the resistance to transfer of heat or water vapour.

Some measure of the relative values of z_T, z_q and z_0 is possible from consideration of the structure of the interfacial sublayer.

4.2.2 Interfacial sublayer relations

The interfacial sublayer is the layer of air adjacent to the surface where the universal surface-layer profile forms and the Reynolds analogy are not valid. The breakdown of the Reynolds analogy may be due to the presence of pressure fluctuations around the roughness elements or to differences in the distribution of sources and sinks of momentum, heat and water vapour on the surface, or within a canopy. For smooth flow the interfacial sublayer is equivalent to the viscous sublayer, and molecular transfer is important. In fully rough flow, the velocity profile in the interfacial sublayer (also called the roughness sublayer) depends on the nature and distribution of the roughness elements.

By analogy with the nondimensional profile forms for the inertial sublayer, profiles within the interfacial sublayer can be represented in terms of the many nondimensional variables likely to affect the flow. These variables include z/h_d, where h_d is the depth of the interfacial sublayer; $Re_+ = u_{*0}h_d/v$ (a roughness Reynolds number); $Pr = v/\kappa_T$ (the Prandtl number); and $Sc = v/\kappa_V$ (the Schmidt number). Additional dependences upon bulk geometry, source and sink positions within the sublayer (e.g. canopy) are to be expected, so clearly any unified treatment covering all surface types cannot be expected. Any meaningful analysis is thus best confined to specific surface types, three being a suitable limit: smooth elements, bluff elements and permeable, well-packed elements. In practice, natural surfaces tend to be intermediate or transitional cases.

In the usual treatment of the interfacial sublayer (e.g. Brutsaert, 1982, Chapter 4), it is necessary to define an interfacial drag coefficient, $C_{D0} = u_{*0}^2/u_d^2$; an interfacial heat transfer coefficient (Stanton number) $St_0 = \theta_{v*0}/(\theta_{vd} - \theta_{v0})$; and an interfacial mass transfer coefficient (Dalton number) $Da_0 = q_{*0}/(q_d - q_0)$. The subscript d here refers to the top of the interfacial sublayer. Although there have been many studies of the form of the velocity profile within the interfacial (viscous) sublayer, this is not the case for the scalar quantities. In order to link the profile forms within the interfacial and inertial sublayers, a region of "overlap" near $z = h_d$ is assumed and the profiles are matched at $z = h_d$. The general bulk relations (Eqs. 3.42, 3.46 and 3.51) can then be decomposed into interfacial and inertial sublayer components, and the inertial sublayer relations are assumed valid at $z = h_d$. This assumption does not seem to be crucial. Equations 3.34, 3.38 and 3.41 for the neutral case can be expressed in terms of differences between a reference level z and $z = h_d$. With C_{D0} set equal to $k^2/[\ln(h_d/z_0)]^2$ (i.e. we assume that $z = h_d$ is high enough for the log law to apply), the unknown variables at level $z = h_d$ (i.e. u_d, θ_{vd} and q_d) can be eliminated, to give

$$(\theta_v - \theta_0)/\theta_{v*0} = B_H^{-1} + k^{-1}\ln(z/z_0) \tag{4.6}$$

$$(q - q_0)/q_{*0} = B_V^{-1} + k^{-1}\ln(z/z_0). \tag{4.7}$$

Here, we have introduced the quantity B^{-1}, used by a number of workers in the past (Owen and Thomson, 1963), with

$$B_H^{-1} = St_0^{-1} - C_{D0}^{-1/2} = k^{-1}\ln(z_0/z_T) \tag{4.8}$$

$$B_V^{-1} = Da_0^{-1} - C_{D0}^{-1/2} = k^{-1}\ln(z_0/z_q). \tag{4.9}$$

For some purposes, it may be preferable to rewrite Eqs. 4.6 and 4.7 in terms of the neutral drag and bulk transfer coefficients defined by Eqs. 3.43 and 3.48, giving

$$C_{HN} = C_{DN}^{1/2}/(B_H^{-1} + C_{DN}^{-1/2}) \tag{4.10}$$

$$C_{EN} = C_{DN}^{1/2}/(B_V^{-1} + C_{DN}^{-1/2}). \tag{4.11}$$

The reader should note that in Eq. 4.6, both θ_0 and B_H^{-1} (or z_T) are undefined as are q_0 and B_V^{-1} (or z_q) in Eq. 4.7. Choosing one specifies the other. In the case of heat transfer, the choice of a suitable surface temperature depends to some extent on convenience; we could take $B_H^{-1} = 0$ and make θ_0 the temperature at z_0. But its measurement would be difficult to make. By identifying θ_0 with a radiative temperature that can be readily measured from an airborne or remote radiometer, and which is a central parameter in the surface energy balance equation (see the next chapter), we transfer the problem to one of determining B_H^{-1}, usually by experiment. The problem of q_0 is less straightforward, although for a saturated surface it is readily identified with the saturated humidity at temperature θ_0. Written in the above form (Eqs. 4.6–4.11), the problem is reduced to determining St_0 (or Da_0) and C_{D0}, in terms of known quantities, for the three surface types to be specified. In the case of C_{D0} this requires the interfacial sublayer depth h_d to be determined.

Smooth surface
Figure 4.3 summarizes experimental data for flow along smooth pipes. The data imply a purely viscous stress for $zu_{*0}/\nu < 5$ (point A), for which range of z the wind profile is linear and satisfies

$$ku/u_{*0} = zu_{*0}/\nu. \tag{4.12}$$

For $zu_{*0}/\nu > 30$ (point B), the stress is purely turbulent and the logarithmic velocity profile is valid. These, and other data for smooth flat plates, suggest taking $h_d = 30\nu/u_{*0}$, whence $C_{D0}^{-1/2} \approx 12$. In contrast, the Stanton and Dalton numbers are likely to depend on Pr and Sc, as has been confirmed in a number of experiments over smooth surfaces and in semi-theoretical or conceptual models. This suggests taking (Brutsaert, 1982, Chapter 4)

$$B_H^{-1} \approx 13.6 Pr^{2/3} - 12 \tag{4.13a}$$

$$B_V^{-1} \approx 13.6 Sc^{2/3} - 12 \tag{4.13b}$$

so that for air in the lower atmosphere, with $Pr = 0.71$ and $Sc = 0.60$ (see

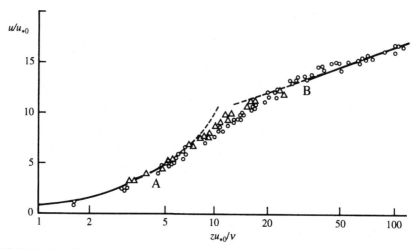

Fig. 4.3 Velocity distribution above a smooth surface, showing the transition from laminar flow (to the left of point A) to fully turbulent flow (to the right of point B). Based on experimental data of Reichardt and Laufer. The curve at small Reynolds number represents a linear velocity distribution (Eq. 4.12), and the logarithmic law is represented by the curve at large Reynolds number. From Hinze (1975, Chapter 7), by kind permission of the McGraw-Hill Book Company.

Appendix 2), $B_H^{-1} \approx -1.7$ and $B_V^{-1} \approx -2.8$. Thus, the use of Eqs. 4.8 and 4.9 suggests $z_0/z_T \approx 0.5$ and $z_0/z_q \approx 0.3$, with z_0 given by Eq. 4.3.

Surface with bluff elements

Bluff elements are roughness obstacles that are impermeable to the wind flow, and whose height h_d is similar in magnitude to the aspect width normal to the mean flow. In the wind tunnel, these obstacles may range from solid spheres, cubes, pyramids, to horizontal rods set normal to the flow. In the atmosphere, examples include densely vegetated vineyards, cabbage plants, pineapple plants and the like, ploughed fields, irregular ice cover. Having discussed the case of smooth flow, where the surface details are not relevant, we turn our attention to aerodynamically rough flow where the roughness Reynolds number, $Re_* = u_{*0}z_0/v$, is generally $\gg 1$.

The effects of the roughness geometry and the height of obstacles on the flow structure and the nature of the wind profile are combined together in the roughness parameter z_0. By introducing z_0, the interfacial drag coefficient is expected to be a function of Re_* only, and thus the coefficients St_0 and Da_0 functions only of Re_*, Pr and Sc. In the case of the drag coefficient, any Re_* dependence is expected to be weak, particularly at high Re_*, when momentum transfer to the surface is primarily due to form drag (related to pressure fluctuations). Observations, however, are few and the behaviour of the drag coefficient is not known in any detail. Experimental velocity profiles in the region of rough walls show that the lower limit of validity of the logarithmic law, i.e. the top of the roughness sublayer, occurs at $z = z_* \approx 10z_0$. In the absence of detailed observations, taking $h_d = z_*$ gives a crude but practical estimate of C_{D0}, corresponding to $C_{D0}^{-1/2} \approx 5$ (Brutsaert, 1982, Chapter 4).

For St_0 and Da_0, observations are not so comprehensive as for smooth surfaces, and are mainly confined to the wind tunnel for $Re_* < 1000$. For Pr and Sc generally in the range 0.5–8, the experimental data lead to (Brutsaert, 1982, Chapter 4)

$$B_H^{-1} \approx 7.3 Re_*^{1/4} Pr^{1/2} - 5 = 6.2 Re_*^{1/4} - 5 \tag{4.14}$$

$$B_V^{-1} \approx 7.3 Re_*^{1/4} Sc^{1/2} - 5 = 5.7 Re_*^{1/4} - 5 \tag{4.15}$$

where we have used $Pr = 0.71$ and $Sc = 0.60$ for air in the lower atmosphere. These two relations are shown in Fig. 4.4 as kB^{-1} versus Re_*, so that values of z_0/z_T and z_0/z_q can be evaluated directly using Eqs. 4.8 and 4.9.

Surfaces with permeable or randomly distributed elements
These surfaces are either of a fibrous nature, with dense packing of individual elements (as with vegetation canopies in general), or fine-grained, solid surfaces (as with soils, sand etc.). In general, detailed experimental data are lacking on the interfacial transfer coefficients. In the case of a canopy, the depth of the interfacial sublayer is approximated as the depth of the canopy itself ($h_d = h_c$) so that, with the logarithmic law assumed valid down to $z = h_c$, $C_{D0}^{-1/2} \approx 2.5$. But this will certainly be an underestimate, since the roughness sublayer probably extends above the canopy top.

Because of uncertainties in the scalar transfer coefficients due to the lack of a firm theoretical model, most relevant experimental data, both in the wind tunnel and in the atmosphere, have included direct measurements of surface temperature or concentration and have utilized the bulk transfer relations to infer transfer coefficients. With suitable stability corrections where necessary (e.g. Eqs. 3.44, 3.49), the neutral coefficients then allow B^{-1} to be inferred from Eqs. 4.10 and 4.11. Results for a range of natural and artificial surfaces are presented in Fig. 4.4, where much of the data relate to heat transfer over natural surfaces, including soil, grass, crops and forest.

The results generally reveal a pronounced dissimilarity between the bulk transfer properties for scalars at permeable (fibrous) surfaces and those at bluff rough surfaces (represented by the curves in Fig. 4.4). For many natural land surfaces, B^{-1} is almost independent of Re_*, for $Re_* > 10$, and a useful approximation for practical application is (Garratt and Francey, 1978)

$$kB_H^{-1} = kB_V^{-1} = \ln(z_0/z_T) = \ln(z_0/z_q) \approx 2. \tag{4.16}$$

Equations 4.14–4.16 show that $z_0 \gg z_T$, z_q over land surfaces, reflecting the more efficient transfer of momentum to the surface compared with heat and mass transfer. In contrast, over the sea where $Re_* \lesssim 1$–10, z_0 is comparable with, or even less than, the scalar roughness lengths (see Section 4.4).

4.3 The vegetation canopy

For tall canopies (e.g. crops, forests) the knowledge and understanding of in-canopy flow and turbulent transfer is of great practical interest, requiring, amongst other things, information on the vertical distribution of sources and sinks of momentum, heat and mass (e.g. in the form of water vapour and

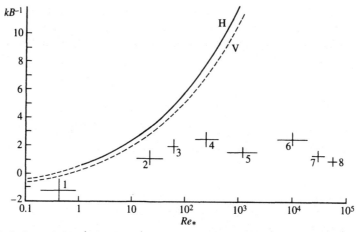

Fig. 4.4 Variation of kB^{-1} ($= \ln(z_0/z_p)$) with roughness Reynolds number. Curves are based on laboratory data and are represented by Eq. 4.14 for heat (H) and Eq. 4.15 for water vapour (V) transfer. Atmospheric data (mainly for heat transfer) are also shown, based on data summarized in Garratt and Hicks (1973) and Brutsaert (1982). Surface types are as follows: 1, sea; 2, vineyard; 3, short grass; 4, medium grass; 5, bean crop; 6, savannah scrub; 7 and 8, pine forests.

carbon dioxide). For example, the distribution of mean velocity, temperature and humidity within the canopy depends on roughness-element structure and density, the penetration of radiation, and the distribution of the above sources and sinks.

In addition, some indication is necessary regarding canopy flow and turbulent structure therein, since these do ultimately relate to the matching of the roughness sublayer to the flow at, or near, the canopy top and to the representation of vegetation in numerical models.

4.3.1 Wind profiles

Knowledge of the actual wind-profile form within the canopy is important, but this profile must match, across the canopy top, the profile form in the roughness sublayer. Within canopies, branches and leaves are usually diffusely distributed and act as a continuum-like sink for momentum. Under such conditions the mean wind speed and shearing stress should decay with depth below the canopy top. Figure 4.5 shows a set of observed canopy wind profiles that illustrate this behaviour. Several profile functions exist, all using the bottom boundary condition that, at $Z = 0$, $u = K_M = 0$. In the simplest approach, the stress divergence in the air, $\partial \tau / \partial z$, is assumed to balance the foliage drag per unit volume of air, $D_f = \rho(A_f/2)C_{Df}u^2$, where A_f is a function of height and equal to the surface area of leaves per unit volume of air, and C_{Df} is a foliage drag coefficient. This simple momentum balance is solved for $u(Z)$ by making suitable assumptions on K_M and $A_f C_{Df}$ as follows.

(i) For constant canopy mixing length l_c, $K_M = l_c^2 \, \partial u / \partial Z$ and $A_f C_{Df}$ is set

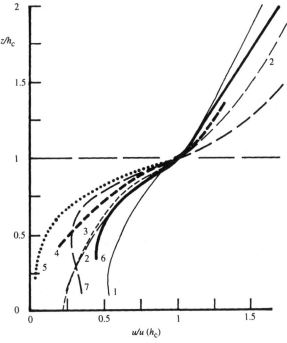

Fig. 4.5 Vertical profiles of normalized wind speed through and above various canopies, for wind-tunnel and atmospheric data. 1, wind-tunnel strips; 2, wind-tunnel artificial wheat; 3, wind-tunnel rods; 4 and 5, corn; 6, eucalyptus forest; 7, pine forest. From Raupach (1989), by kind permission of Springer-Verlag Publishers.

equal to a constant; the result is an exponential wind profile (Inoue, 1963) given by

$$u = u(h_c) \exp(-n_1 \zeta_1) \tag{4.17}$$

where $\zeta_1 = 1 - Z/h_c$.

(ii) If both K_M and $A_f C_{Df}$ are set equal to constants (Thom, 1971), the profile becomes

$$u = u(h_c)(1 + n_2 \zeta_1)^{-2} \tag{4.18}$$

so that both forms have u decreasing as $Z \to 0$ (Eqs. 4.17, 4.18 are not very sensitive to the assumption on K_M).

The coefficients n_1 and n_2 depend upon canopy density and structure, but appear to lie in the range 1–4 for a wide range of canopy types. The above forms for the wind profiles are particularly suitable for canopies with a near-even distribution of foliage with height, but do not reproduce the secondary wind maximum observed in canopies where foliage is concentrated near the top (as in the crowns of some forest types).

Since the flux-gradient assumption, $\tau = \rho K_M \partial u / \partial Z$, has been utilized the exponential wind profile implies an exponential decrease of stress as $Z \to 0$. This is generally supported by observations in uniformly dense canopies (e.g. Finnigan and Mulhearn, 1978).

4.3.2 Temperature and humidity profiles

Typical mean profile forms found in fairly dense canopies are illustrated in Fig. 4.6. These one-hour-averaged profiles were made in a pine forest in the middle of the day, and show several characteristics often observed in plant canopies including

    (i) a temperature maximum near mid-canopy;
    (ii) humidity decreasing upwards, with a large gradient near the ground.

Also shown are time-averaged vertical fluxes illustrating the *countergradient transfer* often observed, particularly for sensible heat, which implies the breakdown of gradient diffusion concepts and flux-gradient relations applied to scalar transfer (Finnigan and Raupach, 1987).

4.3.3 Turbulence

Turbulence in canopies is augmented by the presence of the foliage. In exerting a drag force on the plants, the wind loses momentum and the wind speed decreases rapidly as one descends into the canopy. At the same time, the kinetic energy lost from the mean wind is converted into TKE. Although the absolute TKE generally decreases down into the canopy, the turbulence intensity tends to increase (Raupach, 1988). The evidence for countergradient transfer, briefly mentioned in Section 4.3.2, suggests that the turbulent transfer is dominated by intermittent large structures: "sweeps" and "ejections", as the gust phenomena have been called. Thus, most of the vertical transport is associated with the intermittent, but intense, displacement of warm canopy air by the strong sweeps

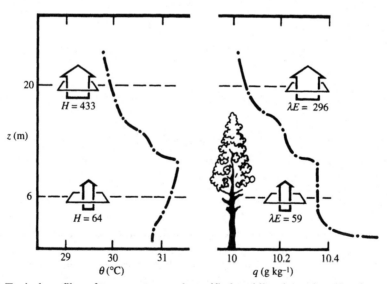

Fig. 4.6 Typical profiles of temperature and specific humidity through a pine forest during daytime with strong solar forcing. Values of sensible and latent heat fluxes in $W\,m^{-2}$ are shown at two levels, illustrating the countergradient (i.e. upwards) transfer deep in the canopy. From Denmead and Bradley (1985); reprinted by permission of Kluwer Academic Publishers.

of cooler, fast-moving boundary-layer air that dominate momentum transfer. They correspond with turbulent eddies of length scale somewhat larger than the height of the canopy, h_c (Finnigan & Raupach, 1987).

In near-neutral conditions, heat transfer and momentum transfer are closely coupled, but in strongly unstable conditions substantial differences occur. Heat transfer becomes dominated by ejections of warm canopy air, identified as rising buoyant plumes, and is presumably dependent on the canopy density and fractional cover of bare ground. With momentum transfer still related to the sweeps or gusts, any simple correspondence of the mechanisms for heat transfer and momentum transfer is no longer observed. In the quiescent periods between sweeps or ejections, comprising much of the averaging period for the mean profiles, very little vertical transfer occurs.

In recent times, many theoretical advances have been made regarding the description of turbulent transfer in plant canopies, using higher-order closure and Lagrangian models (Raupach, 1988). These recognise the problems inherent in the application of K-theory to momentum transfer and scalar transfer within canopies, and attempt to account for the non-Gaussian, intermittent and highly turbulent nature of the transport. Nevertheless, suitably simple but realistic canopy parameterization schemes are required for use in numerical models of the atmosphere, but such schemes must take account of the most important canopy processes. A discussion of this can be found in Chapter 8.

4.4 Flow over the sea

Partly because of the vast extent of the oceans, a major dynamical problem is to determine the wind stress acting on the sea surface. This is important both for the structure of the atmospheric boundary layer over the sea and also in relation to the ocean mixed layer and ocean currents forced by the action of wind at the surface. The wind-stress problem is intimately related to the aerodynamic roughness length and drag coefficient. In fact, an enormous effort has been expended over the years in evaluating drag and heat (mass) transfer coefficients for a wide range of conditions. In essence though, the major differences between flow over land and sea concerns the large heat capacity of the oceans, and the mobile wavy nature of the ocean surface.

4.4.1 Wave properties

The very shortest waves are capillary waves, although most of the spectrum comprises gravity waves, with the longest of these being referred to as swell. Two classical theories of the mechanism of wave growth are the resonance theory of Phillips (1957) and the instability theory of Miles (1957). The Phillips' theory applies to the early stages of wave growth, where energy in each wave component grows linearly with time, whilst Miles' theory applies to the later stages of wave growth, with energy growing exponentially. An important part of these early theories was the proposal that a saturated region of the spectrum exists in the shorter-wave components. Such conditions are attained after the wind has been blowing over a relatively short fetch, with the spectrum of the height variations in the water surface, $\phi(n)$, given by (Phillips, 1958)

$$\phi(n) = \alpha_w g^2 n^{-5}. \tag{4.19}$$

The constant α_w has been derived from observations; and is ≈ 0.0012 if n is in rad s^{-1}.

Wave structure is very much dependent on wind–wave interactions, and depends on such quantities as wave age, fetch and wind speed or u_{*0}. Wave breaking is related to a transfer of energy from the already saturated wave components (ripples) to turbulence, and is associated with air-flow separation at the critical wave components, typically when u_{*0} is greater than the minimum phase velocity of breaking surface gravity waves (≈ 0.23 m s^{-1}). One of the basic problems in connection with wind–wave interaction is to determine the region of the wave spectrum that supports the bulk of the wind stress. It is generally accepted that z_0 is determined principally by the steepest waves rather than the longest waves, the steep short waves being those components from which air-flow separation occurs with phase velocities less than u_{*0}. This criterion for determining air-flow separation from wind waves is supported by both observations and numerical studies.

For the sea surface in general, flow is aerodynamically smooth with $Re_* \approx 0.11$ and fully rough for $Re_* > 2$; for intermediate values of Re_* the flow is referred to as transitional. The transition to fully rough flow at $Re_* \approx 2$ coincides with the onset of wave breaking when $u_{*0} \approx 0.23$ m s^{-1} (the 10 m wind speed ≈ 5.5 m s^{-1}), and is accompanied by a significant increase in the drag coefficient (Melville, 1977). Wave breaking produces *whitecaps*, and the white-cap coverage as a function of wind speed is shown in Fig. 4.7 to illustrate the onset of breaking and its rapid increase with wind speed.

4.4.2 The aerodynamic roughness length

The logarithmic wind profile in neutral conditions, and the general form of the similarity Φ functions in non-neutral conditions, have been broadly confirmed from observations over water surfaces. From such measurements, both z_0 and the transfer coefficients can be determined. It should be noted that, because of the mobile nature of the sea surface, Eq. 3.17 is modified to include the surface drift velocity u_s:

$$k(u - u_s)/u_{*0} = \ln(z/z_0) \tag{4.20}$$

where observations suggest $u_s \approx 0.55 u_{*0}$ (Wu, 1975). In practice, u_s is usually neglected in Eq. 4.20.

In very light winds, evidence suggests that flow closely approximates that over an aerodynamically smooth surface so that z_0 is given by Eq. 4.3. As wind speed, and hence Re_*, increase, flow becomes transitional with evidence of fully rough flow at higher winds. In rough flow conditions, observations support the use of Eq. 4.5. This identifies g, the acceleration due to gravity, as an essential dynamic parameter, and characterizes the equilibrium interaction between wind and waves, with the spectrum of gravity waves acting as roughness elements. Equation 4.5 thus predicts an increase in z_0 and C_{DN} with wind speed; Table 4.1 summarizes a number of estimates of the Charnock constant α_c, and the reader should note that its derivation is sensitive to the value chosen for k. For

Table 4.1. *Values of the Charnock constant α_c from the literature, mainly based on observations made in moderate ($> 5 \, \mathrm{m\,s^{-1}}$) to strong winds.*

Reference	α_c	Comments
1. Charnock (1958)	0.012	
2. Kitaigorodskii and Volkov (1965)	0.035	large lake
3. Wu (1969)	0.0156	oceanic data
4. Clarke (1970)	0.032	based on DW62[a]
5. Smith and Banke (1975)	0.013	oceanic data
6. Garratt (1977a)	0.0144	review; $k = 0.41$
	0.017	review; $k = 0.40$
7. Wu (1980)	0.0185	review
8. Geernaert et al. (1986)	0.0265	limited fetch

[a] DW62 is Deacon and Webb (1962). Note should be made of the following, (i) apart from the reference 8, most recent data sets and analyses give $\alpha_c \approx 0.014$ to 0.019. A value of 0.016 is recommended; (ii) values of α_c depend on the assumed value of k when using measured u, u_{*0} and the log law (e.g. reference 6); (iii) some studies purport to show that α_c is not a constant but a function of water and wave variables, e.g. Melville (1977) and Wu (1982).

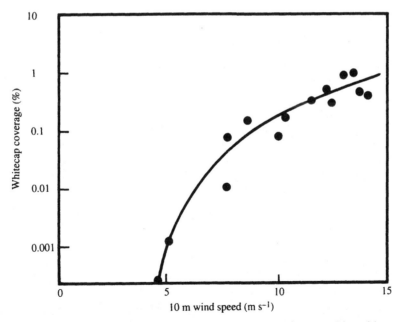

Fig. 4.7 Fraction of the sea surface (expressed as a percentage) covered by whitecaps as a function of the 10 m wind speed. After Kondo *et al.* (1973), *Journal of Physical Oceanography*, American Meteorological Society.

$\alpha_c = 0.016$, Eq. 4.5 gives $z_0 \approx 1\text{–}2$ mm at a wind speed of $20\,\mathrm{m\,s^{-1}}$ at 10 m height, comparable with the roughness of snow or sand surfaces. Two main refinements to Charnock's relation, originally derived from dimensional arguments, are briefly discussed here, since they indicate the large number of physical parameters upon which z_0 may depend. Firstly, it is possible that the nature and properties of the waves are relevant, so that α_c becomes dependent on wave parameters (Melville, 1977). The second refinement concerns the effects of surface tension σ_t and viscosity μ_w (the dynamic viscosity of water) on wave structure at the ocean surface. Oceanic and laboratory data suggest a modified version of Eq. 4.5 (Wu, 1980):

$$z_0 = (\alpha_c u_{*0}^2/g)(\mu_w u_{*0}/\sigma_t)^{m-2} \tag{4.21}$$

with an overall value of $m \approx 7/3$. For most purposes, Eq. 4.5 is an acceptable approximation for winds greater than a few $\mathrm{m\,s^{-1}}$ and for a wide range of conditions generally found over the oceans. Overall, the recommended value of α_c is one lying between about 0.014 and 0.0185 (rather than a value of 0.032 as adopted in many numerical models of the atmosphere).

4.4.3 Drag coefficients

Much effort has gone into evaluating the neutral drag coefficient from observations, and determining the factors upon which it depends (in particular, the wind-speed dependence). For neutral conditions, C_{DN} is given for aerodynamically smooth flow by

$$C_{DN} = k^2/[\ln(u_{*0}z/0.11\nu)]^2 \tag{4.22}$$

and for rough flow by

$$C_{DN} = k^2/[\ln(zg/\alpha_c u_{*0}^2)]^2. \tag{4.23}$$

The variation of C_{DN} with u_{10}, the mean wind speed at 10 m, is shown in Fig. 4.8 where the value of $\alpha_c = 0.016$. The transitional regime corresponds to the wind-speed range $2.5 < u_{10} < 5.5\,\mathrm{m\,s^{-1}}$.

Values of Charnock's constant and numerous expressions describing the function $C_{DN}(u)$ have been derived from extensive observations over both small water bodies (e.g. lakes) and large water bodies (e.g. the open ocean), over a range of fetches and mostly for wind speeds less than about $20\,\mathrm{m\,s^{-1}}$. Critical assessment of many data sets (Garratt, 1977a) has produced the composite data plot shown in Fig. 4.8, which shows a variation consistent with $k = 0.41$ and $\alpha_c = 0.0144$. Results of turbulence sensor comparison experiments suggest that much of the data scatter seen in Fig. 4.8, and much of the systematic differences between data sets, is due to calibration uncertainties associated with sensor performances in the field. The effects of fetch, wind duration, wave properties, unsteadiness in the wind field and water depth are generally obscured by the experimental data scatter. The data do not provide evidence for any discontinuity in the variation of the drag coefficients as the wind increases from light to strong (Wu, 1982).

Observations have usually been represented in algebraic expressions relating drag coefficient to wind speed; for example,

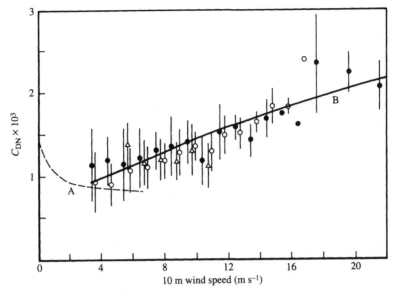

Fig. 4.8 Neutral drag coefficient over the sea as a function of the 10 m wind speed based on individual data taken from the literature. Data points represent many single runs (between 3 and 84) with the root-mean-square variation also shown. Curve A, for smooth flow, represents Eq. 4.3, and curve B, for rough flow, represents Eq. 4.5 ($\alpha_c = 0.016$). After Garratt (1977a), *Monthly Weather Review*, American Meteorological Society.

$$C_{DN} = (a_1 + b_1 u_{10}) \times 10^{-3} \qquad (4.24)$$

$$C_{DN} = a_2 u_{10}{}^{b_2} \times 10^{-3} \qquad (4.25)$$

The wind speed dependences described by Eqs. 4.24 and 4.25 can be related directly to that implied in Eq. 4.5. Thus, the mean variation shown in Fig. 4.8 is closely approximated by $a_1 = 0.75$ and $b_1 = 0.067$, or by $a_2 = 0.51$ and $b_2 = 0.46$.

At wind speeds greater than 20 m s^{-1}, data are scarce but the evidence from several studies suggests that Eq. 4.5 is valid for wind speeds as high as 50 m s^{-1}, with α_c between 0.014 and 0.0185 and C_{DN} is represented by Eq. 4.24 with $a_1 = 0.775$ and $b_1 = 0.066$. Thus, the roughness length concept appears to be useful even for stormy seas where spray is prevalent.

4.4.4 Bulk parameterization for heat and mass transfer
In analogy to the drag coefficient, the bulk coefficients for heat and water vapour transfer are of considerable interest. These bulk transfer coefficients are dependent upon the scalar roughness lengths; the neutral values can also be related to C_{DN} and the parameter B^{-1} through Eqs. 4.10 and 4.11.

For smooth flow over water, the expression for B^{-1} given by Eq. 4.13 should be valid and, for fully rough flow, that given by Eqs. 4.14 and 4.15 for the bluff-rough surface is valid. The expressions for $St_0{}^{-1}$ and $Da_0{}^{-1}$ implied in these equations are not very different to those found with a "surface renewal model" of the interfacial sublayer (Liu and Businger, 1975; Liu *et al.*, 1979). In this, the depths of the interfacial sublayers for momentum (δ_u), heat (δ_T) and water vapour (δ_q) transfer scale with the molecular diffusivities and a time scale, t_*, is identified as the average time of fluid contact with the interface. Thus, $\delta_u = (vt_*)^{1/2}$, $\delta_T = (\kappa_T t_*)^{1/2}$ and $\delta_q = (\kappa_V t_*)^{1/2}$ and the profiles are assumed exponential through the interfacial sublayer (consistent with observations). If t_* is the same as the Kolmogorov time scale for small eddies (Brutsaert, 1975), then $t_* \propto (vz_0/u_{*0}^3)^{1/2}$. For smooth flow, the model predicts $St_0^{-1} \propto Pr^{1/2}$ and $Da_0^{-1} \propto Sc^{1/2}$ with a constant of proportionality ($C_{D0}^{-1/2}$) equal to 16. Reference to Section 4.2.2 for smooth flow over solid surfaces shows small differences in the predicted behaviour, but in practice the results are quite similar. For smooth flow ($u_{10} < 2.5 \text{ m s}^{-1}$), with z_0 given by Eq. 4.3, we take

$$z_T u_{*0}/v \approx 0.2 \tag{4.26a}$$

$$z_q u_{*0}/v \approx 0.3. \tag{4.26b}$$

For rough flow, the surface renewal model gives relations for St_0 and Da_0 that are very similar to those found for bluff solid surfaces (Section 4.2.2). Thus, B^{-1} is given by Eqs. 4.14 and 4.15, so that z_T and z_q can be described by

$$\ln(z_0/z_T) = 2.48 Re_*^{1/4} - 2 \tag{4.27}$$

$$\ln(z_0/z_q) = 2.28 Re_*^{1/4} - 2 \tag{4.28}$$

where we have taken $Pr = 0.71$ and $Sc = 0.60$. In Eqs. 4.27 and 4.28 the roughness length z_0 is given by Eq. 4.5 for fully rough flow, with $\alpha_c = 0.016$. It must be said that Eqs. 4.26–4.28 rely on very few oceanic observations for support, and so must be seen as tentative forms only. They do show that at low winds, in aerodynamically smooth flow, z_T and z_q are greater than z_0, whilst in rough flow the reverse is the case. As with flow over solid surfaces, form drag becomes important at high winds, and momentum transfer to the sea surface is enhanced relative to heat and mass transfer. The neutral transfer coefficients can be derived from these scalar roughness length values using Eqs. 4.10 and 4.11, with z_0 given by Eq. 4.3 (smooth flow) and Eq. 4.5 (rough flow); they are shown in Fig. 4.9. Two interesting features of the predicted behaviour of C_{HN} and of C_{EN} are worth noting: the coefficients tend to a constant value with increasing wind speed in rough flow and increase with decreasing wind speed in smooth flow.

Many observational analyses of the heat and water vapour transfer coefficients, for small and large water bodies, have appeared in the literature (e.g. Large and Pond, 1982). Measurements are notoriously difficult and the most reliable estimates of the neutral coefficients are confined to winds less than 10 m s^{-1}; referred to a reference height of 10 m, values in the range $1–1.5 \times 10^{-3}$ are typical. Figure 4.9 shows a summary of C_{HN} and C_{EN} values based on observations made from ships in the open ocean and from a stable deepwater tower, for winds up to 25 m s^{-1}. At present, these probably represent the best data sets over such a wide wind-speed range; they give

$$C_{HN} = C_{EN} \approx 1.1 \times 10^{-3} \, (\pm 15\%)$$

The predicted larger value of C_{EN} shown in Fig. 4.9, based on $Pr > Sc$, is not

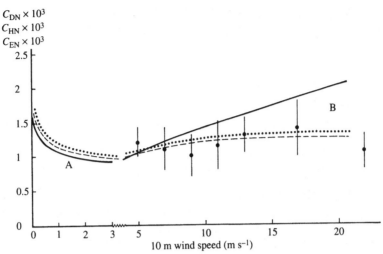

Fig. 4.9 Drag coefficient C_{DN}, heat transfer coefficient C_{HN} and water vapour transfer coefficient C_{EN} as functions of the 10 m wind speed. Curves A are for smooth flow: solid curve C_{DN} (Eq. 4.22); pecked curve, C_{HN} (Eqs. 4.10 and 4.26a); dotted curve, C_{EN} (Eqs. 4.11 and 4.26b). Curves B are for rough flow: solid curve, C_{DN} (Eq. 4.23); pecked curve, C_{HN} (Eqs. 4.10 and 4.27); dotted curve, C_{EN} (Eqs. 4.11 and 4.28). Observational data are from Large and Pond (1982).

apparent in any of the observations to date, but the predicted increase in both neutral coefficients with decreasing wind speed in very light winds is evident in recent observations described by Bradley *et al.* (1991). For wind speeds less than about $1 \, \mathrm{m \, s^{-1}}$, free convection conditions must prevail so that the use of transfer coefficients and Monin–Obukhov theory becomes questionable. As with such conditions over land, a suitable free-convection formulation for sensible and latent heat fluxes is required, though in many numerical models of the atmosphere a minimum wind speed or friction velocity is set and the fluxes are evaluated using standard surface-layer formulations.

Several major problems arise with the measurements of C_{HN} and C_{EN}: firstly, the need to correct for stability effects; secondly, the effects at high wind speeds of spray and airborne evaporation of droplets; thirdly, the correct measurement of sea-surface temperature (Hasse, 1971; Deacon, 1977). Many past observations have used a "bucket" temperature, rather than the radiometric infrared "skin" temperature, so that differences in transfer coefficients between different data sets can be expected where the surface temperature observations are not consistent.

To expand on this last point, let T_0 be the true sea-surface temperature (as measured radiometrically, for example), and let T_w be the "bucket" temperature as measured anywhere from about 10 cm to about 50 cm beneath the surface. Then the temperature difference, $T_0 - T_w$, will arise predominatly because of the presence of the aqueous interfacial sublayer. In direct analogy to the treatment of the interfacial sublayer in the atmosphere, the scaled temperature difference can be identified with an aqueous Stanton number and formulated in a similar way to the atmospheric case, as described above. Observations, however, are inconclusive, though they do show that absolute values of $T_w - T_0$ may reach 1–2 °C at low wind speeds (Saunders, 1967; Hasse, 1971).

4.5 Local advection and the internal boundary layer

Internal boundary layers (IBLs) in the atmosphere are associated with the horizontal advection of air across a discontinuity in some property of the surface. Flow normal to a single, infinitely long discontinuity line represents the simplest of internal boundary-layer problems. Many studies have been carried out, and most can be categorized according to three major features:

 (i) the nature of the surface forcing (this is usually specified in terms of a step change in surface roughness, temperature or humidity, or in the surface flux of heat or moisture);
 (ii) the thermal stratification of the incident (upstream) flow (this is specified as neutral or non-neutral);
 (iii) the horizontal scale or upper limit of the downwind fetch (the growth of the IBL is confined to downwind fetches, of the order of either hundreds of metres (small-scale or microscale) or tens to hundreds of kilometres (mesoscale). In the microscale case, emphasis in the study is on the development of the inner layer whilst mesoscale studies emphasize growth of the whole ABL).

Much of the discussion throughout this book is concerned with ABL flow over a horizontally homogeneous surface of very large extent such that edge effects are not important. In such cases, mean horizontal advection plays no part in determining the local structure of the ABL. In Chapter 3, for example, surface-layer and boundary-layer laws were developed for homogeneous surfaces of specified z_0. Having discussed the physical and aerodynamical basis of z_0, and the analogous scalar roughness lengths, in the earlier part of this chapter we now deal primarily with small-scale characteristics of neutrally stratified flow when differences in surface roughness occur along the flow direction.

4.5.1 Definition of the IBL

Early studies (from the late 1950s to the mid-1970s) were concerned mainly with the problem of neutral flow across a step change in surface roughness. Almost without exception, these studies involved flows confined to small fractions of the upstream boundary-layer depth, i.e. to the wall region in laboratory flows and to the surface layer in the atmospheric case. Our discussion focuses on IBL growth and development in the atmosphere and is confined to neutral conditions and maximum downstream fetches of about 1 km. Discussion of the deeper (thermally stratified) IBL at greater fetches can be found in Chapter 6.

Figure 4.10 shows in schematic form the concept of an internal boundary layer for a step change in roughness, surface temperature, or surface flux, based, in part, on both microscale and mesoscale observations. Typical profiles of wind speed and potential temperature are also shown to illustrate the presence of the

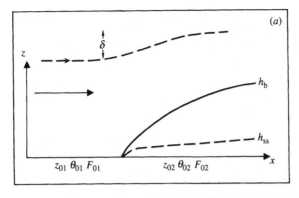

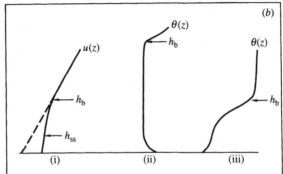

Fig. 4.10 (*a*) Schematic representation of the IBL depth $h_b(x)$ and inner equilibrium layer $h_{ss}(x)$ downstream of a step change in roughness z_0, in temperature θ_0 or in heat or moisture flux F_0. The streamline displacement δ is also shown. (*b*) Vertical profiles at a distance x downstream of the discontinuity: (i) wind profile for neutral flow across a z_0 change (the pecked curve represents the log law in the upstream flow); (ii) θ profile for an unstable IBL; (iii) θ profile for a stable IBL. From Garratt (1990); reprinted by permission of Kluwer Academic Publishers.

IBL. In the case of neutral flow, the IBL can be defined as the region influenced by the surface changes in z_0 and stress τ. In boundary-layer flow over a plate, for example, the IBL is characterized by velocities less than those in the free stream. Thus, the boundary-layer depth h_b was defined by Schlichting (1979, Chapter 7) as the level at which $u = 0.99u_\infty$, where u_∞ is the free-stream velocity (in the atmospheric case, there is no obvious analogue to u_∞). He also defined a displacement thickness and a momentum thickness, the latter being a height scale related to the total loss of momentum due to the boundary layer compared to potential flow. Both these scales are simple fractions of the boundary-layer depth, h_b.

In the case of a response to z_0 changes, in wind-tunnel or atmospheric flow, the IBL can be identified in a variety of ways. In the laboratory case, this may be through use of a velocity criterion (e.g. with the top of the IBL at $z = h_{b1}$ where this is the lowest level at which the velocity lies within one per cent of the upstream velocity at the same level) or a stress criterion (e.g. with the top of the IBL at h_{b2} where τ (at h_{b2}) is within one per cent of the upstream stress at that level). Typically, $h_{b1} < h_{b2}$ since velocity profiles are found to adjust more slowly than stress. In the atmosphere, the top of the IBL can usually be identified by $\partial u/\partial z$ discontinuities or wind-profile "kinks".

In addition to the IBL so defined, an inner layer (depth h_{ss}) exists with $h_{ss} \ll h_b$. This inner, or equilibrium, layer corresponds to the same region found in the unperturbed upstream boundary layer (as described in Chapter 3), in which the profile characteristics are fully governed by the local boundary conditions. Downstream of the leading edge, in neutral conditions, the inner equilibrium layer is characterized by a logarithmic profile form.

For non-neutral flow situations, the above criteria can be used to define h_b, with additional information available from temperature and heat-flux profiles.

4.5.2 Response to roughness changes

Observations of the development of an internal boundary layer downstream of a roughness change reveal the following features.

(i) Above the IBL (defined as described in the previous section) the flow field is characteristic of the upstream conditions, except for a displacement δ of the outer flow field (the streamlines) required by continuity.

(ii) Very near the ground, an inner or "equilibrium" layer exists where the wind profile has completely adjusted to the local boundary conditions.

(iii) Above this inner layer, and within the IBL as a whole, there exists a blending layer (Plate, 1971) in which the velocity distribution gradually changes from the logarithmic form (of the downstream roughness) at low heights to the profile form (of the upstream surface) at heights above the IBL.

(iv) At large distances from either side of the discontinuity, the shear stress at the surface adjusts to that of flow above a uniform surface.

Observations from wind-tunnel, pipe and duct experiments (Garratt, 1990) have concentrated on the development of the IBL and its internal mean and turbulent structure. Atmospheric observations, mainly emphasizing the modification to the low-level wind profiles, have been widely reported based on flow over roughness changes induced by both natural and artificial surfaces. They include simultaneous measurements of velocity profiles and surface stresses at several positions relative to discontinuities separating grass and tarmac, and tarmac and artificial roughness (Bradley, 1968).

The results for surface stress, both for flow from smooth to rough surfaces, and for flow from rough to smooth surfaces, are shown in Fig. 4.11 together with curves to be discussed later in the section. Both sets of data show a trend towards equilibrium stress values (as do the curves), but two features are of particular interest: in the smooth-to-rough case, the stress initially increases to about twice the final (large-fetch) value. In the rough-to-smooth case, the initial stress decreases to about one-half of the final value. This overshooting or undershooting of the downstream stress is evidently a transient response to the instantaneous roughness change and to the evolving velocity profile above. If the velocity profile changed abruptly to its downstream equilibrium form, so would the surface stress and a step change at the surface discontinuity would result.

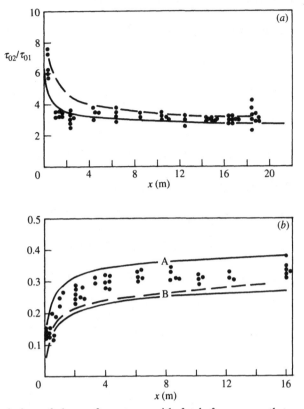

Fig. 4.11 (*a*) Variation of the surface stress with fetch for a smooth-to-rough transition; data points from Bradley (1968). The pecked curve is from Panofsky and Townsend (1964) and the continuous curve is from Rao *et al.* (1974). (*b*) As in (*a*) but for a rough-to-smooth transition. Curves A and B, from Rao *et al.* (1974), have z_{02} equal to 2×10^{-5} m and 2×10^{-6} m respectively. From Garratt (1990); reprinted by permission of Kluwer Academic Publishers.

Results for velocity profiles are shown in Fig. 4.12, together with numerical results from a higher-order closure model to be discussed shortly. Several features are of interest.

(i) There is a systematic shift of the profiles, left or right, as fetch increases and a systematic increase in the upper point of the modified profile as fetch increases – the latter representing the increase in the IBL depth with fetch.

(ii) The mean velocity profile is virtually unchanged above this upper point. This can be interpreted (Plate, 1971) as evidence for only a minor deflection of the streamlines, implying that δ is small (this does depend, however, on the relative change in the magnitude of the roughness).

(iii) At any value of x, there is only a shallow region of the profile in which the velocity distribution deviates from the low-level logarithmic form characteristic of downstream conditions, or the upper logarithmic form above the IBL top.

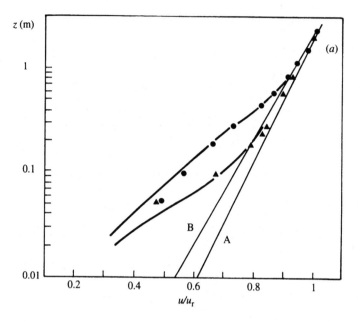

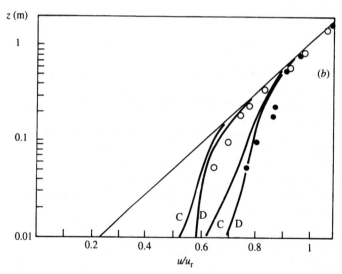

Fig. 4.12 (a) Adjustment of wind profile downwind of a smooth-to-rough transition ($z_{01} = 2 \times 10^{-5}$ m; $z_{02} = 0.0025$ m); data from Bradley (1968). Solid triangles, $x = 2.32$ m; solid circles, $x = 16.42$ m. The continuous curves are from Rao *et al.* (1974). Lines A and B are the upstream profiles assumed by Bradley and by Rao *et al.* respectively. In the abscissa lable, u_r is a reference velocity measured at $z = 2.2$ m. (b) As in (a) but for a rough-to-smooth transition (z_0 values reversed); open circles, $x = 2.1$ m; solid circles, $x = 12.2$ m. Curves C and D have the downstream roughness set at 2×10^{-5} m and 2×10^{-6} m respectively. The straight line is the upstream wind profile assumed by Rao *et al.* In the abscissa lable, u_r is a reference velocity measured at 1.125 m. Based on Garratt (1990).

The governing equations for simple two-dimensional steady mean flow, with rotation and pressure gradient terms set to zero, involve Eqs. 2.21 and 2.41 which can be written as (in reverse order),

$$u\partial u/\partial x + w\partial u/\partial z = \rho^{-1}\partial\tau/\partial z \tag{4.29}$$

$$\partial u/\partial x + \partial w/\partial z = 0. \tag{4.30}$$

To solve for $u(x, z)$ requires an equation for τ and suitable boundary conditions and, in general, numerical methods or approximate techniques are used to find this solution. The latter approach uses analytical theory where the crucial requirement is to represent the relation between stress and the velocity profile so as to solve for $\tau_0(x)$, $u(x, z)$ and $h_b(x)$. The Karman–Polhausen method (Schlichting, 1979, Chapter 10), which involves an integral constraint on the momentum balance of the whole IBL, is based on integration of Eq. 4.29 from the surface to h_b, and substitution for w from Eq. 4.30, to give

$$\partial/\partial x \left(\int_0^{h_b} u^2\,dz \right) - u_h\partial/\partial x \left(\int_0^{h_b} u\,dz \right) = \tau_h/\rho - u_{*2}^2 \tag{4.31}$$

where the subscript 2 refers to downstream flow. Solution of Eq. 4.31 allows $h_b(x)$ and $u_{*2}(x)$ to be evaluated. In this equation, u_h is given by u_1 at $z = h_b - \delta$ (the subscript 1 refers to upstream flow), with τ_h equated to the appropriate upstream stress. This simplistic treatment implies a discontinuity in stress, and is of course physically unrealistic. In order to close the equations the velocity distribution $u_2(z)$ must be specified, and is generally assumed to be given by

$$u_2(z)/u_{*2} = k^{-1}\ln(z/z_{02}) + f(z/h_b) \tag{4.32}$$

where $f(z/h_b) = 0$ for $z/h_b \ll 1$. Various forms for the blending function $f(z/h_b)$ have been suggested (see Plate, 1971), including $f = 0$, with Eq. 4.32 applied not just to the integral approach described above, but to Eqs. 4.29 and 4.30 directly.

Comparison of the theoretical results with Bradley's atmospheric observations of stress variation is shown in Fig. 4.11. This reveals, in the case of smooth-to-rough flow, that at small x/z_0 the data exhibit a more rapid variation of surface stress than that predicted by any theory. In the case of rough-to-smooth flow, the stress variation is well described although absolute values disagree; this is due, in part at least, to the sensitivity to the chosen z_0 value. The comparison of theoretical velocity profiles with observation is shown in Fig. 4.12. Their form and relative displacement have considerable similarity, although the observed profile inflection point appears much more pronounced. The acceleration (deceleration) of low-level air for a rough-to-smooth (smooth-to-rough) transition is very evident in the data.

The crucial shortcoming of many of the theories alluded to above, including diffusion approaches (Philip, 1959) and studies related to temperature changes at the surface (Taylor, 1971), is a failure to represent properly the relation between shear stress and the velocity profile in a transition or non-equilibrium situation. Under such conditions, relations such as $u_* = kz\partial u/\partial z$ are not necessarily appropriate (as evidenced in numerical studies to be discussed shortly), and the local stress may be determined by non-local flow conditions. Put another way, the relation between stress and velocity gradient is not determined solely by mixing length and eddy diffusivity concepts. This problem can be overcome to a great extent by carrying an equation for τ, or for turbulent kinetic energy, and solving the equations numerically.

An equation for τ can be incorporated by utilizing the TKE equation (Eq. 2.72) and the relation between stress and the turbulent kinetic energy, such that $\tau/\rho \propto \bar{e}$ and $\varepsilon = (\tau/\rho)^{3/2}/\lambda$, λ being a length scale having a value of kz close to the surface. The vertical divergence term is parameterized by assuming a flux–gradient relation, with diffusivity identical to that assumed in the stress–velocity-gradient relation (Chapter 8). The problem with this TKE approach is that the relations used between stress and dissipation are probably valid only in constant-stress equilibrium layers rather than in transitional flow regimes.

An alternative approach is to carry the full second-order turbulence equations with suitable parameterization of the third-order terms. Based on this method, the higher-order closure results of Rao *et al.* (1974) are shown in Fig. 4.11 for variation of surface stress with fetch; Bradley's data are given for comparison. Generally the stress distribution is better predicted by this numerical method than by any of the analytical methods. The numerical simulations also provide vertical profiles of stress and other turbulent statistics that are typical of those available only from detailed wind-tunnel experiments (e.g. Mulhearn, 1978). A comparison of velocity profiles is shown in Fig. 4.12. Quite good agreement is found and, in particular, the numerical results, represented by the curves in (*a*) and (*b*), show a transitional velocity profile in the blending region with an inflection point, similar to that actually observed.

The failure of simple mixing-length theory in the region immediately downwind of the roughness discontinuity is well illustrated by reference to the numerical results for the nondimensional gradient Φ, which is given by $\Phi = (kz/u_*)\partial u/\partial z$. Its variation with dimensionless height is shown in Fig. 4.13, where values differing from unity, and hence from equilibrium values, reveal the effect of the transitional flow upon Φ (and other similarity variables) and the failure of mixing-length assumptions based on $l = kz$ (this equation implies $\Phi = 1$).

4.5.3 Growth of the IBL at a roughness change

The growth relation depends to some extent on how the modified region or IBL is defined (see Section 4.5.1). Observations of turbulent flow from smooth to rough surfaces and from rough to smooth surfaces in the atmosphere (e.g. Bradley, 1968) are generally consistent with turbulent boundary-layer growth over a smooth plate, in which case $h_b \propto x^{0.8}$ (Schlichting, 1979, Chapter 21). This result is compared with numerical simulations in Fig. 4.14. Displacement of the curves usually results from different definitions of the IBL. Some studies have attempted to include explicitly the effect of roughness, or roughness change, into the formulation. For example, suggested forms include $h_b \propto x^{0.8}z_0^{0.2}$ (Wood, 1982), with z_0 the greater of z_{01} and z_{02}, and also a more general form for neutral flow (Shir, 1972):

$$h_b = f_1(z_{01}/z_{02})x^{0.8+f_2(z_{01}/z_{02})}. \tag{4.33}$$

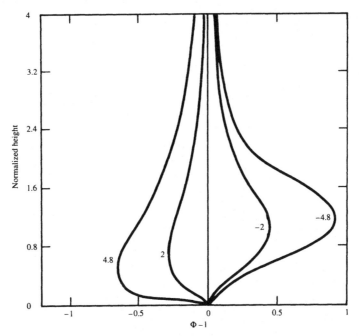

Fig. 4.13 Variation of the nondimensional wind shear $\Phi - 1$ with normalized height, for several values of the roughness-change parameter $M = \ln(z_{01}/z_{02})$, at $x = 2$ m. Results are from the numerical simulations of Rao *et al.* (1974). From Garratt (1990), reprinted by permission of Kluwer Academic Publishers.

According to numerical results, the function f_2 is slightly negative in rough-to-smooth flow, and close to zero for smooth-to-rough flow.

It is possible to utilize a diffusion analogue (i.e. the principle of limited diffusion rate) to evaluate the slope of the IBL and hence obtain a relation between h_b and x (Panofsky, 1973). This uses the concept of zones of influence, with an analogy between the zone influenced by the downstream roughness and the spread of a smoke plume from a ground-level source in uniform roughness (Miyake, 1965). In this, u_* or the vertical velocity variance $\sigma_w{}^2$ is assumed to determine the growth rate, with

$$dh_b/dx \propto \sigma_w/u(h_b) \propto u_*/u(h_b) \propto k/\ln(h_b/z_0) \tag{4.34}$$

and, after integration,

$$(h_b/z_{02})[\ln(h_b/z_{02}) - 1] + 1 = Ax/z_{02} \tag{4.35}$$

with $A \approx 1$.

Analogous relations to Eq. 4.35 can be derived (Pasquill and Smith, 1983) by considering the mean vertical displacement $\bar{z}$ of passive particles diffusing in space at a given time after release from the ground. Using $d\bar{z}/dt = ku_*$, together with Lagrangian similarity arguments (after Batchelor), leads to Eq. 4.35, with h_b replaced by $\bar{z}$ and $A = k^2$. Good correspondence with IBL growth relations, for example, can be achieved with $h_b = 3\bar{z}$. An extension of

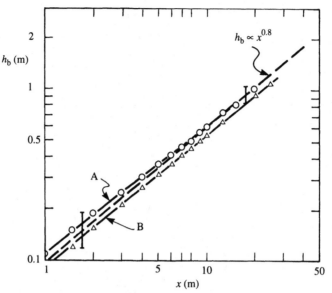

Fig. 4.14 Growth of the IBL based on the model of Rao *et al.* (1974). For comparison, the data of Bradley (1968) are given, represented by the pecked curve (the two vertical bars give an indication of experimental scatter; see Bradley's Fig. 11). The solid curves A and B, which join computed values, correspond to different values of M (A and B have $\ln(z_{01}/z_{02}) = 4.8$ and -4.8 respectively). Based on Garratt (1990).

Miyake's theory (Jackson, 1976) leads to a small correction term (for origin effects) on the left-hand side of Eq. 4.35, with z_{02} replaced by $z_0' = 0.5(z_{01}^2 + z_{02}^2)^{0.5}$. This modified relation agrees satisfactorily with both atmospheric and wind-tunnel measurements.

In terms of the growth of the IBL in the first kilometre or so downwind of the roughness change, Eq. 4.35 allows the fetch–height ratios that are of micro-meteorological relevance to be evaluated. Such data are shown in Table 4.2 for smooth-to-rough flow, with $z_{02} = 0.1$ m, and rough-to-smooth flow, with $z_{02} = 0.001$ m. We have assumed that the inner layer, of depth h_{ss}, is one-tenth of the IBL depth. More rapid growth is demonstrated for the smooth-to-rough case, evidently the result of more intense vertical mixing over the rough surface, with height-to-fetch ratios for the inner layer of about $1/100$ in this case compared with about $1/200$ in the rough-to-smooth case.

The effects of thermal stratification on IBL structure and growth in relation to a roughness change have often been studied in association with the impact of surface heat flux and temperature changes. In the context of the IBL depth, with $h_b \propto x^n$, numerical work (Rao, 1975) has shown that n tends to increase from 0.8 in neutral conditions to about 1.4 in strongly unstable conditions: a much more rapid growth occurs when surface heating is present. Thus, the height-to-fetch ratios quoted above are increased well above $1/100$ in the unstable case (values approach $1/10$) and well below in stable conditions (values approach $1/500$).

Table 4.2. *Calculation of the fetch x required to give an IBL of specified depth h_b, based on Eq. 4.35 with $A = 0.5$, for two values of the downwind roughness length z_{02}, corresponding to smooth-to-rough ($z_{02} = 0.1$ m) and rough-to-smooth ($z_{02} = 0.001$ m) transitions. The depth of the inner equilibrium layer h_{ss} is assumed to be equal to $0.1h_b$*

h_b (m)	h_{ss} (m)	smooth-to-rough		rough-to-smooth	
		x (m)	h_{ss}/x	x (m)	h_{ss}/x
2	0.2	8	1/41	26	1/132
5	0.5	29	1/59	75	1/150
10	1.0	72	1/71	160	1/164
20	2.0	172	1/83	360	1/180
50	5.0	522	1/104	980	1/196
100	10.0	1182	1/118	2100	1/210

Notes and bibliography

Sections 4.1 and 4.2

Early and more recent work on the z_0 and d dependence upon surface characteristics, and descriptions of the interfacial sublayer and scalar roughness schemes, can be found in Chapters 4 and 5 of Brutsaert's book.

As well as identifying d as the level of action of the surface drag on the roughness elements, there have been other attempts to interpret the concept of the zero-plane-displacement physically: (i) by matching in-canopy and atmospheric wind profiles at $z = h_c$, as in

Seginer, I. (1974), Aerodynamic roughness of vegetated surfaces, *Bound. Layer Meteor.* **5**, 383–93;

(ii) by defining d as the vertical displacement of air trajectories passing from a "smooth" reference surface to the rough surface under consideration, as in

De Bruin, H. A. R. and C. J. Moore (1985), Zero-plane displacement and roughness length for tall vegetation, derived from a simple mass conservation hypothesis, *Bound. Layer Meteor.* **31**, 39–49.

For a review of wind-tunnel and atmospheric data relating to the scalar roughness, see

Garratt, J. R. and B. B. Hicks (1973), Momentum, heat and water vapour transfer to and from natural and artificial surfaces, *Quart. J. Roy. Met. Soc.* **99**, 680–7.

A modified form of Eq. 4.14 (with a weaker Re_* dependence) for use at the Earth's surface was proposed by

Zilitinkevich, S. S. (1970), *Dynamics of the Atmospheric Boundary Layer* (trans. C. Long), Hydrometeorol., Leningrad, 239 pp.

Section 4.3

Recent reviews of transfer processes in plant canopies, with emphasis on the breakdown of flux–gradient relations, can be found in

Denmead, O. T. and E. F. Bradley (1985), Flux–gradient relationships in a forest canopy, in *The Forest–Atmosphere Interaction*, eds B. A. Hutchison and B. B. Hicks, pp. 421–42. Reidel, Dordrecht, and in

Finnigan, J. J. and Raupach, M. R. (1987), Transfer processes in plant canopies in relation to stomatal characteristics, in *Stomatal Function*, eds E. Zeiger, G. Farquahar and I. Cowan, pp. 385–429. Stanford University Press, Stanford, CA.

Section 4.4

The values of z_T and z_q given by Eq. 4.26 are similar in magnitude to those of Liu *et al.* (1979), but about one-half of those given by Brutsaert (1982).

Section 4.5

For a recent review on the internal boundary layer and local (as well as mesoscale) advection, the reader should consult

Garratt, J. R. (1990), The internal boundary layer – A review, *Bound. Layer Meteor.* **50**, 171–203,

and Chapter 7 of Brutsaert (1982).

The atmospheric surface-layer data of

Bradley, E. F. (1968), A micrometeorological study of velocity profiles and surface drag in the region modified by a change in surface roughness, *Quart. J. Roy. Met. Soc.* **94**, 361–79,

are ideal for comparison with theoretical and numerical results.

Walmsley, J. L. (1989), Internal boundary-layer height formulae – A comparison with atmospheric data, *Bound. Layer Meteor.* **47**, 251–62,

has considered several IBL depth formulae and compared their predictions with atmospheric data, with fetches confined to less than about 200 m. Equation 4.35 was found to give the best predictions, with $A = 1.25k$.

5

Energy fluxes at the land surface

In this book, the surface is treated primarily as the lower boundary to the atmosphere. For the land surface, the balance of energy fluxes is an important boundary condition, whilst the important elemental surfaces are bare soil and leaves. In the first few chapters, we have not needed to make the distinction between soil or vegetation, and have treated the vertical turbulent fluxes of sensible heat (H_0) and latent heat (λE_0) at the land surface as though they were controlled solely through relations such as Eqs. 3.46 and 3.51, that is, as an interaction of turbulence with established gradients of temperature and humidity. However, the balance of energy fluxes at the surface places a constraint upon the sum of the turbulent fluxes, $H_0 + \lambda E_0$, thus emphasizing the importance of partitioning the *available energy* between the sensible and latent heat fluxes. In addition to this energy balance, the flux of water from the surface into the atmosphere, in the absence of precipitation, must be balanced by the flux of water through the soil towards the surface.

The present chapter discusses the physical significance of several of the individual energy fluxes comprising the energy balance equation, with particular emphasis on the soil heat flux, the radiation fluxes and the moisture flux to and from both vegetated and bare-soil surfaces. Discussion of the moisture flux, and its relation to surface humidity, requires consideration of the concept of surface resistance (vegetation) and soil moisture transfer (bare soil). The treatment provides a useful set of relations upon which to base specific parameterization schemes for the energy fluxes from soil and canopies to be used in numerical models (Chapter 8).

We adopt the sign convention for vertical fluxes, molecular, turbulent and radiative, given in Section 1.7.2: non-radiative fluxes directed away from the surface and radiative fluxes directed towards the surface are taken as positive.

5.1 Surface energy balance and soil heat flux

The primary forcing of the ABL over the land is through solar radiation absorption at the ground which generally results in a diurnal variation in both the surface temperature and the turbulent heat fluxes. The surface temperature itself is constrained by the requirement for a balance of all energy fluxes at the Earth's surface, and the equation describing this energy balance may be used to evaluate the surface temperature for a set of given conditions.

5.1.1 Surface energy balance

In the absence of vegetation, the atmosphere–surface interface is relatively well defined. Conservation of energy at the interface, both instantaneously and averaged in time, requires that

$$R_{N0} - G_0 = H_0 + \lambda E_0 \tag{5.1}$$

which is a surface energy balance (SEB) equation. Here R_{N0} is the net radiative flux density to the surface (i.e. the net radiation) and G_0 is the heat flux into the soil; the sum of the fluxes on the left-hand side of Eq. 5.1 defines the *available energy*. All fluxes have SI units of $W\,m^{-2}$. It is important to note that H_0 not H_{v0}, appears in Eq. 5.1; if H_{v0} is used then the right-hand side becomes $H_{v0} + 0.93\lambda E_0$ for a temperature of 300 K (based on Eq. 2.80). The flux G_0 is to be interpreted as the soil heat flux *at the surface*. The ratio of the turbulent fluxes is represented by the Bowen ratio, $B = H_0/\lambda E_0$. Where the atmospheric fluxes are referred to a level z within the atmospheric surface layer, Eq. 5.1 is exact only when the horizontal advection of heat in the layer between z and the surface is negligible and the vertical fluxes do not vary with height.

For a thin layer of soil of thickness $\Delta z'$, neglecting any horizontal conduction of heat in the soil, the time rate of change in the temperature of the soil layer is governed by the difference in the heat fluxes into and out of the layer. The relation between G_0, the surface soil heat flux, and G_1, the heat flux at a small depth in the soil, can be found using the requirement for heat conservation in this thin layer. With T_s the soil temperature, and z' the vertical coordinate (positive down) in the soil, then

$$\rho_s c_s \partial T_s/\partial t = - \partial G/\partial z' \tag{5.2}$$

where $C_s = \rho_s c_s$ is termed the *volumetric heat capacity* (units $J\,m^{-3}\,K^{-1}$), ρ_s is the soil density and c_s is the soil specific heat. In finite difference form, Eq. 5.2 can be written, for a layer of depth $\Delta z'$ adjacent to the surface,

$$\rho_s c_s \Delta z' \partial T_s/\partial t = G_0 - G_1. \tag{5.3}$$

This allows Eq. 5.1 to be written as

$$R_{N0} - G_1 = H_0 + \lambda E_0 + \partial W_s/\partial t \tag{5.4}$$

where the energy storage is given by $W_s = \rho_s c_s \Delta z' T_s$ and $\partial W_s/\partial t$ is the rate of energy storage per unit area in the soil layer.

For a canopy, or layer of vegetation, of depth h_c, the balance equation at the canopy top can be written

$$R_{N0} - G_0 = H_0 + \lambda E_0 + D_h + \partial W_c/\partial t \tag{5.5}$$

where $\partial W_c/\partial t$ is the rate of energy storage per unit area in the canopy layer, with $W_c = \rho_c c_c h_c T_c$. Here, ρ_c is the canopy density, c_c is the canopy specific heat and T_c is canopy temperature. The term D_h is a horizontal flux of energy due to advection; generally, this term may be neglected, but could be significant near the edges of crops or forests (Thom, 1975). In Eq. 5.5, the turbulent heat fluxes are taken to be the fluxes at the top of the canopy, and are interpreted as the surface fluxes. In many practical situations, both D_h and $\partial W_c/\partial t$ are negligible, and Eq. 5.5 is then identical to Eq. 5.1 for bare soil.

5.1.2 Soil temperatures and heat flux

Values of the soil heat flux at the interface or at a shallow depth depend on many factors, including solar radiation (hence time of day), soil type (hence physical properties) and soil moisture content. On occasions, the flux may be a significant fraction of the net radiation and must be taken into account – for example, when the magnitude of available energy is required or when the surface energy balance equation is to be solved for the surface temperature (see Section 8.2).

The subsurface heat flux G at any level z' can be described by Fourier's law for heat conduction in a homogeneous body: i.e. this flux depends on the temperature gradient as follows:

$$G = -k_s \partial T_s/\partial z' \tag{5.6}$$

where the thermal conductivity $k_s = \rho_s c_s \kappa_s$, with κ_s the thermal diffusivity. Equation 5.6 neglects a latent heat term involving the vertical gradient of soil moisture content because this is usually unimportant. However, the dependence of both k_s and κ_s upon wetness is significant, since, during wetting, air (a relatively poor conductor of heat) in the soil pores is replaced by water (a good conductor). Table A7 in Appendix 4 illustrates the range of values of k_s, ρ_s, c_s and κ_s for several varieties of soil and other surface types. The thermal conductivity shows the greatest variation, varying by an order of magnitude between dry soils and rock (0.25–2.9 W m^{-1} K^{-1}).

Combination of Eqs. 5.2 and 5.6 leads to the heat conduction equation

$$\partial T_s(z', t)/\partial t = \kappa_s \partial^2 T_s(z', t)/\partial z'^2 \tag{5.7}$$

which can be solved for T_s and hence G. Some insight into the dependence of $T_s(z', t)$ on surface forcing and soil thermal properties can be gained by considering the solution to Eq. 5.7 for a purely sinusoidal forcing function. This relates closely to the problem of the penetration of diurnal and annual temperature waves into the ground (Hillel, 1982). Such a solution also allows estimates of κ_s to be made in the field from observations of $T_s(z', t)$ over a period of time.

In the case of diurnal forcing, we take the upper boundary condition (at $z' = 0$) as $T_s(0, t) = T_0(t) = \bar{T} + A_0 \sin \Omega t$, where T_0 is the surface temperature, $2A_0$ is the diurnal temperature amplitude, $\bar{T}$ is the daily average surface temperature and Ω is the angular velocity of the Earth's rotation (7.292×10^{-5} rad s^{-1}). For the lower boundary condition, $T_s(\infty, t) = \bar{T}$, the deep soil temperature. Solution of Eq. 5.7 with these boundary conditions gives

$$T_s(z', t) = \bar{T} + A_0 \exp(-z'/D) \sin(\Omega t - z'/D) \tag{5.8}$$

where D is referred to as the *damping depth* and is given by

$$D = (2\kappa_s/\Omega)^{1/2}. \tag{5.9}$$

It is the depth at which the temperature amplitude has decreased to $1/e$ of the surface amplitude; for the diurnal wave, and for typical soils, $D \approx 0.1$ m. The solution gives an exponential decrease with depth, and a phase shift of z'/D. This is shown in Fig. 5.1 for a complete day and for several depths using assumed soil properties. At a depth $z' = 0.4$ m the peak lags the surface wave by about 12 hours. In the case of the annual cycle, Ω is replaced by $2\pi/(365$ days) so that D increases by a factor of $(365)^{1/2} \approx 19$, and for a typical soil will be about 2 m.

The soil heat flux can now be determined from Eq. 5.6 using the solution given by Eq. 5.8. At $z' = 0$ this reads

$$G_0 = \rho_s c_s (\kappa_s \Omega)^{1/2} A_0 \sin(\Omega t + \pi/4) \tag{5.10}$$

so that the heat flux is $\pi/4$ out of phase with the temperature wave (e.g. maximum G_0 occurs three hours before the maximum surface temperature). Before proceeding, two comments are necessary concerning the practical application of the above theory. Firstly, G_0 cannot easily be measured, so measurement is usually made of the heat flux G_1 at a small depth, z'. The experimental technique involves several heat-flux plates buried horizontally in the soil at a depth of about 10 mm or so. An average of the flux-plate measurements is taken

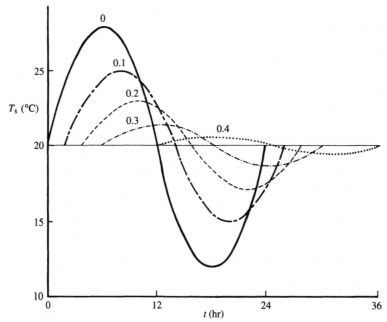

Fig. 5.1 Idealized variation of soil temperature through a diurnal cycle for several depths in the soil (indicated in metres). The curves represent the solutions to Eq. 5.7 for sinusoidal forcing; these are given by Eq. 5.8. A uniform soil is assumed with $\kappa_s = 0.8 \times 10^{-6}$ m^2 s^{-1} and $k_s = 1.68$ W m^{-1} K^{-1}.

so as to give a statistically reliable result. The flux plates actually measure the temperature difference across the plate thickness, and knowledge of soil properties and reliable calibration of the plates then allows evaluation of the soil flux, G_1. To calculate G_0 from G_1 requires use of Eq. 5.3; in this, the storage term $\partial W_s/\partial t$ will be significant only when $\partial T_s/\partial t$ is large, i.e. in late afternoon hours, and around sunrise and sunset. In such situations, with flux plates located at depths of about 10 mm, $\partial W_s/\partial t$ may reach several tens of $W m^{-2}$.

The second comment concerns the solution for soil temperature represented by Eq. 5.8. In practice, the diurnal forcing is not sinusoidal but the sum of a set of harmonics, so this equation is a crude representation of the soil temperature variations. A more appropriate solution is to represent the temperature forcing at the surface as a Fourier series (see Chapter 8). Solutions are obtained numerically in models that include finite difference forms of Eq. 5.7. Figure 5.2 shows solutions for two different soil types (low and high thermal conductivity) in the form of T_s profiles for mid-afternoon and for sunrise, under clear skies. The reduced diurnal amplitude for the wet soil with high conductivity is particularly evident. In the presence of a canopy, diurnal forcing at the ground surface is likely to be much reduced compared to that over a bare soil surface, mainly because the canopy will shade the soil surface.

5.1.3 Heat storage in canopies

For forest canopies in particular, the storage term $\partial W_c/\partial t$ in Eq. 5.5 may be significant at certain times of the day, particularly when canopy temperatures are changing rapidly (Stewart and Thom, 1973). The total rate at which heat energy is stored within a column of unit cross-sectional area extending from the

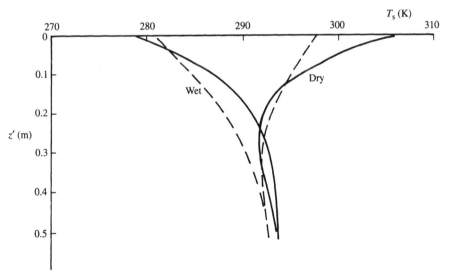

Fig. 5.2 Vertical profiles of temperature within the soil, for saturated and dry sand (physical properties are listed in Table A7). Profiles based on numerical solution of Eq. 5.7 in an interactive soil–atmosphere model are shown for mid-afternoon and for sunrise.

soil surface to the level $Z = h_c$ (the height to which R_{N0}, H_0 and λE_0 refer) is given by

$$\partial W_c / \partial t = S_T + S_q + S_{veg} \qquad (5.11)$$

where

$$(S_T, S_q, S_{veg}) = \int_0^{h_c} [\partial(\rho c_p T, \rho \lambda q, \rho_{veg} c_{veg} T_{veg})/\partial t] \, dZ. \qquad (5.12)$$

In the above, S_T and S_q are respectively the changes of the sensible and latent heat contents of the column of air within the canopy, and S_{veg} is the rate of change of the heat content of the vegetation; ρ_{veg}, c_{veg} and T_{veg} are respectively its density, specific heat and temperature. For a tall forest, the storage term may reach several tens of $W\,m^{-2}$ under extreme conditions.

5.2 Radiation fluxes

Equation 5.1 contains the net radiation which, with the adopted sign convention, is generally positive by day and negative by night, over land. R_{N0} contains a combination of shortwave and longwave fluxes as follows:

$$R_{N0} = R_{s0}(1 - \alpha_s) + \varepsilon_s R_{L0}^d + R_{L0}^u \qquad (5.13)$$

where R_{s0} is the global shortwave radiative flux (positive), R_{L0}^d is the downward longwave flux (positive) and R_{L0}^u is the upward longwave flux (negative), all at the surface. In addition, α_s is the shortwave albedo of the surface and ε_s is its longwave emissivity. In Eq. 5.13, we have allowed for an amount $(1 - \varepsilon_s)R_{L0}^d$ of reflected longwave radiation, which equals zero if the surface is a black body with $\varepsilon_s = 1$.

The flux of radiation emitted by a black body is given by the Stefan–Boltzmann law as σT^4, where $\sigma = 5.67 \times 10^{-8}\,W\,m^{-2}\,K^{-4}$ is the Stefan–Boltzmann constant. Most natural surfaces are "grey" rather than "black". They emit a longwave radiative flux of magnitude $\varepsilon_s \sigma T^4$ given by the modified Stefan–Boltzmann law, where ε_s is less than unity.

5.2.1 Shortwave fluxes

The spectrum of solar radiation received at the top of the atmosphere approximates that of a black body with surface temperature close to 6000 K, and maximum energy emitted at a wavelength of $0.48\,\mu m$ (Wien's Law). For the idealized solar spectrum, the shortwave flux is mainly confined to wavelengths between 0.1 and $4\,\mu m$. At the top of the atmosphere, this shortwave flux is closely related to the *solar irradiance* (formerly the *solar constant*), defined as the flux of solar radiation passing through a plane normal to the solar beam at the top of the atmosphere, with the Earth at its mean annual distance from the Sun. Its value is close to $1367\,W\,m^{-2}$ and will be denoted by S_c. The amount of radiation received at the top of the atmosphere depends mostly on the actual distance of the Earth from the Sun.

The solar flux reduces with depth through the atmosphere, due to scattering, absorption and reflection by clouds and atmospheric turbidity (particles). At the

Earth's surface, the shortwave flux has direct and diffuse components; the diffuse radiation arises from scattering in the atmosphere by molecules (Rayleigh scattering) and suspended particles and, in cloudy conditions, from reflection from clouds. The relation between R_{s0} and S_c can then be written as

$$R_{s0} = \tau_a \sin \Psi S_c^{app} \tag{5.14}$$

where Ψ is the solar elevation angle, S_c^{app} is the apparent solar irradiance taking into account the actual distance of the Earth from the Sun and τ_a is the transmissivity of the atmosphere, or fraction of shortwave radiation reaching the surface when $\Psi = \pi/2$. Typically, with the sun directly overhead, $\tau_a \approx 0.8$ in cloudless skies, and $\tau_a \approx 0.1$ for a complete overcast of low, medium and high clouds. At night (negative Ψ), $R_{s0} = 0$.

A significant fraction of the incident shortwave flux at the surface may be reflected back to the atmosphere; this is related to the albedo of the surface, α_s, defined as the ratio of reflected global shortwave flux to the incident flux. It represents an integral value of the surface reflectivity over the wavelength range $0.1–4 \mu m$, and contrasts with the spectral albedo appropriate to a specific wavelength. The albedos of many natural surfaces are sensitive to the solar elevation angle (water is a good example). For soil in particular, the albedo is strongly dependent on wetness. Table A8 in Appendix 4 gives a summary of mean albedos for a range of surface types, and for several cloud types. They typically range from about 0.1 or less for tropical rainforest to 0.9 for fresh snow.

5.2.2 Longwave fluxes

The absorption and emission of radiation by the "greenhouse gases" in the atmosphere occur in a series of wavelength bands, and are strongly dependent on the vertical distributions of all the relevant gases. Because the Earth–atmosphere system has a much lower temperature than the sun, the emitted radiation has a much longer wavelength than that for the sun. At temperatures near 273 K, the radiation is mainly confined to the wavelength range $4–100 \mu m$ i.e. the infrared region.

Let us consider the general problem of evaluating downwards and upwards longwave fluxes at any height z within the troposphere. In order to avoid radiative transfer models that use numerical solutions of the radiative transfer equations (with line-by-line or band absorption and emission information), an emissivity or grey-body approximation is made. In this, each infinitesimal atmospheric layer (Fig. 5.3) is represented by a single longwave emissivity or transmissivity averaged over all wavelengths, so that the downwards flux at any level z can be written as (Paltridge and Platt, 1976, Chapter 7)

$$R_L^d(z) = \int_z^\infty \sigma T(z'')^4 [\partial \varepsilon(z'', z)/\partial z''] \, dz'' \tag{5.15}$$

$$\approx \sum_{j=1}^\infty \{\sigma T_j^4 [\varepsilon_j(m + dm) - \varepsilon_j(m)]\}.$$

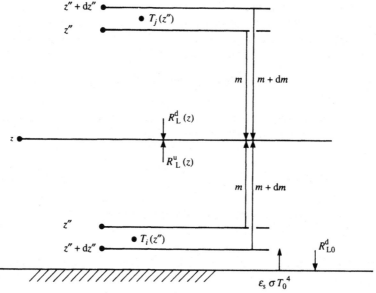

Fig. 5.3 Schematic diagram of the contributions to the downward and upward fluxes at height z from the surface and infinitesimal layers of thickness dz''; m is the mass of absorbing gases, corrected for pressure effects.

Likewise, the upwards flux is given by

$$R_L^u(z) = \int_z^0 \sigma T(z'')^4 [\partial \varepsilon(z'', z)/\partial z''] \, dz'' \tag{5.16}$$

$$+ [R_{L0}^u - (1 - \varepsilon_s)R_{L0}^d][1 - \varepsilon(z, 0)]$$

$$\approx \sum_{i=1}^k - \{\sigma T_i^4[\varepsilon_i(m + dm) - \varepsilon_i(m)]\}$$

$$+ [R_{L0}^u - (1 - \varepsilon_s)R_{L0}^d][1 - \varepsilon(m_k)].$$

In Eq. 5.16, allowance has been made for reflection of the downwards longwave flux at the surface. In addition, the quantity $\varepsilon(z, 0) = \varepsilon(m_k)$ is the effective emissivity of the air between the surface and height z; $\varepsilon(z'', z)$ is the emissivity for the corrected mass m of absorber corresponding to a vertical path from height z to z''. Assuming that in the atmosphere n gases absorb and emit in distinct spectral bands, then the total emissivity ε is the sum of the individual emissivities, less the emissivity due to overlapping of absorption bands e.g. the carbon-dioxide and water vapour bands at 13 μm. Thus, combining the emissivities of water vapour (ε_1), the water vapour dimer in the window region between 8 and 12 μm (ε_2), carbon dioxide (ε_3) and ozone (ε_4) ε is given by

$$\varepsilon = \varepsilon_1 + \varepsilon_2 + \varepsilon_3 + \varepsilon_4 - \varepsilon_1\varepsilon_3. \tag{5.17}$$

In Eq. 5.17, $\varepsilon_1 = \varepsilon_1(m_q)$, $\varepsilon_2 = \varepsilon_2(m_q e)$, $\varepsilon_3 = \varepsilon_3(m_c)$, $\varepsilon_4 = \varepsilon_4(m_O)$, where m_q is the corrected absorber mass of water vapour, m_c and m_O are the absorber amounts of carbon dioxide and ozone respectively, and e is the water vapour pressure in air. In practice, each emissivity ε_n is represented as a set of empirical functions of the absorber amounts, and Eqs. 5.15 and 5.16 are solved numerically in finite difference form (see Fig. 5.3). Such calculations and observations routinely show that contributions to the incoming (downwards) longwave flux are mainly confined to atmospheric levels below about 200 mb, and that in cloud-free, night-time conditions there is a net loss of energy at the surface, with negative R_N.

Turning now to the surface fluxes, the upwards longwave flux R_{L0}^u (terrestrial radiation) results from longwave emission from the surface. For grey-body emission, it is given by

$$R_{L0}^u = -\varepsilon_s \sigma T_0^{\,4} \qquad (5.18)$$

where T_0 is the surface temperature. Most natural surfaces are almost black, with ε_s greater than 0.9; values for a range of surfaces are shown in Appendix Table A8. The downwards longwave flux R_{L0}^d results from emission of infrared radiation from the whole atmosphere, and from clouds. Low clouds, in particular, are dominant since they tend to radiate as black bodies at their cloud-base temperatures. This downwards flux is readily measured, but its computation is by no means straightforward, even in the absence of clouds. Under clear skies, it is strongly dependent on the vertical distributions of temperature and humidity throughout the troposphere and particularly within the ABL.

Several empirical relations exist for the downwards longwave flux R_{L0}^d, both in clear sky and cloudy conditions, which will give values to within about 20 per cent. For example, Swinbank's (1963) relation is based on sea-level night-time data under clear skies, and relates the downwards flux at the surface to the screen temperature T_a,

$$R_{L0}^d \approx 0.94 \times 10^{-5} \sigma T_a^{\,6}. \qquad (5.19)$$

If the surface is assumed to radiate as a black body, an approximate expression for the net longwave flux, R_{N0}^{lw}, in the absence of clouds, utilizes Eq. 5.19, giving

$$R_{N0}^{lw} \approx -\sigma T^4 (1 - 0.94 \times 10^{-5} T^2) \qquad (5.20)$$

where we have assumed $T_0 = T_a$. The function $R_{N0}^{lw}(T)$ is shown plotted in Figure 5.4, suggesting minimum values of net radiation of between -90 W m^{-2} and -100 W m^{-2} that are consistent with field observations, and a maximum diurnal amplitude in the magnitude of the net longwave flux of only 20–30 W m^{-2}. Clouds have a large impact on all the radiative fluxes appearing in Eq. 5.13, and on the net radiation during the night-time. Taking Eq. 5.20 as the cloudless net longwave radiation, a simple correction for clouds (cloud-cover fraction C) is to take (Paltridge and Platt, 1976, Chapter 6)

$$R_{N0C}^{lw} \approx R_{N0}^{lw} + 0.3\varepsilon_c \sigma T_{cb}^{\,4} C \qquad (5.21)$$

where ε_c and T_{cb} are respectively the emissivity and the temperature of the cloud base. The effect of the cloud is seen to be an increase in the downwards longwave flux at the surface, and hence a reduction in the magnitude of the (negative) net radiation. The flux from the cloud assumes that only 30 per cent

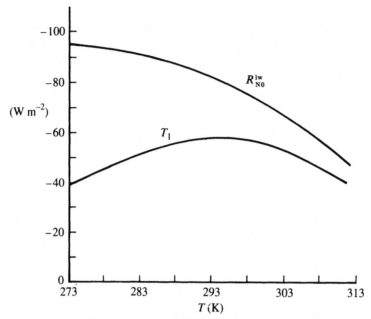

Fig. 5.4 Empirical values of net longwave radiation in cloudless conditions, as a function of absolute temperature, based on Eq. 5.20. The curve labelled T_1 refers to the discussion on dewfall given in Section 5.4, and represents maximum values of dewfall given by the first term on the right-hand side of Eq. 5.50, i.e. ΓR_{N0}, where $\Gamma = s/(s + \gamma)$. Γ is thus dependent solely on temperature, and values of R_{N0} are as given by the upper curve in the figure.

of the total energy in the black body spectrum exists in the window region (8–12 μm), and that the clouds affect the net flux at the surface only via the window region. For low clouds, $\varepsilon_c \approx 1$, for medium-level clouds, $\varepsilon_c \approx 0.9$ and for cirrus clouds, $\varepsilon_c \approx 0.3$.

Illustrative diurnal variations, under cloudless skies, of the net radiation, atmospheric and soil heat fluxes at the surface are shown in Fig. 5.5 for observations and for numerical simulations. Relatively high values of the soil heat flux are observed during the daytime (reaching a maximum of 100–200 $\mathrm{W\,m^{-2}}$ near midday), mainly because the surfaces are bare (*b*) or only sparsely vegetated (*a*). With dense vegetation, soil heat fluxes are usually small because shortwave fluxes at the soil surface are small. The phase lag of 1–2 hours between the sensible heat flux (and surface temperature) and the net radiation is usual for land surfaces, with the maximum near-surface air temperature occurring in late afternoon. At night-time the atmospheric fluxes are usually small due to the strong thermal stability and low turbulent intensities.

Formulations for the sensible and latent heat fluxes have been developed in Chapter 3, using the bulk transfer relations. In the next section, several physical concepts that are useful in the formulation of relations for evaporation will be discussed, and factors that influence evaporation from both saturated and unsaturated surfaces, and evapotranspiration from vegetation, will be identified.

5.3 Evaporation

5.3.1 Introduction

Evaporation is the phenomenon by which a substance is converted from the liquid state into vapour. The reverse process is usually referred to as condensation. Evaporation requires energy to supply the latent heat of vaporization, and involves the diffusion of water vapour away from the evaporating surface. In the natural world, evaporation of water may take place from the surface of water bodies, from the soil surface and from the surface of wetted vegetation. The vaporization of water through the stomata of vegetation is called transpiration, whilst evapotranspiration includes both evaporation and transpiration but makes no distinction between the two. The evaporation of water in the natural environment is one of the main phases of the hydrological cycle, and is an important term in the water budget of a region. The water budget for a soil layer of depth d per unit area of surface is usually written

$$(\rho_w d)\partial \eta/\partial t = \rho_w P - E_0 - F_{wd} \tag{5.22}$$

where ρ_w is the density of water, η is the soil moisture content per unit volume, P is the precipitation rate in units of velocity, F_{wd} is the flux of water from the base of the soil layer in units of $\mathrm{kg\,m^{-2}\,s^{-1}}$ and E_0, the turbulent moisture flux, is positive for evaporation. Terms representing surface runoff of water and snowmelt have been omitted from the right-hand side of the equation. The left-hand side of this equation represents the rate of storage of water in the soil, whilst runoff is assumed to occur if the precipitation rate exceeds a critical value or if the soil layer is saturated during a precipitation or snowmelt episode.

The moisture flux, E_0, has been formulated in terms of surface and ABL parameters (e.g. Eqs. 3.51 and 3.63). From the perspective of an atmospheric scientist, much of the problem of parameterizing evaporation relates to the determination of q_0, the surface humidity, i.e. the air's relative humidity at the humidity roughness height z_q. For a saturated soil, wet canopy or water surface, where the relative humidity is 100 per cent, q_0 equals the saturated value, q_0^*, which is related to the atmospheric pressure and to the saturation vapour pressure, itself a unique and well-known function of surface temperature (see Appendix 2). Note that under these conditions, the surface saturation deficit $\delta q_0 = 0$. For unsaturated or drying soil surfaces, and for canopy surfaces in particular, q_0 is not so readily determined. For the case of dry canopies, the concept of *surface resistance* (or *conductance*) has proved useful, although for drying soils the resistance has no recognizable relationship to the transport processes. This is mainly because transport is by liquid films during the drying stage, while the surface resistance concept is based on vapour transport.

5.3.2 Evaporation from wet surfaces

Combination equation

Let us restrict ourselves to evaporation from saturated surfaces (wet foliage or soil; water), where $q_0 = q^*(T_0)$ at a given atmospheric pressure. The concept of *potential evaporation* (E_P) is closely related to evaporation from a wet surface (McIlroy, 1984). It is defined as the maximum possible evaporation from a given

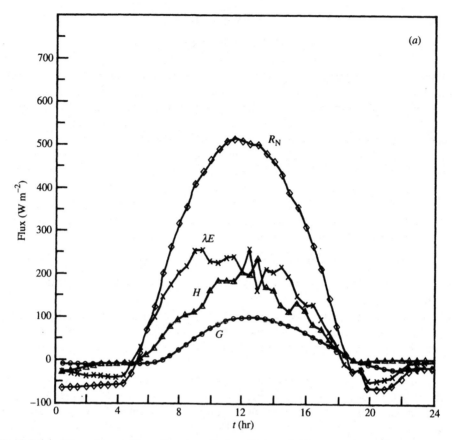

Fig. 5.5 (*a*) Diurnal variation of net radiation R_N, turbulent sensible and latent heat fluxes H and λE and soil heat flux G (at a small depth in the soil) under cloudless skies for a sparse maize crop. Observations are from Noilhan and Planton (1989). (*b*) As in (*a*) but for a bare clay surface, based on numerical simulations in an interactive soil–atmosphere model. Also shown are the time variations of the surface and the 10 m air temperatures. Surface conditions are dry, at the summer solstice and 30 °S; the surface albedo was set at 0.01 so as to generate extreme daytime surface temperatures.

surface, for a given environmental state specified by R_N, G, the aerodynamic resistance r_{aV} (dependent upon wind speed and thermal stability) and the saturation deficit $q^*(T) - q$, abbreviated to δq. Here, T is the temperature at height z, where $T = T(z)$ and $q = q(z)$. The potential evaporation is formally written as (Eq. 3.63),

$$\lambda E_P = \rho \lambda [q^*(T_0) - q]/r_{aV}. \tag{5.23}$$

An alternative expression involves the Penman combination approach (Penman, 1948), which is particularly useful in clarifying the physical constraints on the evaporation. It also eliminates T_0 from the problem, and so is a definite advantage if "standard" observations are used for estimating the potential evaporation. The combination method (so called because energy and aero-

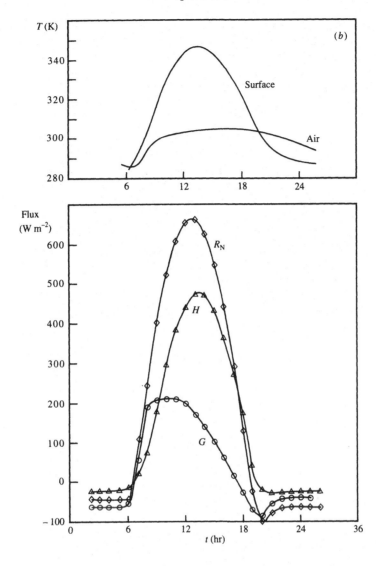

dynamic terms are combined) was originally applied to the case of evaporation from short grass with an ample supply of water, and from water surfaces. The modern formulation is quite general, involving consideration of Eqs. 3.62 and 5.1 together (e.g. Webb, 1975). The crucial step is to write the quantity $q^*(T_0) - q$ as

$$q^*(T_0) - q = s(T_0 - T) + \delta q \qquad (5.24)$$

where $s = \partial q^*/\partial T$ is determined at a temperature $(T_0 + T)/2$, i.e. linearization of the $q^*(T)$ function is assumed. By combining Eqs. 5.23 and 5.24, and eliminating the temperature difference using Eq. 3.62b, the interim relation

$$\lambda E_P = (s/\gamma)H_0 + \rho\lambda(\delta q + sT'')/r_{aV}$$
$$\approx (s/\gamma)H_0 + \rho\lambda\delta q/r_{aV} \qquad (5.25)$$

is derived, where we have assumed $r_{aH} = r_{aV}$ and written $\gamma = c_p/\lambda$. The quantity T'' takes into account the fact that absolute temperature is used in Eq. 5.24, whilst the sensible heat flux is parameterized in terms of potential temperature. Approximately, $T'' = T[(p/p_R)^{-R_d/c_p} - 1] \approx 0.01z$, where z is the height of the atmospheric reference level. Substituting H_0 from Eq. 5.1, with $E_0 = E_P$, gives

$$\lambda E_P = \Gamma(R_{N0} - G_0) + (1 - \Gamma)\rho\lambda\delta q/r_{aV} \qquad (5.26)$$

where we have written $\Gamma = s/(s + \gamma)$. Equation 5.26 has a two-term structure suggesting that the evaporation from a saturated surface has both energy and aerodynamic contributions. This equation is actually a very good approximation even when $T_0 - T$ exceeds several degrees kelvin, so long as s is calculated at a temperature $(T + T_0)/2$. It is applicable to water surfaces, saturated soil and vegetation with wet foliage (whether from intercepted rainfall, dewfall or irrigation spraying), and has been used widely with much success, both for calculating hourly and longer-term evaporation rates (e.g. Thom & Oliver, 1977). From an experimental viewpoint, the main advantage of Eq. 5.26 is that data from only one level in the atmosphere are required to estimate the potential evaporation.

With some restrictions, Eq. 5.26 is a useful form with which to illustrate the dependence of evaporation from a wet surface upon climate and the nature of the surface, i.e. upon net available energy, surface roughness and the saturation deficit in the air. We have used the calculations of Webb (1975), and show the evaporation as a function of available energy $(R_{N0} - G_0)$, for a range of z_0 values, in Fig. 5.6. It should be noted that the aerodynamic term in Eq. 5.26 requires values of temperature, humidity and wind speed in the atmospheric surface layer and, for a well-adjusted air mass, these will depend strongly on the nature of the surface. The calculations upon which the curves in Fig. 5.6 are based do not take all of these considerations into account. For ease of computation, the temperature and humidity are taken to be the same at 10 m over each surface type, 20 °C and 0.0085 respectively. This allows s, γ, ρ, λ and δq to be calculated directly (e.g. Tables A1 and A2). The 10 m wind speed is allowed to vary with z_0 in the calculation of the aerodynamic resistance, $r_{aV} = (C_E u)^{-1}$, with buoyancy effects incorporated. Both u and C_E are calculated as follows.

(i) The geostrophic wind and ABL depth are set at $10\ \mathrm{m\,s^{-1}}$ and 1000 m respectively. The friction velocity u_{*0} is then deduced, for each z_0, from the curves in Fig. 3.12 (iteratively for the non-neutral case), and the 10 m wind speed is inferred from curves in Fig. 3.7(a).

(ii) C_{HN} is calculated from Eq. 3.48, with $z = 10$ m and $z_T = z_0/7.4$, and C_H $(= C_E)$ is deduced from the curves in Fig. 3.7(b).

The two sets of curves, shown in Fig. 5.6, respectively ignore and incorporate the effect of thermal stratification in r_{aV}. The magnitude of the aerodynamic term is given by the intercept of any one curve on the ordinate axis. For large z_0 (e.g. forests), the contribution from the aerodynamic term is relatively large and the effects of thermal stability are considerable. For a real forest wetted by rain, the aerodynamic term (and hence the evaporation rate) might not be as large as

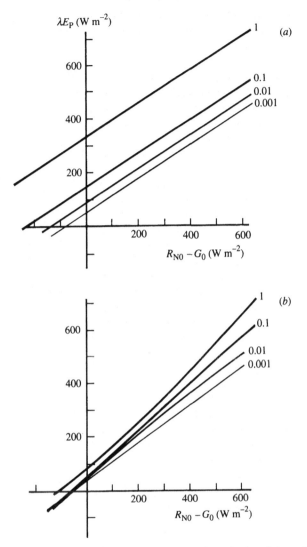

Fig. 5.6 Potential evaporation for different wet surfaces calculated from Eq. 5.26. In (a) neutral conditions have been assumed, and in (b) the full stability correction in r_{aV} is included (see Eqs. 3.47 and 3.57). Note how the effects of thermal stability tend to reduce the direct influence of aerodynamic roughness. Values of z_0 are as follows: 0.001 m, lake; 0.01 m, grass; 0.1 m, scrub; 1 m, forest. Further details of the calculations can be found in Webb (1975).

in this example, simply because the saturation deficit would probably be smaller than used in the calculation. Nevertheless, such large evaporation rates from a wetted forest are not unrealistic, and can only be maintained with a heat flux from the air towards the surface (negative H_0) so that the air is being cooled. This situation is implied in Fig. 5.6 for forests at all values of $R_{N0} - G_0$, and would probably occur, for example, with evaporation of intercepted rain from forests over quite extensive areas (Thom and Oliver, 1977). Negative H_0 corresponds to the condition $R_{N0} - G_0 < \rho\lambda\delta q/r_{aV}$, which can be derived by combining Eqs. 5.1 and 5.26.

The Bowen ratio and energy considerations

Equation 5.26 can be used to introduce the concept of *equilibrium evaporation*. This concept applies to evaporation from a saturated surface into an air flow at very large fetch, where it is assumed that air is saturated, and hence $\delta q = 0$. Evaporation then occurs under advection-free conditions into air having a vertical gradient in q^*, and is equal to $\Gamma(R_{N0} - G_0)$. Thus, the first term on the right-hand side of Eq. 5.26 represents a lower limit to the evaporation from moist surfaces. For the special case of $\delta q = 0$, the Bowen ratio B_1 is readily seen to be given by

$$B_1 = \gamma/s \qquad (5.27)$$

which is a strong function of absolute temperature. Note that Eq. 5.27 can also be inferred from Eqs. 3.46b and 3.51, with

$$B_1 = H_0/\lambda E_0 = c_p(T_0 - T)/\lambda[q^*(T_0) - q^*(T)] = \gamma/s$$

where equal transfer coefficients for heat and water vapour transfer are assumed.

For steady flow over a large area, it is rare for air to be saturated. For the case of minimal advection (large-scale or regional advection only), which is a common occurrence in the real atmosphere, the air is non-saturated ($\delta q > 0$) so that the evaporation rate exceeds the equilibrium rate. The second (aerodynamic) term in Eq. 5.26 can then be seen as a measure of departure from equilibrium in the ABL. In limiting the problem to one of minimal advection, we exclude situations of local advection related to leading-edge effects and internal boundary-layer growth. Under such conditions, the concept of equilibrium evaporation from a saturated surface can be used as the basis for the empirical relationship (Priestley and Taylor, 1972)

$$\lambda E_P = \alpha_{PT}\Gamma(R_{N0} - G_0) \qquad (5.28)$$

where α_{PT} is expected to be greater than unity. Even though Priestley and Taylor intended Eq. 5.28 for application at the large scale over extensive areas of land, determination of α_{PT} has often used observations from local sites. In this sense, reference to Eq. 5.26 shows that no unique value for α_{PT} is to be expected. Even if a strong climatological correlation existed between δq and $R_{N0} - G_0$, variations in roughness and absolute temperature will strongly influence α_{PT}, when the latter is evaluated locally. Under conditions of minimal advection for low-z_0 sites, experimental data show $\alpha_{PT} \approx 1.26$ (Brutsaert, 1982, Chapter 10), though when extended to other surfaces (forests in particular) α_{PT} values are more variable and generally larger. The value of 1.26 for oceans and short vegetation suggests that large-scale advection accounts for about 20 per cent of the evaporation rate. The main shortcoming in Eq. 5.28 is its failure to account in any way for the surface control on potential evaporation, as implied in Eqs. 3.51 (with $q_0 = q^*(T_0)$) and 5.26.

Eq. 5.28 implies that the Bowen ratio B_2 is given by

$$B_2 = (1 - \Gamma\alpha_{PT})/\Gamma\alpha_{PT}, \qquad (5.29)$$

with $B_2 < B_1$. Both parameters are shown in Fig. 5.7 as functions of T, together with several sets of atmospheric observations. These show strong support for the B_2 formulation, though it should be noted that data from wet, large-roughness vegetation are not included.

In the case of tall vegetation, where r_{aV} is small, the status of the surface (wet or dry) is the primary control on the evaporation rate (on the right-hand side in Eq. 5.26, the second term is likely to be greater than the first). When the foliage is wet, E_P can be many times larger than the evapotranspiration rate from dry foliage under conditions of minimal water stress, i.e. where water is freely available to the leaves for transpiration (Monteith, 1981). Large values of E_P for forests are evident in Fig. 5.6, where available energy plays at most a minor role.

When the surface is not wet, determination of the evaporation is less easy since the surface humidity is less than the saturation value and unknown. In addition, the physical differences between a soil surface and a leaf surface suggest that the evaporation from each be treated differently.

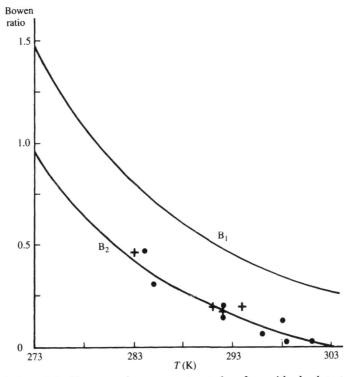

Fig. 5.7 Variation of the Bowen ratio over a saturated surface with absolute temperature. Curves represent $B_1 = \gamma/s$ and B_2 from Eq. 5.29, with experimental data from Priestley and Taylor (1972) (solid circles) and from Garratt and Hyson (1975) (crosses). The observations tend to support B_2, which is consistent with the fact that B_1 gives evaporation rates that are too small, since they are calculated for saturated air.

5.3.3 Evaporation from dry vegetation

Surface resistance

Evaporation is reduced below the potential rate whenever the surface humidity is less than the saturation value (calculated at the existing surface temperature). In the case of a leaf surface, there exists a surface resistance to water vapour transfer through foliage stomata whether the vegetation is under negligible or severe water stress.

The concept of aerodynamic resistance (or conductance) was introduced in Section 3.3.3, with Eq. 3.63 describing E_0 in terms of the humidity difference $q_0 - q$ and aerodynamic resistance r_{aV}. The surface resistance (to be denoted by r_s) applies to unsaturated surfaces and most usually to vegetation – it can be seen as responsible for the reduction of humidity at the "surface" below the saturation value, such that $q_0 < q^*(T_0)$. In the case of vegetation, this resistance represents an effective stomatal resistance to the transfer of water vapour from the internal water sites in the plant to the leaf exterior. The widely applied concept of a simple effective surface, or canopy, resistance is a gross simplification of the real world, making no explicit reference to the detailed canopy structure but, within these limitations, it does provide the best available simple description of the surface control on transpiration.

The use of surface resistance (or conductance) is associated with single-layer and multilayer canopy models of evaporation from plant canopies. Single-layer models are appropriate where the vegetation is essentially seen as a lower boundary condition to the ABL flow, and where length scales of interest are considerably larger than the canopy scale ($\sim h_c$). Thus, they are often used in numerical models of the atmosphere, as well as in hydrological modelling at all scales larger than single stands or crops. In contrast, multilayer models are appropriate when it is necessary to resolve details within the canopy; for example, in the hydrology of small forest stands. Because of our emphasis, a detailed description of multilayer models of a canopy is beyond the scope of the present book, and we will therefore mostly confine the discussion to single-layer models of evaporation from plant canopies. The reader is also referred to Section 8.4 dealing with surface parameterization schemes in numerical models.

A general surface resistance r_s for a partly or wholly vegetated surface can be defined by

$$r_s = \rho[q^*(T_0^{eff}) - q_0]/E_0 \qquad (5.30)$$

where T_0^{eff} is the effective temperature of this surface (and, for a sparse vegetation cover, is most likely to be intermediate between the leaf and soil-surface temperatures). Combination of Eqs. 3.63 and 5.30 allows the unknown q_0 to be eliminated to give

$$\lambda E_0 = \rho\lambda[q^*(T_0^{eff}) - q]/(r_{aV} + r_s) \qquad (5.31)$$

where q is the specific humidity at height z. Eq. 5.30 is mathematically exact, but for a partial canopy cover, the resistance r_s is not readily interpreted physically, since E_0 is not solely due to transpiration but includes contributions from the soil. We can gain some insight into the role of r_s by considering the concept of the *bulk stomatal (physiological) resistance* of a canopy, which is appropriate to a fully vegetated surface.

The sensible heat flux H_0 can be considered to originate on the surface of leaves whose average temperature is T_f. The latent heat flux, however, is driven essentially by the evaporation of liquid water within the leaves, in the many intercellular spaces whose link to the leaf surface is through the stomata. Within the intercellular spaces the humidity can be taken as $q^*(T_f)$. In analogy with r_{aV}, the difference $\delta q_0 = q^*(T_f) - q_0$ is related to the evaporative flux and is proportional to the integrated effect of the physiological resistance of all the leaves. This resistance is almost entirely determined by the stomatal resistance of the leaves – hence by the bulk stomatal resistance – and given by

$$r_{st} = \rho[q^*(T_f) - q_0]/E_0. \tag{5.32}$$

This equation is commonly referred to as the "big-leaf" model, since it basically represents the whole canopy as a single hypothetical "leaf" of resistance r_{st}. Now if r_{sti} is the resistance of a single leaf, then r_{st} is defined as the parallel sum of these leaf stomatal resistances, with

$$1/r_{st} = A^{-1} \sum L_{A,i}/r_{sti} \tag{5.33}$$

where $L_{A,i}$ is the area of the ith leaf and the summation is over all leaves above some representative ground area A. The leaf area index (LAI) is $L_A = \sum L_{A,i}/A$. For many applications, the approximation $r_{st} \approx r_{sti}/L_A$ may be used.

For closed canopies, where E_0 is predominantly transpiration, experimental results support the use of r_{st} in Eq. 5.31. These show that r_{st} depends on several environmental factors, with a strong diurnal variation, as well as varying between species (the reader can find, in Chapter 8, sample empirical relations used in numerical models). Table 5.1 gives a summary of minimum daytime values of r_{st} for several canopy types that are appropriate where there is no water stress. The range of 30–300 s m^{-1} in r_{st} reflects the range encountered in L_A and the differences in r_{sti} between canopy types. Generally, good agreement has been found between r_{st} determined from evaporation measurements in the field (Eq. 5.31) and r_{st} evaluated from Eq. 5.33.

For an open canopy, when the LAI is typically less than unity, the simple model for r_{st} in Eq. 5.32, with its underlying single-source assumption and the physiological nature of surface resistance, is not valid. The total evaporation from an area becomes a combination of foliage and subcanopy evaporation, plus evaporation from the open areas separating the vegetation patches. If Eq. 5.31 is to be used, it can be applied separately to vegetated and bare areas, each with appropriate values of r_s, and the component evaporation rates summed to give the area evaporation. Such an approach must provide a simple but physically plausible description of the transition between bare substrate and a closed canopy. In this transition, the surface resistance must reflect how stomatal control (and the influence of LAI) takes over from soil conditions (in particular, the near-surface moisture content) as vegetation density increases (Shuttleworth and Wallace, 1985).

Table 5.1. *Minimum daytime values of the bulk stomatal resistance r_{st} and the leaf resistance (per unit leaf area) r_{sti} for a range of canopy types. Also shown is the unconstrained surface resistance r_s^+ described in Chapter 8.*

Canopy type	References	r_{sti} (s m^{-1})	r_{st} (s m^{-1})	r_s^+ (s m^{-1})
Forest				
tropical	DS89			50
	S84		125–150	
	S89[a]		80	43
deciduous	DS89			50
	V86		75–160	
	S89		70–100	41
coniferous	DS89			60
	J76[b]	120–2700	90–150	
	S89		70–100	56
	NP89			43
general	P82		100–300	
Other				
savannah scrub	G78		100	
mesophytes[c]	G80	100–200		
grassland	DS89			40
	R80		40	
	P82		100–200	
crops	DS89		30–35	30
	SD87			20–75
	S67	50–320		
	P82		40–130	
	NP89			20–150
citrus	P82		250	

[a]Shuttleworth (1989) gives a number of estimates of minimum r_{st} values for several forest sites, and quotes an overall value of 100 ± 30 s m^{-1} for forests.
[b]The reader is referred to Table 12 of Jarvis *et al.* (1976), where values for a range of conifer and other tree types, from numerous references, are given.
[c]Mesophytes are tropical and temperate plants needing plentiful water for survival.
Sources: DS89, Dorman and Sellers (1989); G78, Garratt (1978); G80, Gates (1980); J76, Jarvis *et al.* (1976); NP89, Noilhan and Planton (1989); P82, Perrier (1982); R80, Russell (1980); S67, Slatyer (1967); S84, Shuttleworth *et al.* (1984); S89, Shuttleworth (1989); SD87, Sellers and Dorman (1987); V86, Verma *et al.* (1986).

Modified combination equation
Using the surface saturation deficit defined as $\delta q_0 = q^*(T_0) - q_0$, the analogous combination equation to Eq. 5.26 for E_0 is

$$\lambda E_0 = \Gamma(R_{N0} - G_0) + (1 - \Gamma)\rho\lambda(\delta q - \delta q_0)/r_{aV} \qquad (5.34a)$$

$$= \lambda E_L - (1 - \Gamma)\rho\lambda\,\delta q_0 r_{aV}^{-1}, \qquad (5.34b)$$

where λE_L is, by inspection, the sum of the two terms $\Gamma(R_{N0} - G_0)$ and $(1 - \Gamma)\rho\lambda\,\delta q/r_{aV}$. That is, E_L is the non-potential equivalent to E_P as given by Eq. 5.26 (the distinction is made, since only for wet surfaces is E_P given simultaneously by Eqs. 5.23 and 5.26).

For a canopy surface, an alternative form of Eq. 5.34 introduces surface resistance through the relation between δq_0 and r_{st} (assumed for present purposes to be the same as r_s) given by Eq. 5.32. The result is (Thom, 1972)

$$\lambda E_0 = \Gamma^*(R_{N0} - G_0) + (1 - \Gamma^*)(\gamma/\gamma^*)\rho\lambda\,\delta q/r_{aV} \tag{5.35}$$

where $\Gamma^* = s/(s + \gamma^*)$ and

$$\gamma^* = \gamma(1 + r_s/r_{aV}). \tag{5.36}$$

Equation 5.35 is often referred to as the Penman–Monteith relation (Monteith, 1981). It may also be written as

$$E_0/E_L = [1 + (1 - \Gamma)r_s/r_{aV}]^{-1} \tag{5.37}$$

where

$$\lambda E_L = \Gamma(R_{N0} - G_0)_{dry} + (1 - \Gamma)\rho\lambda\,\delta q/r_{aV}. \tag{5.38}$$

In Eq. 5.38, the available energy is appropriate to a dry surface (cf. Eq. 5.26). In contrast, it can be seen that the ratio of actual to potential evaporation (that is, the apparent potential evaporation that would occur from a hypothetical wet surface at the temperature of the existing dry surface), given by combining Eqs. 5.23 and 5.31, is

$$E_0/E_P = (1 + r_s/r_{aV})^{-1}. \tag{5.39}$$

Implicit in Eq. 5.39 is the assumption that surface temperature does not change between the wet and dry surfaces, which is quite unrealistic, and limits the practical application of the equation. In fact, for soils as well as vegetation, many workers calculate E_P using data taken under dry conditions, leading to unrealistically high values near midday when surface temperatures are relatively high. In contrast, the use of Eq. 5.37, which does allow an adjustment in the surface temperature, gives a more realistic assessment of the factors affecting the decrease in evaporation as the foliage dries and r_s becomes important. Figure 5.8 shows E_0/E_L as a function of r_s for low and high roughness canopies (and for two temperature values). When r_{aV} is large (small z_0) even significant values of r_s do not reduce the evaporation much below E_L. In contrast, where r_{aV} is small (large z_0) finite values of r_s may reduce E_0 well below E_L – as in the case of forest evapotranspiration under conditions of minimal water stress.

5.3.4 Evaporation from drying soils and soil moisture relations

The structure of the ABL and the partitioning of energy at the surface are very much dependent on the wetness of the surface, and on the amount of moisture in the soil near the surface. For example, the soil heat flux depends on the thermal and physical properties of the soil, which themselves are highly dependent on soil water status (Appendix equation A24). In addition, the soil albedo, which is a critical parameter affecting the surface energy balance, depends on surface wetness (Appendix equations A25, A26).

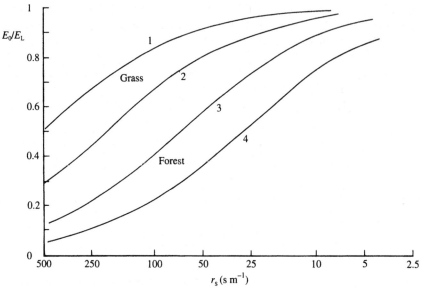

Fig. 5.8 Variations of E_0/E_L (Eq. 5.37) with surface resistance. Values of r_{aV} have been calculated for neutral conditions, with $z_q = z_0/7.4$. For short grass ($z_0 = 0.0025$ m): curve 1, $T = 303$ K; curve 2, $T = 278$ K. For forest ($z_0 = 0.75$ m): curve 3, $T = 303$ K; curve 4, $T = 278$ K.

The evaporation is determined, in part, by the surface humidity and hence by the soil moisture status. For a drying soil surface, internal forces within the soil restrict the flow of water or vapour to the surface and, in the absence of precipitation, the evaporation must be balanced by an equal flux of water from the soil, F_w.

For a drying soil, the evaporation is given by either a bulk transfer relation (Eq. 3.63) or a combination relation (Eq. 5.34b, with E_L defined by Eq. 5.38). By defining a soil-surface relative humidity, $r_h = q_0/q^*(T_0)$, these two equations can be rewritten as

$$E_0 = \rho[r_h q^*(T_0) - q]/r_{aV} \tag{5.40}$$

or

$$E_0 = E_L - (1 - \Gamma)\rho(1 - r_h)q^*(T_0)/r_{aV}. \tag{5.41}$$

For relative humidities less than unity, both relations quantify the reduction of E_0 below the evaporation from a saturated soil surface (where $r_h = 1$). They also show that the relative humidity at the surface does not have to be zero before the evaporation rate becomes zero. For example, according to Eq. 5.40, evaporation is zero when $r_h = q/q^*(T_0)$, that is, when vertical humidity gradients in the air disappear.

In Eqs. 5.40 and 5.41, the evaporation rate is primarily controlled by the atmospheric conditions when r_h is unity or just below unity (first stage of drying), but is controlled by the soil conditions when r_h is small and tending to zero (second stage of drying). For bare soil surfaces, we wish to explore the soil moisture relations that will allow r_h, and hence q_0 and E_0, to be evaluated. The relative humidity is a dominant factor determining the evaporation, and is closely related to the soil moisture content and other soil properties.

The *soil volumetric moisture content* η and its saturation value η_s quantify the amount of water in the soil. This may exist in liquid or vapour form within the soil pores, and coexist with air spaces around the soil particles. Generally speaking, the vertical flux of water in a soil will depend on vertical differences in the hydraulic head of water, comprising both pressure-head and gravitational-head components. In the absence of any pressure head, the effects of gravity on the viscous vertical flow of water are represented by the *hydraulic conductivity*, K_η. The presence of a hydrostatic pressure head, or suction, is represented by the *moisture* or *matric potential*, ψ (units of length). Basically, ψ equals the work required to extract water from the soil against capillary (surface tension) and adsorption forces. When the soil is saturated ψ has a minimum value, but not all air spaces will be filled with water (the porosity is a measure of the maximum amount of water that soil can hold). For many soils η_s has a value of about $0.4 \, \text{m}^3 \, \text{m}^{-3}$.

Darcy's law, originally conceived for saturated soils, can be extended to the unsaturated case if allowance is made for the hydraulic conductivity to be a function of ψ (Hillel, 1971). It states

$$F_w = - \rho_w K(\psi) \partial(\psi + z')/\partial z' \tag{5.42}$$

where ρ_w is the density of liquid water. Equation 5.42 is a simplified form of the vertical equation of motion for water flow in the soil (Pielke, 1984, Chapter 11). For present purposes, we treat the one-dimensional case only (for vertical transfer), but even with this restriction the formulation of moisture transfer can be very detailed. Some simplifying assumptions are necessary, but the important elements of the transfer equations, which affect the evaporation process and ABL structure in general, are retained. Now ψ can be written as a function of η (see later), and with a diffusivity defined by

$$D_\eta = K_\eta \partial \psi/\partial \eta, \tag{5.43}$$

Eq. 5.42 can be written

$$F_w = - \rho_w D_\eta \partial \eta/\partial z' - \rho_w K_\eta. \tag{5.44}$$

It is possible for the variables K_η and D_η, as well as ψ, to be related to the soil moisture content. Also, the diffusion coefficient has both liquid and vapour contributions, though as a first approximation it is permissible to ignore vapour transfer. Certainly, in isothermal conditions, vapour-pressure differences through the body of the soil are likely to be very small, and soil air is likely to be nearly vapour-saturated most of the time. When significant temperature gradients exist, vapour differences will be large so that vapour transfer may be non-negligible.

Finally, conservation of soil moisture requires

$$\rho_w \partial \eta/\partial t = - \partial F_w/\partial z'. \tag{5.45}$$

Under conditions of high surface evaporation, F_w will be highly negative at small z', decreasing towards zero as z' increases. Equation 5.45 thus implies a decrease in η with time. In order to solve Eqs. 5.44 and 5.45 for $\eta(z', t)$, the variables K_η and D_η, together with ψ, need to be specified as functions of η. Empirical functions do exist, based on field data for a range of soil types. Suitable relationships can be written as

$$\psi = \psi_s(\eta/\eta_s)^{-b} \tag{5.46}$$

$$K_\eta = K_{\eta_s}(\eta/\eta_s)^{2b+3} \tag{5.47}$$

$$D_\eta = -(bK_{\eta_s}\psi_s/\eta)(\eta/\eta_s)^{b+3} \tag{5.48}$$

based on the work of Clapp and Hornberger (1978). Table A9 in Appendix 4 gives values of several relevant variables for a number of soil types showing, in particular, how rapidly K decreases as the soil dries.

The surface humidity still has to be related to the soil moisture. To do this, consider that the surface moisture potential, which is a measure of the soil wetness, is related to water vapour at equilibrium by the relative humidity, r_h (Philip, 1957):

$$r_h = \exp\left(-g\psi_0/R_v T_0\right). \tag{5.49}$$

Here, ψ_0 and T_0 are surface values of ψ and T, and R_v is the gas constant for water vapour. Equation 5.49 is derived from a combination of the hydrostatic equation and the ideal gas law for water vapour within the soil, with the resultant differential equation integrated between a saturated and an unsaturated state. Thus, with η determined from the solution of Eqs. 5.44 and 5.45 with appropriate boundary conditions, extrapolation to $z' = 0$ gives η_0; use of Eq. 5.46 then allows ψ_0 to be evaluated, and hence r_h from Eq. 5.49. The surface humidity, q_0, can then be determined from $q_0 = r_h q^*(T_0)$.

Analytical solutions of Eqs. 5.44 and 5.45 are not readily obtained so, for illustration, we consider some numerical solutions obtained using a multi-level soil model interactive with an atmospheric model. For example, Fig. 5.9(a) shows the relative humidity as a function of η/η_s for three soil types, using the data in Table A9, and Fig. 5.9(b) gives the corresponding variation of E_0/E_L (Eq. 5.41) as a function of the fractional moisture content, calculated for a range of climate situations (i.e. values of net radiation, humidity deficit and absolute temperature). The important feature of these two diagrams is the rapid decrease in evaporation when η falls below a critical value, called the wilting point η_w, which depends upon soil type. Values of η_w for all soil types are included in Table A9. Physically, η_w represents the point at which evaporation is limited by the soil moisture transport to the surface, and where the soil moisture diffusivity becomes a strong function of moisture content.

For practical purposes, whether in the field or in a numerical model of the atmosphere, less formal methods for evaluating evaporation than those based on the above soil-moisture relations may be appropriate. These are discussed in Chapter 8.

This section has dealt with the problem of evaporation from bare soil and vegetated surfaces. However, the E_0 term appearing in the energy-balance equation actually represents the moisture flux in general, and so can also be interpreted as the condensation. In the case of condensation, the surface is

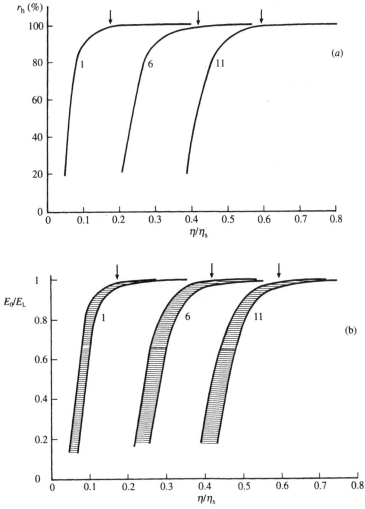

Fig. 5.9 (*a*) Relative humidity r_h as a function of relative soil moisture content η/η_s, based on Eq. 5.49 and data in Table A9 for soil types 1 (sand), 6 (loam) and 11 (clay). Calculations are for a temperature T_0 of 303 K. The vertical arrows indicate the wilting points. Note that combining Eqs. 5.46 and 5.49 allows r_h to be calculated from $\ln r_h = -(g/R_v T_0)\psi_s(\eta/\eta_s)^{-b}$. (*b*) E_0/E_L as a function of the relative soil moisture content, based on numerical simulations in an atmospheric model for a range of climate conditions (mid-latitude summer) represented by the shaded regions (the temperature range is 283–303 K and $q = 0.005$).

treated as saturated so that many of the relationships developed for potential evaporation apply.

5.4 Condensation

When the surface temperature cools sufficiently to allow the surface humidity to fall below the humidity in the air ($q_0 < q$), $\partial q/\partial z$ close to the surface becomes positive, and the flux of water vapour into the surface results in *condensation at the surface* or *dewfall*. Under such conditions, and depending upon the nature of the surface, the effective q_0 can be taken as $q^*(T_0)$. This saturated condition is related either to a continuous film of water on the surface or to water drops distributed over the surface with saturated vapour in contact with the surface.

Such conditions for dewfall occur regularly around the world, usually at night-time and under clear skies. The onset of dewfall occurs when the surface cools to the dew-point temperature, and will coincide with a reversal in the sign of the humidity gradient. Dewfall, as an atmospheric source of water, in fact is only one of two important contributions to dew formation; the other has been termed *distillation* by Monteith (1957). In distillation, soil is the additional source of dew formation on canopies (no matter how sparse). Observations of dew amounts tend to give values less than 0.5 mm per night (Garratt and Segal, 1988). Upon what external factors does dewfall in particular depend?

To answer this important question, use can be made of the combination-type equation (Eq. 5.26) discussed previously for evaporation from wet surfaces. This can be applied to the idealized cases of a closed canopy and a bare soil surface as sketched in Fig. 5.10. In the canopy case, it is necessary to define "the surface" as the zero-plane displacement, i.e. at $z = 0$. This means that sub-canopy fluxes up from the ground to the foliage (denoted by subscript U) are negative (they are in a negative z region); e.g. both H_U and E_U in Fig. 5.10 are negative (as are H_D and E_D, the subscript D denoting downwards fluxes to the canopy). The total dew $E = E_U + E_D$, with E_D given by Eq. 5.26. If we assume a saturated soil, this equation can be simplified by use of Eq. 5.27, which implies $G_0 = \lambda E_U(1 + \gamma/s)$. This allows the canopy dewfall E^{can} to be written

$$\lambda E^{\text{can}} = \lambda E_D + \lambda E_U = \Gamma R_{N0} + (1 - \Gamma)\rho\lambda\delta q/r_{aV}^{\text{eff}} \qquad (5.50)$$

where r_{aV}^{eff} is an effective canopy aerodynamic resistance that accounts for atmospheric and in-canopy heat fluxes. In the bare soil case, total dew is E_D (dewfall), so

$$\lambda E^{\text{soil}} = \Gamma(R_{N0} - G_0) + (1 - \Gamma)\rho\lambda\delta q/r_{aV}. \qquad (5.51)$$

Apart from factors that influence dew formation, Eqs. 5.50 and 5.51 allow maximum rates of condensation to be evaluated. With our chosen sign convention for fluxes, both E (for condensation) and R_{N0} are negative at night, so maximum negative values of E will occur when the air is saturated and $\delta q = 0$. In Eqs. 5.50 and 5.51 these maximum values are given by the first term on the right-hand side and, for example in the canopy case, are mainly a function of absolute temperature. Using Eq. 5.20 to estimate R_{N0}, peak values of this term of about $-60 \, \text{W m}^{-2}$ occur at $T = 293 \, \text{K}$ (Fig. 5.4) corresponding to dew formation of $0.08 \, \text{mm hr}^{-1}$, or nearly 1 mm for a 12-hour night. Field observations tend to lie below this, particularly for bare soil where G_0 (towards the surface) in Eq. 5.51 reduces the condensation rate E further. In the canopy case, the ratio of dewfall (E_D) and distillation (E_U) is a strong function of wind speed and absolute temperature. Thus, as $u \to 0$ and hence $K_w \to 0$, $E_D \to 0$ and dew will be solely the result of distillation.

In the above, we have assumed saturated air. Under non-saturated air

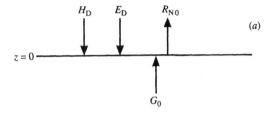

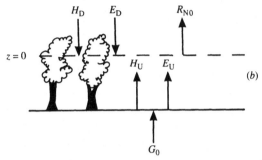

Fig. 5.10 Schematic representation of surface energy balance terms for (*a*) bare soil and (*b*) a canopy for the special case of dewfall in stably stratified conditions.

conditions condensation will be less than the above maximum possible values (given by the first terms on the right-hand side of Eqs. 5.50 and 5.51), and will be zero when

$$\Gamma R_{N0}^{can} = - (1 - \Gamma)\rho\lambda\delta q/r_{aV}^{eff} \qquad (5.52)$$

$$\Gamma(R_{N0} - G_0)^{soil} = - (1 - \Gamma)\rho\lambda\delta q/r_{aV}. \qquad (5.53)$$

For given values of the humidity deficit in the air, the right-hand sides of Eqs. 5.52 and 5.53 are a strong function of wind speed, thermal stability and z_0. For illustration, consider Eq. 5.50 for the canopy, for which Eq. 5.52 is the special case of $E = 0$. The variation of E^{can} with wind speed is shown in Fig. 5.11 for two values of absolute temperature and several values of δq. Values of r_{aV} have been determined using Eq. 3.57, and the relation between wind speeds at 0.5 and 10 m height based on the wind relation described by Eq. 3.34. As an example, the curves suggest that at an air relative humidity of 70–80 per cent, condensation is replaced by evaporation for $u_{10} > 6.5 \text{ m s}^{-1}$, or $u_{0.5} > 2.25 \text{ m s}^{-1}$.

In the ideal one-dimensional horizontally homogeneous situation, dewfall at the surface is balanced by the local time rate of change of water vapour in the column above and, for turbulent transfer, within the vertical column comprising the nocturnal boundary layer (assuming that the dew formation is occurring at night). Taking the simplified form for moisture conservation in the lower atmosphere (refer to Eq. 2.47), this requires, when integrated across the ABL, that

$$E_D/\rho = \int_0^h \partial q/\partial t \, dz \qquad (5.54)$$

where we have assumed that the water vapour flux is zero at the ABL top

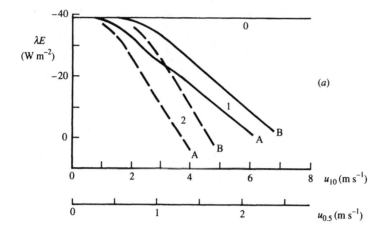

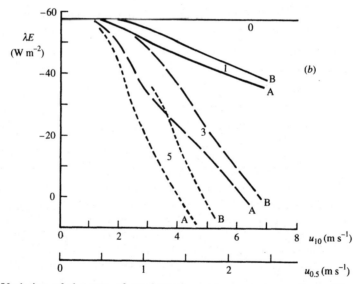

Fig. 5.11 Variation of the rate of condensation onto a canopy surface with wind speed, based on Eq. 5.50 for (*a*) $T = 273$ K and (*b*) $T = 293$ K. The first term on the right-hand side of Eq. 5.50 is taken from Fig. 5.4 (curve T_1). In (*a*), the two sets of curves correspond to $\delta q = 1$ and 2 g kg^{-1} as indicated; in (*b*), the three sets of curves correspond to $\delta q = 1$, 3 and 5 g kg^{-1}, with the horizontal line in both diagrams representing $\delta q = 0$ (note that the value of λE in this case is the value of T_1 in Fig. 5.4 at the corresponding temperature). Each pair of curves for specific δq values has values of $\Delta \theta$ of 0.625 K (A) and 2.5 K (B), where $\Delta \theta$ is an air–surface temperature difference.

In (*a*), values of $\delta q = 0$, 1 and 2 g kg^{-1} correspond to relative humidities in the air of 100%, 74% and 47% respectively. In (*b*), relative humidities are 100%, 93%, 79% and 66% for $\delta q = 0$, 1, 3 and 5 g kg^{-1} respectively. From Garratt and Segal (1988): reprinted by permission of Kluwer Academic Publishers.

$(z = h)$. With significant dewfall, where E_D is negative, q decreases with time. Thus, for dewfall amounting to 0.5 mm over a 12-hour period $(E_D = -29\,\mathrm{W\,m^{-2}})$, the average q in a column 100 (250) m deep decreases by 4 (1.7) $\mathrm{g\,kg^{-1}}$ approximately. Observations show that such time changes do not occur (Garratt and Segal, 1988) and that Eq. 5.54 is generally not satisfied. Rather, much of the water supplied to the surface derives from horizontal advection into the vertical column.

Notes and bibliography

Section 5.1

A detailed discussion on the components of the surface energy balance applied to a canopy can be found in

Thom, A. S. (1975), Momentum, mass and heat exchange of plant communities, in *Vegetation and the Atmosphere*, Vol. 1, *Principles*, ed. J L. Monteith, pp. 57–109. Academic Press, London.

Calculations of canopy storage for a real forest canopy are given in

Stewart, J. B. and A. S. Thom (1973), Energy budgets in pine forest, *Quart. J. Roy. Met. Soc.* **99**, 154–70.

Section 5.2

For details on longwave schemes, and empirical relations for the downwards flux at the surface, see Chapters 6 and 7 in

Paltridge, G. W. and C. M. R. Platt (1976) *Radiative Processes in Meteorology and Climatology*. Elsevier, Amsterdam, 318 pp.

Section 5.3

The reader will find a coherent and organized introduction to the theory of evaporation and associated relationships, together with an outline of their history and application, in

Brutsaert, W. (1982), *Evaporation into the Atmosphere*. Reidel, Dordrecht, 299 pp.

For a discussion on the relation between evaporation and surface temperature, with many useful references, see

Monteith, J. L. (1981), Evaporation and surface temperature, *Quart. J. Roy. Met. Soc.* **107**, 1–27.

The dependence of evaporation upon radiant energy and the aerodynamic characteristics of the underlying surface is dealt with at length in the following two papers:

Priestley, C. H. B. and R. J. Taylor (1972), On the assessment of surface heat flux and evaporation using large-scale parameters, *Mon. Wea. Rev.* **100**, 81–92;

Webb, E. K. (1975), Evaporation from catchments, in *Prediction in Catchment Hydrology*, eds T. G. Chapman and F. X. Dunin pp. 203–236. Australian Academy of Science, Canberra.

The following two papers are recommended for their discussions of the Penman–Monteith (combination) equation in the context of plant water use and the extrapolation of stomatal concepts from the leaf to regional scales:

Jarvis, P. G., Edwards, W. R. N. and H. Talbot (1981), Model of plant and crop water use, in *Mathematics and Plant Physiology*, eds D. A. Rose and D. A. Charles-Edwards, pp. 151–94. Academic Press, London:

Jarvis, P. G. and K. G. McNaughton (1986), Stomatal control of transpiration: Scaling up from leaf to region, *Adv. Ecol. Res.* **15**, 1–49.

Section 5.4

The classical paper on dew is

Monteith, J. L. (1957), Dew, *Quart. J. Roy. Met. Soc.* **83**, 322–41.

The following paper gives a review of the dew literature, and relates dewfall to ABL properties.

Garratt, J. R. and M. Segal (1988), On the contribution of atmospheric moisture to dew formation, *Bound. Layer Meteor.* **45**, 209–36.

6

The thermally stratified atmospheric boundary layer

In earlier chapters, we dealt with the application of physical laws to describe the dynamics and thermodynamics of the ABL, and developed scaling laws for wind and other properties. A main emphasis was on the surface fluxes, and how these depend on surface and atmospheric conditions. In the present chapter we look more closely at specific features of the ABL, features that are closely related to turbulent and radiative characteristics in the presence of buoyancy. The intention is to deal with the convective and stable boundary layers separately, with an emphasis on simple analytical and conceptual models that address a range of physical phenomena (e.g. entrainment and mixed-layer growth; the surface inversion; the nocturnal jet). A distinction is made between the continental and marine ABL, mainly because of the impact of diurnal forcing on the former, and for many purposes we assume the absence of cloud. The chapter finishes with a section on the growth of the boundary layer in flow across a coastline, extending the discussion on the small-scale structure of the IBL in Chapter 4 to mesoscale aspects.

6.1 The convective boundary layer

6.1.1 Introduction

In the absence of complete cloud cover, the convective boundary layer (CBL) over land shows a strong diurnal development, and in mid-latitudes in summertime typically reaches a height of 1–2 km by mid-afternoon. From sunrise on, and throughout the daylight hours, this development (see Fig. 6.1) includes the following:

(i) the breakdown of the nocturnal inversion through heating, and the development of a shallow, well-mixed layer (A in Fig. 6.1);

(ii) the subsequent development of a deep, well-mixed boundary layer, possibly with a strong capping inversion; this elevated inversion layer atop the CBL is called the interfacial layer and should not be confused with the interfacial sublayer adjacent to the surface (Chapter 4);

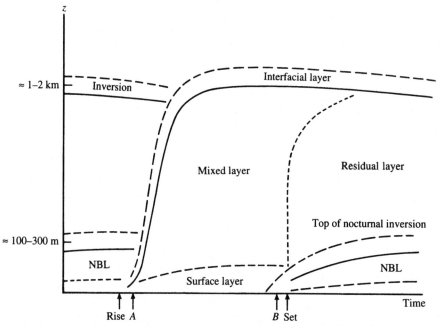

Fig. 6.1 Schematic representation of ABL evolution throughout the diurnal period over land under clear skies.

(iii) evidence of stable stratification in the upper regions of the CBL, apparently related to entrainment processes across the inversion;
(iv) prior to sunset, development of a surface inversion related to surface cooling (B in Fig. 6.1).

The mean structure of the quasi-stationary CBL, as exists in the late morning or in the afternoon over land, or over the sea during a cold-air outbreak for example, can be summarized as follows:

(i) The surface layer: this is limited to depths $z < |L|$, within which Monin–Obukhov theory is valid. Mean profiles, spectra and integral statistics, when suitably nondimensionalized, depend upon $\zeta \,(= z/L)$, except that for horizontal velocity turbulence components h/L is the relevant scaling. Here h is the CBL depth.
(ii) The free convection layer: this is confined to $|L| < z < 0.1h$, in which height range u_{*0} is no longer an important velocity scale, but z scaling still is. In fact u_{*0} never goes to zero in practice, but the condition of "local free convection" determines the scaling for turbulence structure. This involves the scales u_f and θ_f (Eq. 3.30).
(iii) The mixed layer: this comprises most of the CBL, with $0.1 < z/h < 1$. With strong turbulent mixing present, characterized by large values ($> 1\ \mathrm{m\,s^{-1}}$) of the turbulent velocity scale w_*, vertical gradients of θ_v and mean wind are small even in the presence of large geostrophic wind shear. The whole layer tends to warm at a uniform rate, implying that the heat flux H decreases approximately linearly with height. This can be seen by

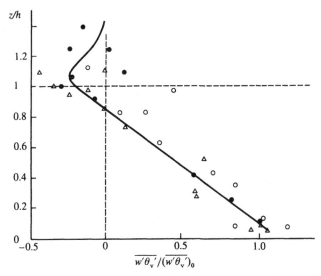

Fig. 6.2 Experimental data on the vertical variation of the virtual heat flux, normalized by its surface value; h is the depth of the mixed layer. Data are for three days from the 1983 ABL experiment; see Stull (1988, Figs. 3.1, 3.2 and 3.3). See also Fig. 6.23 of this volume.

taking the partial derivative with respect to z of Eq. 2.46 without the radiation term and noting that $\partial\theta_v/\partial z$ remains constant during the heating process. Thus,

$$\partial/\partial t(\partial\theta_v/\partial z) = -(\rho c_p)^{-1}\partial/\partial z(\partial H_v/\partial z) \approx 0,$$

whence $\partial H_v/\partial z$ equals a constant. Observations support the linear decrease of H_v with height; those summarized in Fig. 6.2 have the measured heat fluxes normalized by their surface values, and the height normalized by the ABL depth. When moisture fluxes are present, H and H_v decrease at different rates (H faster than H_v, when $\overline{w'q'}$ is upwards throughout the ABL), though the difference is not so large as to be readily detected in experimental data (see Section 6.1.3). The well-mixed assumption for scalars is not always met and, on many occasions, θ_v increases and q decreases in the upper part of the mixed layer. These features are partly related to the process of entrainment (see Section 6.1.5). Throughout most of the daytime, the CBL depth increases with time due to entrainment and encroachment processes.

Within the mixed layer, z and u_{*0} are no longer relevant to the structure and properties of turbulence, with h, w_* and T_* emerging as dominant scales.

(iv) The *inversion* or *interfacial layer*: this is dominated by local entrainment effects and by the properties of the capping inversion and stable region above. Acoustic sounder observations reveal evidence of an undulating inversion, with hummocks associated with thermal convective penetration from below. In the atmosphere, the mean thickness of this layer may be as much as $0.5h$.

So far as the mean structure of the CBL is concerned, the main problems of interest concern (i) the mean properties of the mixed layer; (ii) the growth of the CBL, $\partial h/\partial t$, and the depth itself; (iii) the entrainment fluxes of momentum, heat and mass at h; (iv) the description of turbulence fluxes as a function of height within the CBL. Before considering each of these in more detail, we describe some aspects of the physical structure of the CBL.

6.1.2 Physical aspects

The turbulence field within the CBL is dominated by coherent physical structures, or large eddies. The convective circulations, whose upward arm comprises buoyantly driven updrafts, extend vertically over the whole CBL. They have lifetimes of order h/w_*, ≈ 500 seconds, with $h = 1000$ m and $w_* = 2$ m s^{-1}, and a horizontal scale of about $1.5h$. From observations of gull soaring, and from the experience of glider pilots, it has long been known that the structure of thermal upcurrents in the CBL is of two basic types. In strong winds, the upcurrents are arranged in lines approximately along the mean wind direction, corresponding to horizontal roll vortices. In light to moderate winds, the upcurrents occur as isolated axisymmetric thermals or plumes. In recent times, such a picture has been confirmed and extended by field observation (Webb, 1977) and numerical simulation (Schmidt and Schumann, 1989).

Thermal and plume structures

Convective updrafts have their origin near the surface in the superadiabatic layer, and occur as small-scale sheets or thermals of buoyant air. They are readily identified with the characteristic ramp structure seen in a temperature time series, with the maximum updraft and temperature usually located just inside the upwind edge of the ramp (Fig. 6.3). Higher up, these ramps merge into larger-scale plumes or columns with typical widths of several hundred metres and maximum updraft speeds of 2–5 m s^{-1} at $z/h \approx 0.75$. In the sinking environmental air through which the plumes rise, the stratification is usually slightly stable, even more so if there is entrainment of warmer air into the top of the CBL. Consistent with this environmental stability, the temperature excess of plumes relative to the environment decreases with height, passing through zero near $z/h \approx 0.75$. Thermals rising from the surface remote from strong convergence (updraft) regions do not merge but decay whilst rising through the larger-scale downdrafts. At the top of the boundary layer, rising air spreads laterally producing hummocks and disturbances at the interface (penetrative convection), and returns as descending air around the updraft regions. The downdrafts are closely related to the entrainment process. They occur as small-scale sheets in the entrainment zone, becoming more diffuse, larger-scale downdrafts at lower levels in the CBL. If the updrafts reach the lifting condensation level of the air within them, cumulus clouds will form, thus providing a new source of buoyant energy to aid in penetration of the capping inversion. According to LES results (Schmidt and Schumann, 1989), the entrainment heat flux at the top of the CBL is carried by cold updrafts and warm downdrafts in the form of wisps at scales comparable to h. These wisps

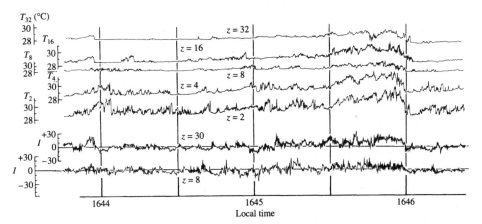

Fig. 6.3 Examples of recorded fluctuations of temperature T (in °C) and inclination angle I over flat, dry grassland (at Hay, Australia) at heights from 2 m to 32 m as shown. The mean wind speed at 2 m was $4.5\,\mathrm{m\,s^{-1}}$ with $L = -18.5\,\mathrm{m}$. The traces illustrate the characteristic ramp structure for temperature associated with passing thermals (indicated also by positive inclination, or upwards motion). From Webb (1977).

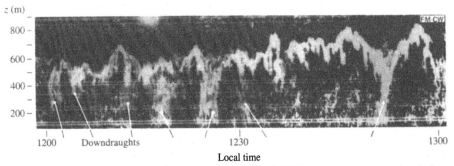

Fig. 6.4 FM-CW radar time–height presentation of CBL structure over land for a one-hour period. The bright vertically aligned regions correspond to downdraught structures (or wisps), and the bright undulating continuous region to the interfacial (entrainment) layer. From Rowland and Arnold (1975), by kind permission of the authors.

are evident in laboratory photographs of turbulent entrainment across a density interface, and in radar and acoustic sounder records of the CBL (Fig. 6.4).

The whole convective circulation system, comprising columnar updraft and diffuse downdraft arms, travels in the mean wind direction at a speed about that of the wind averaged through the boundary layer. In plan form, the thermal plumes or columns in light to moderate winds tend to be in the form of walls, and arranged in an irregular polygon pattern of horizontal dimension about 1–3 km. Figure 6.5 shows an excellent LES representation of the cellular structure of the convection in the CBL, in terms of the fluctuation vertical velocity and temperature fields. Cells of 1–2 km diameter are clearly evident, although the structure and dimensions do tend to vary with height.

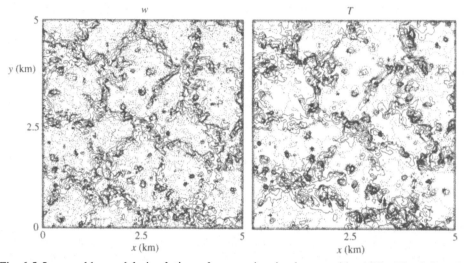

Fig. 6.5 Large-eddy model simulation of convection in the unstable ABL. The left and right panels show contour plots of normalized vertical velocity and normalized temperature fluctuations respectively. Each panel has a 5 km side, and the fields are shown for $z/h = 0.25$. The pecked curves correspond to negative contour values (i.e. cool, subsiding air). Thus, the patterns demonstrate an irregular polygon structure with warm, rising motion confined mostly to the "thin" walls of the 1–2 km-wide columns. From Schmidt and Schumann (1989).

Waves in the inversion or interfacial layer

Strong wind shear is likely across the stable layer at the top of the CBL so that Kelvin–Helmholtz waves are generated. These waves form because the flow becomes dynamically unstable in the presence of shear. They are also likely at a smaller scale, near the tops of penetrating updrafts where large local shears are also likely. The larger-scale waves in particular, after breaking and decaying into turbulence, contribute to the entrainment process across the interface.

Penetrative convection within the stable, overlying region also generates internal gravity waves, which may propagate in many directions away from the CBL. Their influence on the energy or momentum budget of the CBL, and on the entrainment rate, appears to be very small under most conditions of interest.

Entrainment

Entrainment is the process whereby miscible fluid is exchanged across a density interface bounding a region of turbulent flow (Turner, 1973); it is of central importance in governing the structure of boundary-layer clouds. In the exchange process, relatively quiescent fluid is engulfed by turbulent motions penetrating across the mean density interface and is subsequently mixed into the turbulent region. Smaller-scale motion is rapidly damped by the interfacial density gradient so that a sharp interface is maintained which advances into the quiescent layer causing the turbulent layer to thicken. The existence of the sharp temperature step at the inversion is an indication that some of the energy released by gravitational instability at the heated boundary has been used to produce entrainment across the inversion interface.

With entrainment, air is transferred across the capping inversion from above to within the CBL at the expense of the turbulent kinetic energy. This is usually the means by which any turbulent layer adjacent to a non-turbulent region increases its depth with time. So far as the CBL is concerned, relevant mechanisms include shear-stress driven entrainment and the buoyancy driven entrainment associated with penetrative convection. In contrast, *encroachment* is growth of the CBL arising from surface heating alone, and takes place in the absence of a capping inversion in conditions of non-penetrative convection.

Horizontal roll vortices
The formation of these rolls under combined surface heating and strong wind conditions (i.e. slight to moderate thermal instability only), appears to be related to thermal and inertial instabilities of the mean flow in the CBL (Stull, 1988, Chapter 11). These instabilities are associated with weak, horizontal helical circulations that may concentrate convective thermals and plumes into rows, resulting in along-wind convergence zones. If moisture is readily available, cumulus cloud streets can form along the convergence lines. Strong air-mass modification during cold-air outbreaks provides the optimum conditions for rolls and the associated cloud streets, and at larger fetches out to sea may be transformed into mesoscale cellular convection.

6.1.3 Mean mixed-layer properties

General
The mean structure of the CBL can be represented in varying degrees of complexity depending on the problem at hand, the main difference between model versions usually concerning the assumed structure of the inversion layer. Two versions are represented schematically in Fig. 6.6: in the first version, shown in (*a*) and (*b*), the thin surface and free-convection layer are incorporated as a lower layer, so that the CBL becomes a two-layer model incorporating zero-order jumps at the top. For many purposes, the heat capacity and momentum deficit of this thin lower layer relative to the mixed layer as a whole can be ignored, and the CBL can be considered as a single slab. This version is known as the "zero-order jump" or "slab" model of the CBL. The second version, (*c*), incorporates an inversion layer of finite thickness, Δh, and is known as a "first-order jump" model. Both zero-order and first-order models are used frequently in many studies of CBL growth and entrainment, though more refined versions that incorporate a more detailed description of the inversion layer do exist (Deardorff, 1979). A most important quantity appearing in the problem here concerns the entrainment heat flux at h, H_{vh}, shown in Fig. 6.6 in association with the heat-flux profile.

The conservation equation for any property s (e.g. as represented by Eqs. 2.44 to 2.47) can be integrated with respect to height z, between the surface and h, to yield an equation for the vertical average, or slab, property, s_m. For the zero-order model, the upper limit of integration is set at $h - \varepsilon$ and the limit

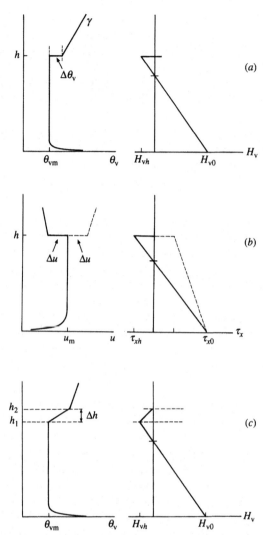

Fig. 6.6 Schematic representation of the vertical structure of the CBL, for wind and temperature. In (*a*) and (*b*) the inversion layer is assumed to be infinitesimally thin, and represented by zero-order jumps in properties. The CBL is assumed to be well mixed, except for a thin surface layer. The corresponding flux profiles are also shown. In (*c*), the inversion layer is more realistic, of finite thickness Δh, and this version is referred to as the first-order jump model.

$\varepsilon \to 0$ taken after integration; for the first-order model no such procedure is necessary. The result is

$$\partial s_m/\partial t = [\overline{(w's')}_0 - \overline{(w's')}_h]/h \qquad (6.1)$$

where the slab value of s is defined as

$$s_m = h^{-1} \int_{z_0}^{h} s \, dz. \qquad (6.2)$$

We should note that the result in Eq. 6.1 utilizes the Leibnitz rule of integration, i.e.

$$\int_{z_0}^{h} (\partial s/\partial t)\,dz = \partial/\partial t \left(\int_{z_0}^{h} s\,dz \right) - s(h)\,\partial h/\partial t$$

$$= h\,\partial s_m/\partial t - [s(h) - s_m]\partial h/\partial t \qquad (6.3)$$

with $s(h) = s_m$. Equation 6.1 describes the time rate of change for mixed-layer temperature or humidity, with radiation effects, for example, ignored. For momentum, the analogous result includes pressure and rotation terms, with (cf. Eqs. 2.44, 2.45)

$$\partial u_m/\partial t = f(v_m - v_{gm}) + [(\overline{u'w'})_0 - (\overline{u'w'})_h]/h \qquad (6.4)$$

$$\partial v_m/\partial t = -f(u_m - u_{gm}) + [(\overline{v'w'})_0 - (\overline{v'w'})_h]/h. \qquad (6.5)$$

These equations for slab properties contain entrainment fluxes at h. They are written formally for the horizontally homogeneous case, with the winds at and above the boundary-layer top assumed to be geostrophic. Although the vertically-averaged geostrophic wind components are given in Eqs. 6.4 and 6.5, Eq. 6.1 for θ_v and Eqs. 6.4 and 6.5 taken together are strictly valid for barotropic conditions only. For baroclinic cases, where u_g and v_g vary with height (implying that $\partial\theta_v/\partial x$ and $\partial\theta_v/\partial y$ are non-zero), the θ_v gradient terms need to be included in Eq. 6.1.

Equations 6.1, 6.4 and 6.5 are associated with linear turbulent flux profiles, since the time change of any mean property must be uniform throughout the CBL. Taking Eq. 6.1 with $s = \theta_v$, we see that the heat flux at the surface will heat the CBL, as will the entrainment heat flux if it is of the opposite sign, i.e. downwards across the inversion. In the case of humidity, with $s = q$, the surface moisture flux or evaporation tends to increase the slab humidity whilst an upwards (entrainment) moisture flux, with dry air above the ABL, will tend to dry out the CBL. For winds, if $u_m > 0$ (Eq. 6.4) we expect $(\overline{u'w'})_0 < 0$, since surface friction will remove momentum from the layer. If the wind component u above the CBL ($z > h$) is larger than u_m, entrainment of overlying air causes acceleration, and is manifest as a negative value of $(\overline{u'w'})_h$. If the wind aloft has u smaller than u_m, the entrainment flux causes deceleration in the CBL and so $(\overline{u'w'})_h$ is positive (Fig. 6.6(b)).

With $\Delta s_h = s(h) - s_m$, the entrainment fluxes can be formally related to the jumps Δs_h at $z = h$ by use of the zero-order model. We take the one-dimensional conservation equations (2.44–2.47) with a vertical advection term included and integrate between limits of $h - \varepsilon$ and $h + \varepsilon$. By applying Eqs. 6.3, 6.4 or 6.5 as appropriate, and taking $\varepsilon \to 0$, the following relations for the properties result:

$$(\overline{u'w'})_h = -\Delta u_h(\partial h/\partial t - w_h) \qquad (6.6)$$

$$(\overline{v'w'})_h = -\Delta v_h(\partial h/\partial t - w_h) \qquad (6.7)$$

$$(\overline{w'\theta_v'})_h = -\Delta\theta_{vh}(\partial h/\partial t - w_h) \qquad (6.8)$$

$$(\overline{w'q'})_h = -\Delta q_h(\partial h/\partial t - w_h). \qquad (6.9)$$

Here the mean vertical velocity at $z = h$ is represented by w_h. From the above discussion, it follows that entrainment always leads to an increase in h ($\partial h/\partial t > 0$) in the absence of subsidence.

As the slab properties change, so do the respective jumps across the top of the CBL. We consider the zero-jump model only, and note that $\partial s(h)/\partial t = (\partial s/\partial z)_+(\partial h/\partial t - w_h)$ where $(\partial s/\partial z)_+$ is the vertical gradient of property s above h. Then the jump equations can be written, by identity,

$$\partial(\Delta s_h)/\partial t = (\partial s/\partial z)_+(\partial h/\partial t - w_h) - \partial s_m/\partial t \qquad (6.10)$$

with $\partial s_m/\partial t$ given by Eq. 6.1 for θ or q, and by Eqs. 6.4 and 6.5 for the wind components. In the case of temperature (with $(\partial\theta/\partial z)_+ = \gamma_\theta$), for example, we conclude from Eq. 6.10 that $\Delta\theta_{vh}$ changes in three ways. Since $\gamma_\theta > 0$, it tends to increase both when the mixed-layer height increases and when there is subsidence ($w_h < 0$), as described by the first term on the right-hand side. It tends to decrease as the mixed layer warms, as described by the second term on the right-hand side.

A simple mixed-layer model
This is based on the zero-order jump scheme (Fig. 6.6(a)). The basic equations that govern the growth and temperature of the one-dimensional horizontally homogeneous mixed layer are based on Eqs. 6.1 and 6.10 for θ_v, and on Eq. 6.8, i.e.

$$\partial\theta_{vm}/\partial t = [(\overline{w'\theta_v'})_0 - (\overline{w'\theta_v'})_h]/h \qquad (6.11)$$

$$\partial(\Delta\theta_{vh})/\partial t = \gamma_\theta(\partial h/\partial t - w_h) - \partial\theta_{vm}/\partial t. \qquad (6.12)$$

These two equations, together with Eq. 6.8, need to be solved for h and θ_{vm} in particular, but they contain seven unknowns, five if we assume that w_h and γ_θ are known: θ_{vm}, $\Delta\theta_{vh}$, h and the two fluxes. A separate equation for $(\overline{w'\theta_v'})_0$ can be readily incorporated or we can assume that the surface heat flux is specified as a lower boundary condition. This leaves four unknowns, and closure of the set is usually achieved by utilizing a parameterized form of the TKE equation (see Section 6.1.5). The above set of equations can be extended to include analogous quantities in the velocity components, and the specific humidity, but at the expense of extending the closure requirements. The closure problem involves a parameterized form for the entrainment heat (and other) flux, but before considering this further we digress and look at several special cases of CBL growth.

6.1.4 CBL growth
In order to study mixed-layer growth, the growth rate $\partial h/\partial t$ has to be related to known properties of the boundary layer, including surface heat flux, the CBL depth, $\Delta\theta_{vh}$ and γ_θ.

Equations 6.11 and 6.12 can be combined to give

$$\partial(\Delta\theta_{vh})/\partial t = \gamma_\theta(\partial h/\partial t - w_h) - [(\overline{w'\theta_v'})_0 - (\overline{w'\theta_v'})_h]/h \qquad (6.13)$$

where $(\overline{w'\theta_v'})_h$ can be replaced using Eq. 6.8. Equation 6.13 immediately yields the encroachment rate, for which $\Delta\theta_{vh} = \overline{(w'\theta_v')}_h = 0$. Thus, with $w_h = 0$,

$$(\partial h/\partial t)^{\text{enc}} = \overline{(w'\theta_v')}_0/\gamma_\theta h. \tag{6.14}$$

For typical midday conditions under clear skies in summer, we can take $h = 1000$ m, $\gamma_\theta = 5$ K km^{-1} and $\overline{(w'\theta_v')}_0 = 0.2$ m s^{-1} K to give a growth rate of about 0.04 m s^{-1} or 144 m hr^{-1}. To determine the growth rate in the presence of an entrainment heat flux it is necessary to make a crucial closure assumption, i.e. that the entrainment flux is a constant fraction β of the surface flux (see Eq. 6.31). This assumption thus neglects any contribution mechanical turbulence makes to CBL growth. We have

$$\overline{(w'\theta_v')}_h = -\beta\overline{(w'\theta_v')}_0 \tag{6.15}$$

so that, with $w_h = 0$, we can rewrite Eq. 6.13, utilizing Eq. 6.8, as

$$\beta^{-1}(1 + \beta)\Delta\theta_{vh}\partial h/\partial t = \gamma_\theta h\,\partial h/\partial t - h\,\partial(\Delta\theta_{vh})/\partial t. \tag{6.16}$$

This is a homogeneous differential equation with solution (e.g. Betts, 1973 and 1974)

$$\Delta\theta_{vh} = \gamma_\theta\beta h/(1 + 2\beta), \tag{6.17}$$

and indicates the connection between the presence of a temperature jump or inversion and a non-zero value of β. If $\Delta\theta_{vh} = 0$, then $\beta = 0$ and we have encroachment. Substituting into Eq. 6.8, and using Eq. 6.15, yields

$$\partial h/\partial t = (1 + 2\beta)\overline{(w'\theta_v')}_0/\gamma_\theta h. \tag{6.18}$$

There are several points of interest that are worthy of discussion, as follows.

(i) Comparison of Eqs. 6.14 and 6.18 shows that entrainment of heat and encroachment together are greater than encroachment alone, by a factor of $1 + 2\beta \approx 1.4$. Thus entrainment typically contributes $2\beta/(1 + 2\beta) \approx 30$ per cent to the total growth rate.

(ii) If we accept Eq. 6.15 for the buoyancy flux, Eq. 2.79 implies that the ratio for the sensible heat fluxes will differ from β when moisture fluxes are present, and that the decrease of $\overline{w'\theta'}$ with height will be more rapid than that of $\overline{w'\theta_v'}$, for upwards moisture fluxes. Thus

$$\overline{(w'\theta')}_h/\overline{(w'\theta')}_0 = -\beta - 0.61\theta[\beta\overline{(w'q')}_0 + \overline{(w'q')}_h]/\overline{(w'\theta')}_0 \tag{6.19}$$

which gives a ratio more negative than $-\beta$. In fact, assuming $\overline{(w'q')}_0 = \overline{(w'q')}_h$, $\beta = 0.2$, $\theta = 300$ K and a Bowen ratio of unity, we find that the right-hand side is equal to -0.29.

(iii) Returning to Eq. 6.18, rearrangement gives

$$\partial/\partial t(h^2/2) = \gamma_\theta^{-1}(1 + 2\beta)\overline{(w'\theta_v')}_0. \tag{6.20}$$

If γ_θ is taken as constant this can be integrated with respect to time to give

$$(h^2 - h_{00}^2)/2 = \gamma_\theta^{-1}(1 + 2\beta)\int_0^t \overline{(w'\theta_v')}_0\,dt' \tag{6.21}$$

where h_{00} is the height at $t = 0$. For encroachment only, with $h_{00} = 0$ we obtain

$$h^2 = (2/\gamma_\theta)\int_0^t (\overline{w'\theta_v'})_0 \, dt'. \tag{6.22}$$

which represents simple growth and heating of the CBL, as illustrated in Fig. 6.7(*a*).

(iv) Equation 6.22 is a simple representation of growth of the shallow morning CBL through the surface inversion. Thus, let the surface inversion, of depth h_i, have a linear θ_v profile, such that $\gamma_\theta \approx \Delta\theta_v/h_i$, where $\Delta\theta_v$ is the initial temperature change across the inversion. Take the heat flux as a linear function of time, $(\overline{w'\theta'})_0 = (t/T)(\overline{w'\theta_v'})_n$, where $(\overline{w'\theta_v'})_n$ is the midday value and $T \approx 3$ hours. Equation 6.22 then gives an estimate of the time required to completely heat up and dissipate, or "breakdown", the inversion, i.e.

$$t = (Th_i\Delta\theta_v / (\overline{w'\theta_v'})_n)^{1/2}. \tag{6.23}$$

For a maximum heat flux of $500 \, \mathrm{W\,m^{-2}}$, inversion depth of $250 \, \mathrm{m}$ and $\Delta\theta_v = 10 \, \mathrm{K}$, which are all typical of mid-summer conditions under clear skies in mid-latitudes, Eq. 6.23 suggests a breakdown time of about 2.5 hr. In mid-winter, with much smaller heat fluxes, this could be in excess of 5 hr. Such values are consistent with observations, though this simple model does not take into account entrainment and dynamic effects.

(v) Mechanical turbulence is excluded from the above analysis by specifically linking the entrainment heat flux to the surface heat flux (Eq. 6.15) or setting it to zero (yielding Eq. 6.14). The introduction of frictional effects requires a more general closure assumption for $(\overline{w'\theta_v'})_h$ in order that the growth equation can be solved. An interesting case involves a wind-driven mixed layer with zero surface heat flux, so that growth is by mechanical mixing only. For an initially stable θ_v profile with lapse rate γ_θ, mixing after time t produces the modified profile shown in Fig. 6.7(*b*). Simple considerations show that the temperature jump is given by $\Delta\theta_{vh} = \gamma_\theta h/2$ so that the original stratified air has been mixed, with no net heating of the layer below h. The reader should note that this value of $\Delta\theta_{vh}$ is the solution of Eq. 6.13 with $(\overline{w'\theta_v'})_0 = 0$ and $(\overline{w'\theta_v'})_h$ given by Eq. 6.8, and is equivalent to setting $\beta = \infty$ in Eqs. 6.16 and 6.17. Substitution of $\Delta\theta_{vh} = \gamma_\theta h/2$ into Eq. 6.8 with $w_h = 0$ gives the growth rate of the layer as

$$\partial h/\partial t = -2(\overline{w'\theta_v'})_h/\gamma_\theta h, \tag{6.24}$$

which also results from Eq. 6.18 with $\beta = \infty$.

A suitable closure for the wind-mixed layer utilizes an analogous relation to Eq. 6.15 for buoyancy-driven entrainment. In this case, we use the fact that the entrainment flux is the result of the surface stress acting over a layer of depth h. If we set, with c_1 an unknown constant,

$$(\overline{w'\theta_v'})_h = -c_1 u_{*0}^3/h, \tag{6.25a}$$

which is consistent with TKE considerations (see next section), the growth rate can be written

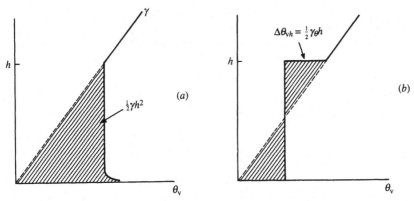

Fig. 6.7 Schematic representation of the heating of a convective boundary layer for (*a*) encroachment, and (*b*) mechanical entrainment only. In (*a*) $\Delta\theta_{vh} = 0$ and in (*b*) $\overline{(w'\theta_v')}_0 = 0$. The shaded area in (*a*) represents the time-integrated surface heat flux (see Eq. 6.22); in (*b*), the areas are equal, since there is no net heat input but, rather, a vertical redistribution of heat in the vertical. From Tennekes and Driedonks (1981): reprinted by permission of Kluwer Academic Publishers.

$$\partial h/\partial t = 2c_1 u_{*0}{}^3/\gamma_\theta h^2. \qquad (6.25b)$$

Integration gives $h \propto u_{*0} t^{1/3}$, which contrasts with the behaviour for buoyancy-driven entrainment where $h \propto t^{1/2}$, based on Eq. 6.22 with constant heat flux (Tennekes and Driedonks, 1981).

The zero-order jump, or slab, model is useful for many applications, but is necessarily limited when the interfacial thickness, Δh, is involved. This may occur, for example, when the closure relations for the entrainment fluxes are based on the TKE equation with ε scaled to Δh^{-1} (sse Section 6.1.5). In such circumstances, higher-order jump models are used, which reflect the fact that $\Delta\theta_{vh}$ occurs over a finite thickness. In the first-order approach (Fig. 6.6(*c*)), h_1 is the relevant CBL depth, whilst h_2 represents the mean height to which penetrative convection, in the form of energetic thermals, reaches. In this model, the minimum negative heat flux occurs at h_1 though there is a modified form that places this minimum flux midway between h_1 and h_2. In contrast with the zero-order model, the entrainment heat flux can be eliminated in the first-order approach, with a closure assumption required for Δh. The possible need for the first-order model is apparent, since many observations on CBL growth due to buoyancy-driven entrainment suggest $\Delta h/h \approx \Delta h/h_1 \approx 0.3$.

6.1.5 Entrainment fluxes

These are formally given by Eqs. 6.6–6.9, and are related to the jumps in properties across the top of the CBL. For the heat flux, the simplest closure assumption is that the entrainment flux is proportional to the surface heat flux (Eq. 6.15). However, since entrainment involves a downwards heat flux across a stable region, this would seem to be at the cost of the expenditure of energy, since warmer air that is being entrained would prefer to rise. This energy comes from the surface heat flux, or from mechanical generation due to wind shear.

Closure schemes generally are based upon, or utilize, the TKE equation so as to formulate a relationship between entrainment and the TKE balance. For buoyancy-dominated layers, an important hypothesis (e.g. Ball, 1960; Randall, 1980a) states that the rate of entrainment is determined by the condition that the ratio of the total buoyant dissipation of TKE (call this N_B) in the mixed layer to the total buoyancy production (P_B) is a constant. This requires

$$N_B/P_B = \text{constant}. \tag{6.26}$$

We consider the TKE equation (Eq. 2.74) integrated across the CBL. With $\partial \bar{e}/\partial t$ and shear production terms assumed small, this reduces to

$$(g/\theta_v)\int_{z_0}^{h} \overline{w'\theta_v'}\,dz = \int_{z_0}^{h} \varepsilon\,dz \tag{6.27}$$

$$= N_B + P_B.$$

Here, the left-hand side is comprised of opposing positive (production, P_B) and negative (dissipation, N_B) contributions. For the assumed heat-flux profile shown in Fig. 6.6(a),

$$\overline{w'\theta_v'} = (\overline{w'\theta_v'})_0(1 - z/h) + (\overline{w'\theta_v'})_h(z/h), \tag{6.28}$$

the two component terms in Eq. 6.27 can be written as

$$P_B = (g/\theta_v)\int_{z_0}^{h}(1 - z/h)(\overline{w'\theta_v'})_0\,dz \tag{6.29}$$

$$= (g/\theta_v)h(\overline{w'\theta_v'})_0/2,$$

$$N_B = (g/\theta_v)\int_{z_0}^{h}(z/h)(\overline{w'\theta_v'})_h\,dz \tag{6.30}$$

$$= (g/\theta_v)h(\overline{w'\theta_v'})_h/2.$$

Equation 6.26 then becomes

$$N_B/P_B = (\overline{w'\theta_v'})_h/(\overline{w'\theta_v'})_0 = -\beta. \tag{6.31}$$

Two extreme cases are of interest here; firstly, when viscous dissipation is negligible (or is balanced by any shear production), all buoyant generation goes into the entrainment flux. In this case, $\beta = 1$ (Ball, 1960). Secondly, when the interface is marginally stable ($\Delta\theta_{vh} \to 0$), boundary heating produces a profile that just encroaches on the initial profile, i.e. it changes directly from the initial form to a well-mixed layer with no temperature discontinuity above (Carson and Smith, 1974). In this case, $\beta = 0$ (see schematic temperature profile in Fig. 6.7(a)).

An alternative approach to Eq. 6.26 compares the energy used for entrainment with that produced by the destabilizing flux at the ground. In this (Manins and Turner, 1978), the relevant energy ratio is β^* which equals the ratio of the increase in potential energy due to the redistribution of the initial profile to the potential energy made available by heating from below. For present purposes, we take as given

$$\beta^* = (2\beta + 1)/3 \tag{6.32}$$

so that $\beta^* = 1$ when $\beta = 1$ (heat fluxes are equal but of opposite sign) and all energy released by heating is used to redistribute any property s. For $\beta = 0$ (encroachment), $\beta^* = 1/3$, implying that one-third of the energy released by heating from below is used for redistribution (of the fluid from the stable environment), whilst two-thirds has been dissipated.

Theoretical and experimental approaches have suggested values of β between 0 and 1, though the bulk of experimental data for the daytime CBL over land gives $\beta \approx 0.2$. With a value of β of about 0.2–0.25, the entrainment flux contributes on average about 20 per cent to the CBL heating, and about 30 per cent to the growth (Eq. 6.18).

Equation 6.15 for the entrainment flux is also consistent with a closure scheme based on the budget of TKE at the inversion base. In the case of buoyancy-driven entrainment, this can be seen by assuming that the buoyancy loss of TKE is balanced by a flux convergence due to penetrative convection. Viscous dissipation and shear-stress terms are assumed negligible. Then

$$(g/\theta_v)(\overline{w'\theta_v'})_h = \partial/\partial z(\overline{w'e}) \approx -w_*^3/h$$

whence Eq. 6.15 results, using the definition of w_*. In the case of friction-driven entrainment, Eq. 6.25a follows by assuming that the buoyancy loss of TKE is balanced by a flux convergence due to shear-driven mixing. Then

$$(g/\theta_v)(\overline{w'\theta_v'})_h = \partial/\partial z(\overline{w'e}) \approx -u_{*0}^3/h.$$

In the presence of significant velocity jumps at h, viscous dissipation and shear-stress terms would need to be included in any treatment of this closure problem, since it can be assumed that shear-driven entrainment will be important.

The consequences of entrainment are very apparent in observed profiles of the sensible heat flux (Fig. 6.2). In this figure, an extensive zone between 0.8 and $1.3h$ is seen to be affected by entrainment, with maximum negative values of $\overline{w'\theta_v'}$ reaching 40 per cent of the surface value. In the case of entrainment fluxes of momentum, these are expected to be significant whenever there are large velocity jumps in the presence of a growing mixed layer (Eqs. 6.6 and 6.7). The jumps for a quasi-steady CBL, as might exist over land in the afternoon, are closely given by

$$\Delta u_h \approx u_g(h) - u_m \tag{6.33}$$

$$\Delta v_h \approx v_g(h) - v_m \tag{6.34}$$

if the wind above the CBL is assumed to be geostrophic. In contrast, velocity jumps across the growing CBL during the morning over land can be much larger, due to the nocturnal jet (Section 6.2) overlying the shallow mixed layer and remnants of the surface inversion. In such situations, observations have revealed entrainment stresses to be as high as twice the surface values (Deardorff, 1973).

6.1.6 Representation of fluxes throughout the CBL

For modelling purposes, vertical fluxes of momentum, heat and water vapour and other scalars can be evaluated using either local or non-local closure schemes (Chapter 8). It is apparent that convective mixed-layer turbulence is one of the more difficult regimes to model using local closure, because the important eddy length scales are comparable with the thickness of the CBL. Rather, the validity of K-theory and mixing length assumptions requires eddies that are much smaller than the scale of interest. In the presence of large eddies and intense mixing, the CBL can often be asumed to be well mixed (See Figs. 3.10 and 6.6), though on some occasions over land and sea it is distinctly unmixed in one or more of the properties. This lack of mixing is apparent in the profiles of potential temperature and specific humidity shown in Fig. 6.8.

The application of K-theory to the θ_v profiles in particular has obvious limitations, since the heat flux may often appear to be countergradient on average, and therefore will require K to be negative, and in parts of the profile (in mid-regions of the CBL usually) may even be singular. Local closure has problems because it relates the flux of a quantity at a given level to properties at that level. The alternative approach uses non-local closure in the form of integral or transilient methods. Nevertheless, many numerical models of the atmosphere, both small-and large-scale, use local closure with suitable modifications to bypass the countergradient transfer problem (see Chapter 8).

The nature of the scalar mean profiles in the CBL determines whether local closure will be appropriate. For the ideal CBL considered here, entrainment profoundly influences the mean temperature and scalar mixing ratio profiles (θ_v and q, for example) through what is referred to as "top-down diffusion". This type of diffusion differs from the familiar "bottom-up" diffusion because the buoyant forcing of the convective turbulence occurs with vertical asymmetry. Numerical simulations reveal that the flux-gradient relationship (involving an eddy diffusivity) for a passive scalar in a mixed layer depends on the boundary from which the flux originates (e.g. Wyngaard and Brost, 1984). To see this, and how it relates to the K-problems, let us consider a passive, conservative scalar whose mean concentration c satisifies

$$\partial c/\partial t = -\partial(\overline{w'c'})/\partial z \tag{6.35}$$

where $\overline{w'c'}$ is the scalar flux. If temporal changes in the scalar gradient $\partial c/\partial z$ are assumed negligible then the flux profile is linear and the vertical profiles can be represented as in Fig. 6.9. The flux profile can be written as

$$\overline{w'c'} = (\overline{w'c'})_0(1 - z/h_1) + (\overline{w'c'})_h(z/h_1) \tag{6.36}$$

so that the flux at any level has contributions from bottom-up diffusion (through $(\overline{w'c'})_0$) and from top-down diffusion (through $(\overline{w'c'})_h$).

Next consider two extreme situations, where either $(\overline{w'c'})_h$ is zero (bottom-up diffusion only) or $(\overline{w'c'})_0$ is zero (top-down only). In the first case the associated eddy diffusivity, K_b, is given by

$$K_b = -\overline{w'c_b'}/(\partial c_b/\partial z)$$
$$= -(\overline{w'c'})_0(1 - z/h_1)/(\partial c_b/\partial z) \tag{6.37}$$

where Eq. 6.36 has been used and where subscript b refers to bottom-up transfer only. In the second case, the associated eddy diffusivity, K_t, is given by

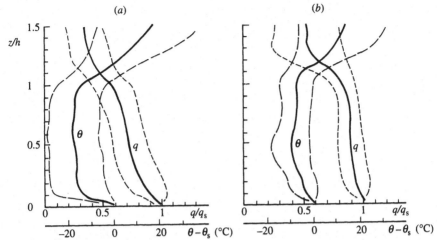

Fig. 6.8 Mean CBL profiles of θ and q observed in (*a*) north-east Colorado and (*b*) south-eastern Australia. Before averaging, q data were normalized by the screen-level value, and θ data were reduced by the screen-level value. In (*a*), mean $h = 1710$ m, mean $q_s = 0.0093$; in (*b*), mean $h = 915$ m and mean $q_s = 0.0041$. Pecked lines denote standard deviations about the mean profiles. After Mahrt (1976), *Monthly Weather Review*, American Meteorological Society.

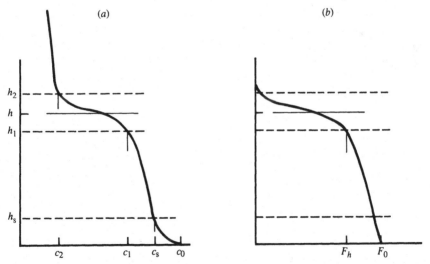

Fig. 6.9 Schematic representation of profiles of (*a*) mean scalar concentration and (*b*) scalar flux through, and above, the CBL. Levels h_s, h_1 and h_2 correspond to the tops of the surface layer, nominal mixed layer and inversion (entrainment) region with associated concentration values of c_s, c_1 and c_2 respectively. F_h is the entrainment flux at h_1. After Wyngaard and Brost (1984), *Journal of Atmospheric Sciences*, American Meteorological Society.

$$K_t = -\overline{w'c_t'}/(\partial c_t/\partial z)$$
$$= -(\overline{w'c'})_h(z/h_1)/(\partial c_t/\partial z) \qquad (6.38)$$

where again Eq. 6.36 has been used and where subscript t refers to top-down transfer only. With fluxes at both boundaries, a crucial step is to assume superposition of properties appropriate to the single-flux situations, so that

$$c = c_b + c_t \qquad (6.39)$$
$$c' = c_b' + c_t'. \qquad (6.40)$$

In this general case, it then follows that

$$\partial c/\partial z = -[(\overline{w'c'})_0(1 - z/h_1)/K_b + (\overline{w'c'})_h(z/h_1)/K_t] \qquad (6.41)$$

which shows that

  (i) if both fluxes are of the same sign, the vertical gradient of c will tend to be large; this is characteristic of the humidity profile;
 (ii) if the fluxes are of opposite sign, the vertical gradient may be close to zero; this occurs with the temperature profile.

We also have for K, using Eq. 6.41,

$$K = -\overline{w'c'}/(\partial c/\partial z)$$
$$= \frac{K_b K_t \overline{w'c'}}{K_t(\overline{w'c'})_0(1 - z/h_1) + K_b(\overline{w'c'})_h(z/h_1)}. \qquad (6.42)$$

Since K depends on the fluxes, this shows that the closure is inherently non-linear with $0 < K < \infty$. The poor behaviour in K can occur even when K_b and K_t are well behaved, but different (see Holtslag and Moeng, 1991).

For the determination of the c-profile, Eq. 6.41 requires knowledge of the eddy diffusivities before integration can be made. The form of the diffusivities can be obtained from large-eddy simulations (e.g. Moeng and Wyngaard, 1984), with

$$K_t = 1.4w_*z(1 - z/h_1)^2 \qquad (6.43)$$
$$K_b = w_*h_1(1 - z/h_1)/g_b. \qquad (6.44)$$

The top-down diffusivity is well behaved and positive throughout the convective boundary layer. In contrast, the mean gradient function g_b is positive in the lower regions of the mixed layer and near the surface, but changes sign near $z/h_1 = 0.6$ and is negative above (Moeng and Wyngaard, 1984). This is a clear illustration of the breakdown of the K-model, with K_b having a singularity in the mid-regions of the mixed layer. The result for K_b shows that the bottom-up process generates a countergradient flux in the upper mixed layer. Substitution into Eq. 6.41 and integration gives

$$c = -[(\overline{w'c'})_0/w_*h_1]\int g_b dz - 0.7[(\overline{w'c'})_h/w_*](1 - z/h_1)^{-1} + \text{constant}. \quad (6.45)$$

With suitable bottom boundary conditions, the c profile described by Eq. 6.45 for various values of the ratio $R = (\overline{w'c'})_h/(\overline{w'c'})_0$ is shown in Fig. 6.10. Small

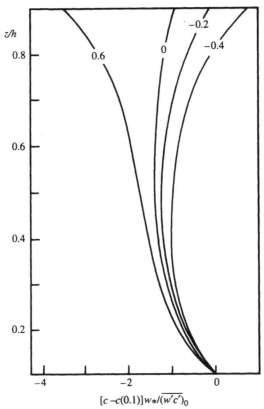

Fig. 6.10 Scaled concentration profiles based on Eq. 6.45, for four values of the parameter $R = \overline{(w'c')}_h / \overline{(w'c')}_0$ and the g_b profile taken from Fig. 1a of Moeng and Wyngaard (1984). Here, $c(0.1)$ is the concentration at $z/h = 0.1$ and is calculated from Eq. 6.45. Negative R gives temperature-like profiles, positive R gives humidity-like profiles. Based on Wyngaard and Brost (1984) and Moeng and Wyngaard (1984).

negative R values give a profile form similar to that for potential temperature (e.g. with $R = -\beta = -0.2$) and illustrate the countergradient heat flux problem in the upper regions of the CBL. Positive R values yield distinctly "unmixed" profiles, as occurs with humidity at times (Fig. 6.8).

As mentioned earlier, an alternative to K or higher-order local closure involves a non-local closure approach. This can readily reproduce the results of bottom-up/top-down diffusion theory and allow heat fluxes, for example, in conditions of apparent zero-gradient or countergradient flow (Fiedler, 1984).

6.1.7 Decay of the CBL over land

Decay of the CBL sets in when turbulence aloft cannot be maintained against viscous dissipation. Over land, under clear skies, this normally happens in late afternoon and towards sunset, when the surface buoyancy flux decreases rapidly towards zero (and may change sign). Under these conditions, the main source of TKE is removed, with consequent decay in TKE and other turbulent properties in the deep, near-adiabatic remnant of the daytime boundary layer. At the same time, the CBL becomes markedly non-stationary, in contrast with the quasi-stationary behaviour throughout most of the afternoon. After sunset, turbulence in the upper part of the old daytime boundary layer continues to decay, whilst at low levels both a surface inversion and a shallow, nocturnal boundary layer develop with the new surface conditions.

Studies and observations of the CBL decay are very few, and most of our understanding has emerged recently from large-eddy simulation studies (e.g. Nieuwstadt and Brost, 1986). With H_{v0} changing instantaneously to zero with a fully-developed CBL, decreases in both TKE and temperature variance appear to scale (to a first approximation) with the dimensionless time tw_*/h. Temperature variance decays rapidly, though with TKE there is a time $t \sim h/w_*$ over which it remains nearly constant. From then on, the w variance decreases as t^{-2} and the horizontal velocity variances as t^{-1}. The heat flux appears to decay from the surface upwards, with a tendency for the negative entrainment flux to propagate downwards from the inversion. The decay period (about one hour for the TKE to decrease by an order of magnitude) represents the transition between the CBL and the stable, shallower, nocturnal boundary layer.

6.2 The stable (nocturnal) boundary layer

6.2.1 Introduction

The study of the stable nocturnal boundary layer (NBL) over land, from an observational standpoint, is inherently more difficult than that of the CBL. The reason is that buoyancy forces act to suppress turbulence, so that the boundary layer is much shallower, and turbulence levels lower. In addition, wave motions can exist simultaneously with turbulence and so complicate NBL structure and the interpretation of data, and radiative effects (longwave) are relatively much more important. In particular, the latter play an important role in the development of the surface inversion, which can exist in windless conditions, and thus in the absence of any significant turbulence. Thus, it is necessary to distinguish between the NBL, defined in terms of the depth of turbulence, and the surface inversion, defined usually in terms of the temperature profile characteristics. Figure 6.11 gives some sample mean θ profiles from two field experiments, with the top of the two regions indicated. In general, the surface inversion is deeper than the NBL, as determined from both model simulations and observations. The NBL can generally be defined as the shallow, turbulent layer above which the mean shear stress and heat flux are negligibly small. This also implies that the layer is one in which the gradient Richardson number is sub-critical.

The NBL, of depth h, is likely to be well defined under clear skies and in moderate to high wind speeds over horizontally homogeneous terrain. On other occasions, particularly at low wind speeds, the NBL may be shallow and intermittent, with waves and intermittent patchy turbulence aloft. The structure of the NBL is actually quite sensitive to sloping terrain, and is closely associated with katabatic, or drainage, flows. In addition, development of the NBL and surface inversion often occurs simultaneously with the formation of the nocturnal jet.

In the present section, we describe the characteristics of the NBL, surface

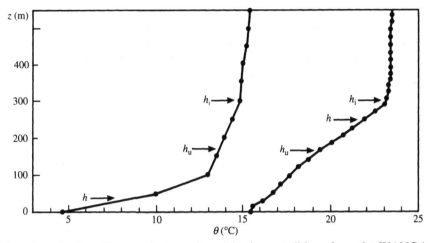

Fig. 6.11 Sample θ profiles under clear sky, night-time conditions from the WANGARA (left) and VOVES (right) experiments. Arrows indicate the respective heights of the surface inversion (h_i), low-level wind maximum (h_u) and NBL (h). After André and Mahrt (1982), *Journal of Atmospheric Sciences*, American Meteorological Society.

inversion and nocturnal jet, and discuss appropriate scaling laws for the NBL. Because of the time-evolving nature of the low-level flow at night, the mid-latitude NBL may never reach steady state (i.e. a quasi-stationary condition as occurs with the CBL in the afternoon). It is possible that the steady state may occur over the Antarctic continent during the long winter "night-time".

6.2.2 Stationary NBL relations

We consider the simplified thermodynamic and momentum equations (Eqs. 2.46, 2.52, 2.53) with the radiative flux divergence assumed zero and fluxes constant in time (so defining a *stationary* NBL). In order to derive solutions for a wide range of NBL characteristics, an analytical approach is adopted (see Nieuwstadt, 1985) that uses a number of simplifying assumptions and a crucial closure statement. This provides a very useful framework against which to compare the behaviour of the real-world, evolving NBL. In a stationary NBL of depth h_e, the cooling rate $\partial\theta_v/\partial t$ can be constant, as must then be the vertical gradient of θ_v. The equations are then written in terms of the vertical gradients, so that

$$\partial^2\overline{w'\theta_v'}/\partial z^2 = 0 \tag{6.46}$$

$$-\mathrm{i}\rho f \partial\mathbf{V}/\partial z + \partial^2\boldsymbol{\tau}/\partial z^2 = 0. \tag{6.47}$$

Here, we revert to complex vector notation, with $\mathbf{V} = u + \mathrm{i}v$ and $\boldsymbol{\tau} = \tau_x + \mathrm{i}\tau_y$, and u, v in surface-layer coordinates. The atmosphere is assumed barotropic with the stress parallel to the shear. Solutions that require suitable closure hypotheses are sought for the fluxes and velocity in particular. The basis of the closure is the assumption that turbulence obeys local scaling (Nieuwstadt, 1984), as discussed in Section 3.5.3. A consequence of local scaling is that dimensionless combinations such as Rf and Ri (Eqs. 2.81, 2.83) approach constant values as height increases. This seems to accord with observations (see Figs. 3.24 and 6.12) and large-eddy numerical simulations, at least well away from the surface, but cannot be valid within the surface layer because of the nature of the profiles. In addition, there is evidence (from model results in particular) that the layer-averaged flux Richardson number, $\overline{Rf}$, is approximately 0.2 in stationary conditions, implying the existence of a critical value (Nieuwstadt and Tennekes, 1981). This parameter, the ratio of the vertically averaged buoyancy destruction of TKE and vertically averaged shear production of TKE, is defined formally as

$$\overline{Rf} = -N_B/P_S \qquad (6.48)$$

where

$$N_B = (g/\theta_v)\int_0^h \overline{w'\theta_v'}\,\mathrm{d}z \qquad (6.49)$$

$$P_S = \rho^{-1}\int_0^h \boldsymbol{\tau}\cdot\partial\mathbf{V}/\partial z\,\mathrm{d}z. \qquad (6.50)$$

In contrast, the bulk ABL Richardson number, Rb $(= gh(\theta_{vh} - \theta_0)/\theta_v V_h^2)$, is usually observed to be in the range 0.25–0.5, even in quasi-stationary conditions, and typical of the range in model values for stationary conditions. Such small values of $\overline{Rf}$, for example, mean that only a small fraction of the TKE is destroyed by buoyancy. The remaining part is lost by viscous dissipation, mainly because other terms in the TKE budget are negligible.

Before solving Eqs. 6.46 and 6.47, there are advantages to be gained by an appropriate scaling using the following nondimensional variables: $\sigma = \tau/\rho u_{*0}^2$, $Y = \overline{w'\theta_v'}/(\overline{w'\theta_v'})_0$, $\xi = z/h_e$, $\mathbf{s}_v = (L/u_{*0})\partial\mathbf{V}/\partial z$ and $s_\theta = (L/\theta_{v*0})\partial\theta_v/\partial z$. Two length scales have been introduced, h_e and L (the Obukhov length), so that the velocity and temperature gradients can be scaled by L instead of by h_e. One reason for this is to allow consistency with surface-layer theory. In scaled form, the two above equations become

$$\partial^2 Y/\partial\xi^2 = 0 \qquad (6.51)$$

$$\partial^2\sigma/\partial\xi^2 - i\gamma_c^2\mathbf{s}_v = 0 \qquad (6.52)$$

whilst Ri and Rf can be written

$$s_\theta/|\mathbf{s}_v|^2 = k\,Ri \qquad (6.53)$$

$$Y/(\sigma\cdot\mathbf{s}_v) = k\,Rf, \qquad (6.54)$$

where k is the von Karman constant. With the equations written in the above form, we immediately find that a nondimensional parameter γ_c appears (in Eq. 6.52), defined by

$$h_e = \gamma_c(u_{*0}L/|f|)^{1/2} \qquad (6.55)$$

which is consistent with Eq. 3.73 which describes the influence of buoyancy on the nondimensional boundary-layer depth under idealized conditions, with the stability function G equal to $(u_{*0}/|f|L)^{-1/2}$. This provides a firm dynamical basis for the depth of the stationary NBL. Equation 6.55 was first derived by Zilitinkevich (1972) using scaling considerations; modelling studies suggest that $\gamma_c \approx 0.4$ for horizontal terrain (Garratt, 1982).

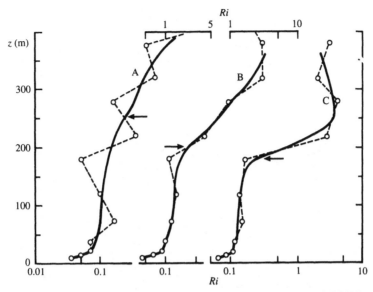

Fig. 6.12 Vertical profiles of gradient Richardson number from the KOORIN experiment at 2200 (A), 0000 (B) and 0300 (C) local time. The horizontal arrow gives the estimated height where $Ri = 0.25$. The pecked curves join the observed values, the continuous curves are drawn by eye to give the smoothed variation. From Garratt (1982): reprinted by permission of Kluwer Academic Publishers.

The set of equations 6.51–6.54, together with the boundary conditions $|\sigma| = 1$, $Y = 1$ for $\xi = 0$ and $|\sigma| = 0$, $Y = 0$ for $\xi = 1$ form a closed set and allow an explicit solution. We consider first the vertical heat flux, and Eq. 6.51; it follows that

$$\overline{w'\theta_v'}/(\overline{w'\theta_v'})_0 = 1 - \xi. \tag{6.56}$$

Such a linear form is supported by observations shown in Fig. 6.13 for the Cabaow site in Holland. The linear form seen in this set of data suggests that radiative effects are small, though closer to the surface this is unlikely to be the case. For the turbulent momentum flux, Eq. 6.56 combined with Eqs. 6.52 and 6.54 gives an equation for σ, the dimensionless stress, i.e.

$$\sigma^* \cdot \partial^2\sigma/\partial\xi^2 - i(\gamma_c{}^2/k\,Rf)(1 - \xi) = 0 \tag{6.57}$$

where σ^* denotes the complex conjugate. This has a solution, *for Rf constant*,

$$\sigma = (1 - \xi)^{(3+i\sqrt{3})/2} \tag{6.58}$$

only if $\gamma_c{}^2/(k\,Rf) = \sqrt{3}$. Thus, if we set $k = 0.4$ and $Rf = 0.2$, the parameter $\gamma_c \approx 0.37$, which is close to observed and modelled values. That is, the scaled height of the NBL is closely related to the stress distribution and the value of the critical flux Richardson number. From Eq. 6.58 the vertical distribution of stress magnitude is given by

$$|\tau|/\rho u_{*0}{}^2 = (1 - \xi)^{3/2} \tag{6.59}$$

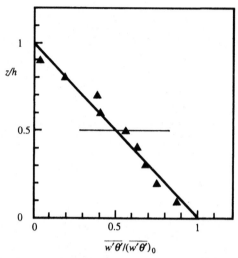

Fig. 6.13 The normalized heat flux as a function of normalized height, showing the Cabaow observations and the predictions of Eq. 6.56, where it is assumed that $\overline{w'\theta'} \approx \overline{w'\theta_v'}$. From Nieuwstadt (1985), by permission of the Oxford University Press.

and is compared with Cabaow data in Fig. 6.14. In this case, the flux distribution does not seem to be sensitive to the unsteadiness of the wind field.

Consider now the velocity $\mathbf{V}(\xi)$ and temperature $\theta(\xi)$ profiles. First, we can derive expressions for s_v and s_θ by substituting Eqs. 6.56 and 6.58 into Eqs. 6.53 and 6.54 to give

$$s_v = (k\,Rf)^{-1}(1 - \xi)^{(-1+i\sqrt{3})/2} \tag{6.60}$$

$$s_\theta = [Ri/(k\,Rf^2)](1 - \xi)^{-1}. \tag{6.61}$$

From the definition of s_v, integration of Eq. 6.60 gives the velocity profile

$$(\mathbf{V}_g - \mathbf{V})/u_{*0} = (h_e/L)(k\,Rf)^{-1}\exp(-i\pi/3)(1 - \xi)^{(1+i\sqrt{3})/2} \tag{6.62}$$

where $\mathbf{V}_g = u_g + iv_g$ (i.e. the value of $\mathbf{V}$ at $\xi = 1$). Equation 6.62 must also satisfy $\mathbf{V} = 0$ at $\xi = 0$, giving

$$|\mathbf{V}_g|/u_{*0} = (h_e/L)(k\,Rf)^{-1} \tag{6.63}$$

$$\alpha_g = \pi/3 \tag{6.64}$$

where α_g is the angle between the surface and geostrophic winds. Note that Eq. 6.63 is comparable to a resistance law (cf. the relevant equations and discussion in Section 3.4), with the geostrophic drag coefficient $\to 0$ as $h_e/L \to \infty$. Equation 6.63 has the following remarkable feature: using the definition of L and 6.55, together with the fact that $\gamma_c^2/(k\,Rf) = 3^{1/2}$, reduces to

$$(g/\theta_v)(\overline{w'\theta_v'})_0 = -|\mathbf{V}_g|^2\,Rf\,|f|/3^{1/2}. \tag{6.65}$$

This shows that the buoyancy flux has an upper limit that is independent of u_{*0} and h_e, and depends only on the external factors $|\mathbf{V}_g|$ and $|f|$. With $|\mathbf{V}_g| = 10\,\mathrm{m\,s^{-1}}$, $|f| = 10^{-4}\,\mathrm{s^{-1}}$ and $Rf = 0.2$, we get a realistic value of the heat

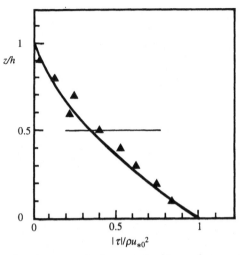

Fig. 6.14 The normalized stress magnitude as a function of normalized height, showing the Cabaow observations and the predictions of Eq. 6.59. From Nieuwstadt (1985), by permission of the Oxford University Press.

flux of around $-40\,\mathrm{W\,m^{-2}}$. The prediction of a specific value of the surface heat flux is somewhat surprising, but is the result of the strong stability limit $h_e/L \to \infty$ (see Derbyshire, 1990) and the assumption of local scaling.

In addition, for $\xi \to 0$ in the surface layer, Eqs. 6.62–6.64 give

$$u/u_{*0} = (z/L)(k\,Rf)^{-1} \tag{6.66}$$

$$v/u_{*0} = 0, \tag{6.67}$$

i.e. a linear profile consistent with Monin–Obukhov scaling, However, the log-linear form described by Eq. 3.37 is not obtained because of the unrealistic assumption of constant Rf near the surface. Substituting Eq. 6.63 into Eq. 6.62 gives

$$(\mathbf{V}_g - \mathbf{V})/|\mathbf{V}_g| = \exp\left(-\mathrm{i}\pi/3\right)(1 - \xi)^{(1+\mathrm{i}\sqrt{3})/2} \tag{6.68}$$

where scaling by $|\mathbf{V}_g|$ has removed any h_e/L dependence. This dependence can be included by replacing $|\mathbf{V}_g|$ with u_{*0} using Eq. 6.63, to give consistency with the predictions discussed in Section 3.4. Plots of the wind speed and direction predicted by Eq. 6.68 are shown in Fig. 6.15, and observations made under quasi-steady, barotropic conditions are also given, for comparison. Good agreement is found in respect of (i) the linear nature of the wind profile through a large part of the NBL and (ii) restriction of the turning of the wind to the upper part of the NBL. However, observations show much less turning than the 60 degrees predicted by the theory.

Integration of Eq. 6.61, *with Ri a constant*, gives a temperature profile

$$(\theta_v - \theta_{v0})/\theta_{v*0} = -(h_e/L)[Ri/(k\,Rf^2)]\ln(1 - \xi), \tag{6.69}$$

which is also shown in Fig. 6.15 together with observed values. In the

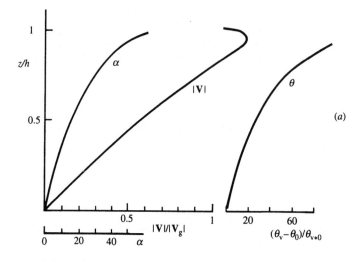

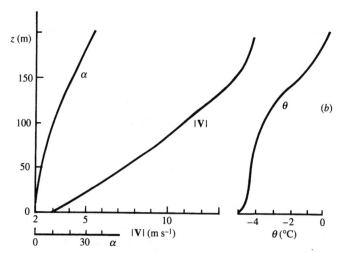

Fig. 6.15 (*a*) Predicted values of cross-isobar flow and normalized wind speed (Eq. 6.68) and of normalized temperature difference (Eq. 6.69) as functions of normalized height. (*b*) Observations from Cabaow of cross-isobar flow angle, wind speed and temperature as functions of height in the NBL. From Nieuwstadt (1985), by permission of the Oxford University Press.

atmosphere, the temperature profile is affected by radiative effects, so that the present theory can only be expected to perform well where turbulent cooling dominates over radiative cooling, i.e. under moderate- to high-wind conditions. At small heights, Eq. 6.69 suggests a linear profile and therefore is inconsistent with the log-linear form. At greater heights, the curvature is positive and consistent with the absence of radiative effects (see Section 6.2.4). One other point about Eq. 6.69: its form gives a singularity at the top of the NBL, which is an undesirable feature of the simple theory. This seems to be related to the fact that constant values of Ri and Rf, which are consistent with a local balance of TKE production and dissipation, cannot apply near $z = h_e$. In addition, the theory requires a constant cooling, and so cannot satisfy a constant temperature at $z = h_e$ (which must occur, in the absence of turbulence and radiative effects).

The assumptions made in the above theory are somewhat limiting in a number of respects (for example, the theory is valid only in the large stability limit $h_e/L \to \infty$), yet its predictions for the stationary NBL act as an important link with our understanding of the real-world NBL. In general, the profile forms are not valid (i) in the surface layer, (ii) under conditions where radiative effects dominate, (iii) in the near-neutral limit (the assumption of constant Ri and Rf is consistent with large h_e/L only) or (iv) under unsteady conditions.

The remainder of the section is concerned with the NBL and surface-based inversion where the restrictions of stationarity are relaxed.

6.2.3 Growth and the NBL height

The NBL is rarely in a steady state, but evolves slowly in time. The evolution of the depth can be described by a linear relaxation equation whose solution is forced towards an equilibrium value given by Eq. 6.55. This rate equation is of the form (Nieuwstadt and Tennekes, 1981)

$$\partial h/\partial t = (h_e - h)/T \qquad (6.70)$$

where T is a time scale to be defined shortly. A number of relations for $\partial h/\partial t$ have been deduced over the years (e.g. see Zeman, 1979), and Eq. 6.70 is an example of one such relation that is based on TKE considerations for the whole layer. The approach revolves around the use of the layer-averaged flux Richardson number, $\overline{Rf}$, and thus involves the shear production and buoyancy destruction averaged across the NBL. With suitable simplifying assumptions, the time scale in Eq. 6.70 is found to be $T = -\Delta\theta_{vB}(\partial\theta/\partial t)^{-1}$, with $\Delta\theta_{vB} = \theta_{vh} - \theta_0$. Thus, the time scale governing the approach of h towards equilibrium is dependent on the surface cooling rate, which typically decreases with time, so that T increases with time. Early in the night, over land under clear skies, the surface cooling rate (SCR) $\approx -1 \, \text{K} \, \text{hr}^{-1}$ giving a value of T of a few hours. By early morning, before sunrise, when $\Delta\theta_{vB} \approx 5\text{–}10 \, \text{K}$ and the SCR $\approx -0.5 \, \text{K} \, \text{hr}^{-1}$, T may reach 10 hours or more. This implies that h will change only slowly in response to changes in external conditions though, apart from boundary-layer changes within a few hours of sunset, h values can be close to the equilibrium value h_e. Overland, under clear skies and with a moderate geostrophic wind, the NBL depth is typically close to 100 m. Inserting such a value of h, and typical values of T, into Eq. 6.70, shows that $|\partial h/\partial t|$ varies by $1\text{–}10 \, \text{m} \, \text{hr}^{-1}$ during the course of a night.

6.2.4 Influence of infrared radiation

In the case of horizontally homogeneous conditions, the thermodynamic equation (Eq. 2.46) represents the contributions of turbulence and radiation to the total cooling, with the temperature structure $\theta_v(z)$ being very dependent on the presence of radiation. The cooling rate at the surface is dominated by radiative effects, and controlled by the surface energy balance equation. Likewise, the cooling rate at the top of the NBL and above, where turbulence is negligible, is mainly determined by radiative cooling.

In modelling studies, calculations of R_N can be made using formulations discussed in Section 5.2. Results of NBL simulations where radiative cooling is included (e.g. Garratt and Brost, 1981) give an indication of the importance of the radiative term, in the presence of strong turbulent activity. They show that the cooling at h (predominantly due to radiative effects) can be a significant fraction of the SCR, and that the layer-averaged turbulent and radiative cooling are both significant fractions of the total cooling. Of course, in calm conditions, turbulence is non-existent and any cooling is then the result of radiative effects alone (in non-homogeneous conditions, advection and subsidence may also be significant).

Studies (e.g. Garratt and Brost, 1981) suggest that the NBL has a three-layer structure so far as cooling is concerned:

(i)   the first, or bottom, layer coincides with the surface layer ($z < 0.1h$) and is dominated by radiation;

(ii) throughout most of the NBL ($0.1h < z < 0.8h$), turbulent cooling dominates and $\theta_v(z)$ is nearly linear;

(iii) the uppermost part of the NBL ($0.8h < z < h$) is dominated by radiative cooling.

A tendency towards constancy of $\partial\theta_v/\partial t$ with height is a natural requirement of the NBL's evolution towards a steady state. With no radiation present (Eq. 2.46), this implies a linear $\overline{w'\theta_v'}$ profile. In the presence of radiation, when longwave flux divergence cannot be ignored, constancy of total cooling with height, particularly near the surface where radiative cooling decreases rapidly upwards, may actually require turbulent warming near the surface. The result is a low-level maximum in $\overline{w'\theta_v'}$, as revealed in both model simulations and observations (Fig. 6.16).

6.2.5 Profile curvature

When conditions are quasi-steady, the curvature of the θ_v profile across the NBL and throughout the surface inversion (see next subsection) depends mostly on the relative contributions of radiative and turbulent cooling. This can be seen as follows. With strong winds and intense turbulent mixing, profiles tend to be well mixed (with θ_v constant with height through the mid-sections of the NBL) and generally concave to the x-axis when plotted in the traditional way. With weak winds, radiative cooling tends to be greatest at the surface, decreasing rapidly with height, and so yielding a θ_v profile convex to the x-axis. Simple empirical expressions for $\theta_v(z)$ are useful for practical applications and can be written as

$$(\theta_v - \theta_0)/\Delta\theta_{vB} = (z/h)^n \qquad (6.71)$$

or alternatively as

$$(\theta_v - \theta_{vh})/\Delta\theta_{vB} = -(1 - z/h)^m. \qquad (6.72)$$

A convenient way of quantifying the temperature profile structure is to introduce a scaled profile curvature parameter defined as (André and Mahrt, 1982)

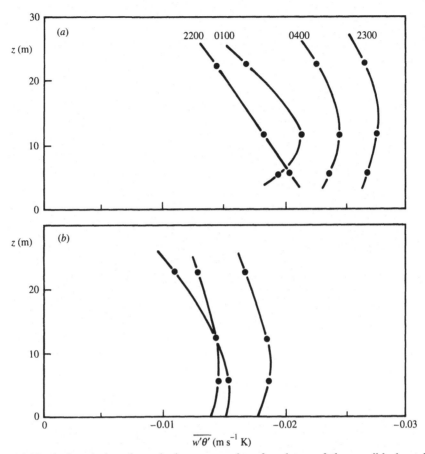

Fig. 6.16 Vertical variation through the nocturnal surface layer of the sensible heat flux: (a) from the Kansas observations (Izumi, 1971) with local time indicated; (b) from model simulations of the NBL, at 3, 6 and 12 hours after sunset. After Garratt and Brost (1981), *Journal of Atmospheric Sciences*, American Meteorological Society.

$$\delta = 1 - 2(\theta_{v,h/2} - \theta_0)/\Delta\theta_{vB} \qquad (6.73)$$

whence for $n = 0.5$, 1 and 2, δ has values of -0.41, 0 and 0.5 respectively. Figure 6.17 shows two sets of observations for which the relative magnitudes of radiative and turbulent cooling are quite different. For the Wangara data, the NBL was extremely shallow (h here is equated with the surface inversion depth) and the curvature is significantly negative (radiation dominant). For the second case, that of a stable internal boundary layer over the sea (see Section 6.4), the curvature is positive and turbulent cooling dominates (the tendency here is for the development of a well-mixed layer beneath a sharp elevated inversion).

6.2.6 Growth of the surface inversion

The surface inversion over land represents the time-integrated cooling due to turbulent, radiative and other effects, i.e. it reflects the cooling history from near sunset to any time during the night, as determined by integrating Eq. 2.46

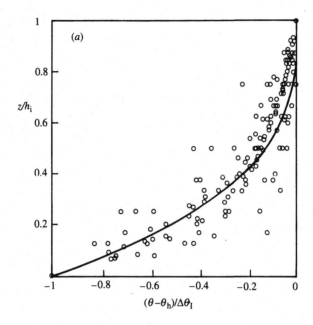

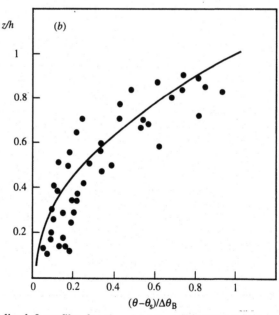

Fig. 6.17 (*a*) Normalized θ profiles for the night-time WANGARA data, where h_i is the surface inversion height. The solid curve represents Eq. 6.72 with $m = 3$. After Yamada (1979), *Journal of Applied Meteorology*, American Meteorological Society. (*b*) Normalized θ profiles for a stably stratified internal boundary layer (of depth h) over the sea; the sea-surface temperature is denoted by θ_s. The solid curve represents Eq. 6.71 with $n = 2$. From Garratt and Ryan (1989): reprinted by permission of Kluwer Academic Publishers.

with respect to time. Since such an inversion can develop in the absence of turbulence, its depth is generally found to be greater, under most conditions, than the NBL depth. The inversion depth is somewhat arbitrary, but can be taken as the depth h_i at which $\partial \theta_v / \partial z$ decreases sharply towards zero.

A rate equation for the depth h_i can be readily derived for the horizontally homogeneous case, by intergrating Eq. 2.46 across the layer and assuming that the heat flux at the inversion top is zero (Yamada, 1979). This gives, with the aid of the Leibnitz rule,

$$\partial/\partial t\left[\int_0^{h_i} \theta_v \, dz\right] - \theta_v(h_i)\partial h_i/\partial t = (\overline{w'\theta_v'})_0 + \int_0^{h_i} (\partial \theta_v/\partial t)_r \, dz \qquad (6.74)$$

where the second term on the right-hand side represents the radiative cooling. Substituting Eq. 6.71 into the left-hand side of Eq. 6.74 and assuming that the radiative cooling decreases linearly with height (from the SCR at the surface to zero at h_i) yields the following result:

$$\Delta\theta_{vI}\partial h_i/\partial t = - a(\overline{w'\theta_v'})_0 - bh_i\partial\theta_{v0}/\partial t. \qquad (6.75)$$

Here, $a = (n + 1)/n$ and $b = (1 - n)/2n$. The right-hand side gives the contributions of turbulence (through the surface heat flux) and radiation (through the SCR) to the inversion evolution. When turbulence dominates, we take $n \approx 2$, and

$$\partial h_i/\partial t \approx - 1.5(\overline{w'\theta_v'})_0/\Delta\theta_{vI} \qquad (6.76a)$$

≈ 15–$20 \, \mathrm{m \, hr^{-1}}$. In contrast, when turbulence is weak, $n \approx 1/2$, and

$$\partial h_i/\partial t \approx - (h_i/2\Delta\theta_{vI})\partial\theta_{v0}/\partial t \qquad (6.76b)$$

$\approx 15 \, \mathrm{m \, hr^{-1}}$, showing that both turbulent and radiative processes have comparable effects on the growth of the surface inversion.

At the end of a 12-hour night under clear skies, the surface inversion may be several hundreds of metres deep with $\Delta\theta_{vI}$ as large as 10–15 K. Its breakdown occurs after sunrise due to surface heating, as described in Section 6.1. Within the surface inversion, there exists a shallower turbulent layer (NBL), and probably also internal gravity waves in the upper part of the inversion. With moderate to strong geostrophic winds a pronounced wind maximum near the top of the NBL, and usually well within the surface inversion, may exist; this is often referred to as the *nocturnal jet*.

6.2.7 The nocturnal jet

A low-level wind maximum, called the nocturnal jet if it is sufficiently pronounced, generally forms at night-time overland under clear sky conditions. The winds above the surface layer may be significantly supergeostrophic, and persist for much of the night. The wind maximum may be more or less intense depending upon a range of factors, including baroclinity, sloping terrain, radiative cooling in the air, surface cooling rates, conditions at sunset, frictional decoupling of the air aloft from the surface, and time changes in the geostrophic wind. Its importance lies in the potentially large wind shears that may be generated by its formation and the potential for significant horizontal advection (and hence horizontal transport) at several hundred metres above the surface at night. Figure 6.18 presents observations, from several sites, that reveal evidence of the formation of a low-level wind maximum during the night. There have been numerous studies of this phenomenon, involving theory, numerical simulation and observational analyses.

An appropriate dynamical explanation and description of the jet, and of known or observed wind variations in the lower atmosphere throughout the diurnal cycle, involves the theory of damped and undamped inertial oscillations. First, consider that just before sunset over horizontal terrain the winds in the CBL above the surface layer will be given approximately by Eqs. 6.4 and 6.5, with $\partial/\partial t = (u'w')_h = (v'w')_h = 0$. At this time, radiative flux divergence effects become increasingly important, leading to temperature gradients aloft increasing with time. In regions of small vertical wind shear (above the surface layer), Ri will increase until $Ri \gg Ri_c$ leading to significant turbulent stress decay. If the stress terms in Eqs. 6.4 and 6.5 become zero, an undamped inertial oscillation will result (this is likely at heights of several hundred metres), otherwise the wind undergoes a damped oscillation. The above is basically the classical treatment of Blackadar (1957) in which the ageostrophic wind, in balance with the stress field in late afternoon, goes through an undamped oscillation when released of all frictional constraint after sunset.

From our previous discussions, we can anticipate damped oscillatory behaviour of the wind within the shallow NBL (by definition, the stress divergence will be non-zero there), and possible undamped behaviour above the NBL near the top of the surface inversion, and above. Consider Eqs. 2.52 and 2.53, and denote the ageostrophic wind vector as $\mathbf{V}_{ag} = \mathbf{V} - \mathbf{V}_g$. With stress divergence set to zero, the momentum equation in vector form becomes

$$\partial \mathbf{V}_{ag}/\partial t = -if\mathbf{V}_{ag} \tag{6.77}$$

with the assumption that the geostrophic wind vector is constant in time. The solution of Eq. 6.77 is

$$\mathbf{V}_{ag} = \mathbf{V}_{ag}^i \exp(-ift) \tag{6.78}$$

where $\mathbf{V}_{ag}^i$ is the ageostrophic wind at $t = 0$. This solution exhibits the following interesting features.

(i) It is governed by the inertial period $2\pi/|f|$, and hence depends on the latitude.

(ii) The amount by which the wind goes supergeostrophic is partly determined by the magnitude of the length of night relative to the inertial period, and partly by the initial conditions.

(iii) $\mathbf{V}_{ag}^i$ is affected by the roughness of the surface; the greater z_0, the greater is the ageostrophic wind magnitude (refer to Eqs. 3.75, 3.76).

Figure 6.19 illustrates the solution given by Eq. 6.78 for low- and high-roughness surfaces in the southern hemisphere. For a latitude of 40 degrees, the maximum wind occurs for $\pi/2 < |f|t < \pi$, i.e. $t \approx 6$ hours. For a latitude of 15 degrees, the maximum is found at $t \approx 18$ hours. In this case, with $t = 0$ corresponding with sunset, the maximum cannot be achieved since night-time lasts for only 12 hours or so.

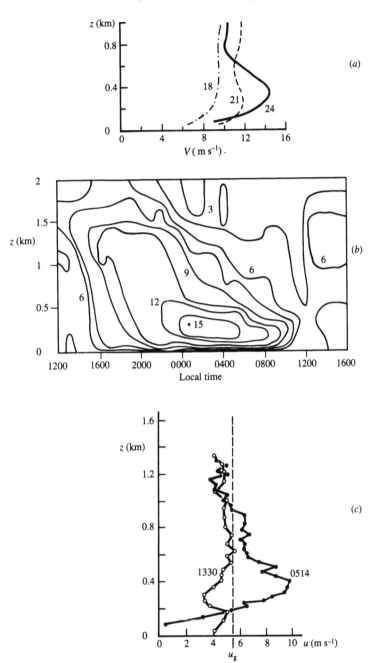

Fig. 6.18 Observations illustrating the formation of the nocturnal jet. (*a*) Wind-speed profiles on day 13 of WANGARA, local times indicated. (*b*) Height–time cross-section of wind speed (in m s^{-1}) on days 13/14 at WANGARA. Isopleths of wind speed are drawn at 1.5 m s^{-1} intervals. (*c*) Profiles of the *u*-component of the wind velocity, with the *x*-axis along the geostrophic wind direction, for mid-afternoon (1330 UT, 6 August, 1974) and early morning (0514, 7 August, 1974) near Ascot, England. After Thorpe and Guymer (1977), *Quarterly Journal of the Royal Meteorological Society*.

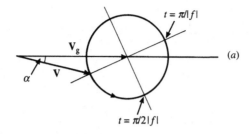

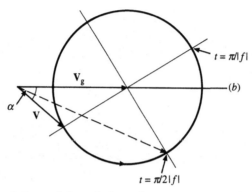

Fig. 6.19 Illustrated solutions of the unbalanced momentum equation (Eq. 6.77) for (*a*) a low-roughness surface and (*b*) a high-roughness surface; undamped inertial oscillations are shown for the southern hemisphere in the form of anticlockwise rotation of the wind vector (**V**) about the geostrophic wind vector (**V**$_g$).

The variation of **V**$_{ag}$ with height depends not only on the stress field, but also crucially on geostrophic wind shear. For example, the jet phenomenon represented by Eq. 6.78, applied over a range of heights, is enhanced when $|\mathbf{V}_g|$ decreases with height, and is diminished when $|\mathbf{V}_g|$ increases with height (Zeman, 1979). In addition, temporal variations of the large-scale geostrophic wind (i.e. the pressure field) are of potential importance. Thus, the jet is strengthened if **V**$_g$ increases in magnitude during the night, or if it rotates so as to increase the radius of the inertial circle shown in Fig. 6.19 (clockwise and anticlockwise in the northern and southern hemispheres respectively).

To expand on the above discussion, let us use a simple two-layer model of the lower atmosphere (Thorpe and Guymer, 1977). This consists of a well-mixed single slab during daytime, and a two-layer system at night, with the upper (deeper) layer becoming decoupled from the lower (representing the NBL); see Fig. 6.20. With coordinate axes set so that $\mathbf{V}_g = (u_g, 0)$, the wind field **V** for each layer can be written as

$$\partial \mathbf{V}/\partial t = -\,\mathrm{i}f(\mathbf{V} - u_g) - \boldsymbol{\tau}_0/\rho h \qquad (6.79)$$

noting that $\boldsymbol{\tau} = 0$ in the night-time upper layer. During the daytime, h is the CBL depth, and at night it is the NBL depth. To solve Eq. 6.79, the stress must

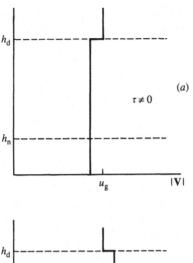

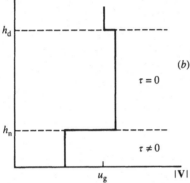

Fig. 6.20 Simple two-layer model of the ABL at night and during the day; h_d is the daytime ABL depth, h_n the night-time depth. During the day (a) the ABL is well mixed with a subgeostrophic uniform wind and non-zero stress; at night (b), the ABL is characterized by light winds and non-zero stress, whilst in the layer above the stress is zero and the winds supergeostrophic. Based on Thorpe and Guymer (1977).

be parameterized and suitable boundary conditions used. If a drag coefficient is used, the equation must be solved numerically. Alternatively, an analytical solution that is useful for illustrative purposes results if we set $\tau_0 = \rho k_s \mathbf{V}$. Denoting day and night velocities as $\mathbf{V}_d$ and $\mathbf{V}_n$, and upper and lower slab velocities by $\mathbf{V}^U$ and $\mathbf{V}^L$, suitable boundary conditions are $\mathbf{V}_n^U = \mathbf{V}_d$, and $\mathbf{V}_n^L = \mathbf{V}_d$ at sunset, and, $h_d \mathbf{V}_d = h_n \mathbf{V}_n^L + (h_d - h_n)\mathbf{V}_n^U$ at sunrise for momentum conservation (these determine the constants A_d, A_n^L and A_n^U below). If we assume k_s is constant in time, the solution for the daytime is

$$\mathbf{V}_d = if(k_s/h_d + if)^{-1}u_g + A_d \exp\left[-(if + k_s/h_d)t\right]. \tag{6.80}$$

For the night-time lower layer (NBL),

$$\mathbf{V}_n^L = if(k_s/h_n + if)^{-1}u_g + A_n^L \exp\left[-(if + k_s/h_n)t\right]; \tag{6.81}$$

and for the night-time upper layer (no turbulence),

$$\mathbf{V}_n^U = u_g + A_n^U \exp(-ift). \tag{6.82}$$

Equation 6.82 is identical to Eq. 6.78 for the undamped oscillatory solution; Eqs. 6.80 and 6.81 are the damped solutions. They are shown in Fig. 6.21(a) in hodograph form, for $k_s = 0.04\,\mathrm{m\,s^{-1}}$, $h_n = 200\,\mathrm{m}$, $h_d = 1000\,\mathrm{m}$ and for mid-latitudes in the southern hemisphere. After sunset, the wind in the lower layer is rapidly modified by friction and decreases significantly in magnitude, together with rotation towards a cross-isobar flow of about 64 degrees. In the upper layer, supergeostrophic flow results after a few hours with rotation along an inertial circle analogous to that shown in Fig. 6.19. A numerical solution, using Eq. 3.42 with $C_{DN} = 0.008$, gives similar results, as seen in Fig. 6.21(b).

6.2.8 Other influences

The simple evolution of the surface inversion, NBL and nocturnal jet implied in the idealized treatments described previously is probably rare in the real world. There, factors such as changes in external conditions, heterogeneous and sloping terrain, advection and subsidence, mesoscale influences and gravity waves combine to produce more complicated behaviour. When winds are light and turbulence weak, the existence of internal waves within the stably stratified environment probably plays an important role in the budget of TKE, in the evolution of the very shallow NBL and in vertical momentum transfer. Nevertheless, observations are available where these influences have been minimized, whilst numerical experiments allow them to be excluded completely.

The presence of sloping terrain can influence the structure of the NBL considerably, and can result in local slope flows (also called katabatic or drainage flows) even in the absence of any large-scale flow (see e.g. Brost and Wyngaard, 1978). The slope surface and air are cooled by longwave radiation into space, and there is likely to be a turbulent heat flux from the air into the ground once flow is established. This flux, and the radiative cooling in the air immediately above the surface, increase the density of the air relative to the environment at the same level. The height to which the denser air extends is no more than a few hundred metres, over slopes of less than 10 degrees. Referring to Fig. 6.22, let the terrain slope be α, the potential temperature deficit in θ for the katabatic flow, relative to the ambient field, be d' (thus $d' = \theta_v - \theta_v^{amb}$), and take the x-axis downslope and the z-axis normal to the slope. The simplified u-momentum equation (Eq. 2.44) becomes

$$\partial u/\partial t = f(v - v_g) - (g/\theta_v)d' \sin\alpha - \partial(\overline{u'w'})/\partial z, \tag{6.83}$$

with the v equation (Eq. 2.45) remaining unaltered. For a stable atmosphere, $d' < 0$ so that the buoyancy force on the right-hand side of Eq. 6.83 results in acceleration down the slope. This buoyancy, or terrain-slope, term , when added to the pressure-gradient term, gives an effective pressure gradient that depends on z, since d' is a strong function of z. This situation is analogous to that existing in the baroclinic, horizontal case in the presence of a horizontal temperature gradient (thermal wind). The existence of a height-dependent effective pressure gradient in the lower atmosphere over sloping terrain is well known. It is an important component of the explanation of the nocturnal jet over the Great Plains in the USA (e.g. Lettau, 1967), and must be considered in any description of winds over Antarctic and Greenland slopes.

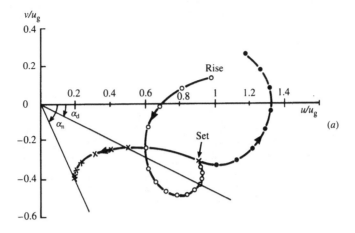

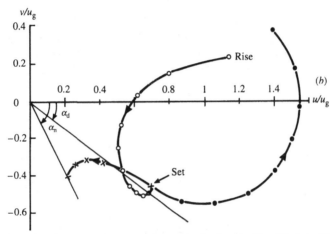

Fig. 6.21 Solutions of the momentum equations for the layers identified in Fig. 6.20: (*a*) analytical solution for a linear stress law (Eqs. 6.80–6.82); (*b*) numerical solution for the more realistic quadratic stress law. The time evolution of the wind components in the single daytime layer between sunrise and sunset, and for the two layers between sunset and sunrise, is illustrated. Typical daytime (α_d) and night-time (α_n) cross-isobar flows are also indicated. Note that at the end of the night, winds in the lower and upper layers are assumed to be instantaneously mixed, with momentum conservation, to give the initial mixed-layer wind at sunrise. After Thorpe and Guymer (1977, *Quarterly Journal of the Royal Meteorological Society*.

Mixing between the ambient and cooled air layers appears to be of great importance to the spatial growth of the katabatic flow, and to the generation of a retarding stress on the flow (e.g. Manins and Sawford, 1979). This retarding stress can be identified with the interfacial stress due to the mixing, and may be dominant when the surface stress is small under the influence of strong stable stratification close to the ground. In the presence of sloping terrain, and hence

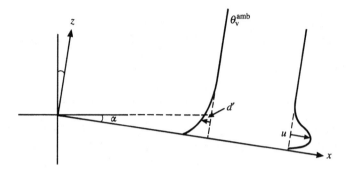

Fig. 6.22 Schematic representation of the downslope flow typical of night-time flow under light wind, clear sky conditions. Here, α is the slope angle and d' is the θ deficit of the flow relative to the ambient field.

of slope flows and interfacial mixing, NBL characteristics are modified compared with flow over a horizontal surface. Both numerical models and observational studies have been used to study the effects of slope. As an example, Table 6.1 summarizes observations and model results of the nondimensional NBL depth for a range of slope angles. Values of the parameter γ_c appearing in Eq. 6.55 are sensitive to both the slope magnitude and the angle between the wind direction and the slope fall vector. The impact of sloping terrain in the Koorin case is clearly evident.

The chapter, thus far, has dealt with characteristics of the continental ABL during the day and night. For completeness, some discussion is appropriate on the marine ABL, and on the transition between the marine and continental ABL at the coastline.

6.3 The marine atmospheric boundary layer

6.3.1 Introduction

So far as theoretical formulations are concerned, those developed in Chapters 2, 3 and 4 apply equally to land or sea surfaces. The spectral and integral properties of turbulence described in Chapter 3, derived mainly from land data, are supported broadly from the marine observations that have been published in the literature.

The major feature that distinguishes the marine ABL from its continental counterpart is its wet, mobile, lower boundary. The presence of water results in a number of consequences.

   (i) Air is usually moister over the sea, compared with the land.
   (ii) In the marine case, large excursions from near-neutral conditions are restricted to relatively small regions of the global oceans. Thus, over much of the ocean, the surface heat flux does not play a large role in determining the boundary-layer structure.

Table 6.1. *Observed values of the Zilitinkevich constant* γ_c *(Eq. 6.55) for four sites, together with numerical values based on one-dimensional NBL simulations, illustrating the effects of slope (magnitude α)*

Model calculations incorporate the observed wind direction relative to the sloping terrain, with appropriate values of latitude (ϕ) and roughness length (z_0)

				γ_c	
Site	ϕ	z_0 (m)	α	Observed	Calculated
Koorin	− 15	0.1	0.002	0.13	0.14
Minnesota	45	0.01	0.001	0.37	0.38
Wangara	− 30	0.001	0	0.39	0.37
Cabauw	45	0.1	0	0.42	0.39

Source: Garratt (1982).

(iii) Diurnal variations in the sea-surface temperature, and hence diurnal forcing of the marine ABL, are small mainly because of the large heat capacity of the oceanic mixed layer, of depth h_w. The ratio of the heat capacities for atmospheric and oceanic mixed layers, $(\rho c_p h)_{air}/(\rho c_p h)_{water} \approx 1/350$, for mixed-layer depths of 100 and 1000 m for water and air respectively.
(iv) Over large areas, the sea surface is relatively uniform.
(v) Dynamic interaction occurs between water waves and surface-layer turbulence.

6.3.2 Observations and the ABL depth

Observational validation of ABL theory, and details of ABL structure, have relied heavily on data obtained over land surfaces, usually in clear to partially cloudy skies, with significant diurnal influences. In spite of the considerable advances in observational capabilities in the last three decades, our knowledge of the structure and kinematics of the ABL rests on a relatively sparse observational database. This is particularly true of the marine ABL, where experimental studies (Table 6.2) have tended to focus on three boundary-layer types:

(i) the tropical and sub-tropical marine ABL, usually associated with cumulus (Cu) and stratocumulus (Sc) clouds;
(ii) the mid-latitude marine ABL (open ocean), associated with Cu and Sc clouds;
(iii) the sub-tropical and mid-latitude marine ABL generated during cold-air outbreaks to the east of the continental land masses, associated with Sc and Cu cloud formation.

It is evident from the above that the marine ABL can often be classed as a cloud-topped boundary layer, and detailed discussion is delayed until Chapter 7. It is apparent both from observations (usually involving aircraft) and modelling studies that, where clouds are present, and particularly fair-weather cumulus,

Table 6.2. *Field experiments related to investigations of the marine ABL*

Name	Location	Reference
AMTEX 1974, 1975	East China Sea	Lenschow and Agee (1976)
ATEX 1973	Tropical Atlantic	Augstein *et al.* (1973)
BOMEX 1969	Caribbean	Kuettner and Holland (1969)
COAST 1983	Dutch coast	Weill *et al.* (1985)
FGGE 1979	Tropical Pacific	Firestone and Albrecht (1986)
GATE 1974	Tropical Atlantic	Keuttner and Parker (1976)
JASIN 1978	North-east Atlantic	Royal Society (1983)
MASEX 1983	Atlantic coast, USA	Atlas *et al.* (1986)

the moisture flux near the cloud base is a significant fraction of the surface evaporation. In contrast, entrainment moisture fluxes near the mixed-layer top in clear skies are usually small. This is evident in Fig. 6.23, which also shows virtual heat fluxes tending to small values at cloud base irrespective of cloudiness. For this particular marine situation, the mixed-layer or ABL top is defined at the level, near cloud base, where θ_v begins to increase significantly with height. We shall see that this may not be the most appropriate definition, irrespective of whether clouds are present or not. The mean heat flux profile for these GATE data, normalized as in Section 6.1, reveals a small but negative entrainment flux at the ABL top, similar to that found for the convective ABL over land. These data, in fact, illustrate the slightly unstable conditions often found over tropical and sub-tropical open ocean areas, under fair-weather conditions, where h/L is typically in the range -10 to -1. The surface sensible heat flux and virtual heat flux are small, but the evaporation rates are relatively large; the radiative cooling near cloud base tends to maintain a sub-cloud layer cooler than the sea (Betts and Ridgway, 1989).

In contrast to the above, the marine ABL during cold-air outbreaks can be highly unstable (e.g. the AMTEX data shown in Fig. 3.11 lie in the range $-200 < h/L < -20$) with latent heat fluxes well in excess of the sensible heat fluxes. Absolute values of the evaporation rates can reach $1000\,\mathrm{W\,m^{-2}}$ in extreme conditions.

The definition of the marine ABL, if it is to be consistent with that used throughout, must relate to the structure of small-scale turbulence, and so the top may not necessarily correspond with either a low-level inversion or cloud top. In the literature, because the marine ABL is often associated with clouds, several definitions of the ABL top can be found. Thus, the marine ABL may be taken to be that part of the lower atmosphere that is directly coupled to the surface by turbulent transfer, including fair-weather cumulus and stratocumulus clouds (Stull, 1988). In this case, $z = h$ corresponds with the cloud top and a pronounced inversion; the boundary layer tends not to be well mixed (as deduced from mean temperature profiles) above cloud base. In contrast, the ABL top may be defined to be at or near cloud base. This usually corresponds both to the top of a relatively well-mixed layer (based on mean θ, θ_v and q profiles) and to the height at which turbulent fluxes of momentum and heat

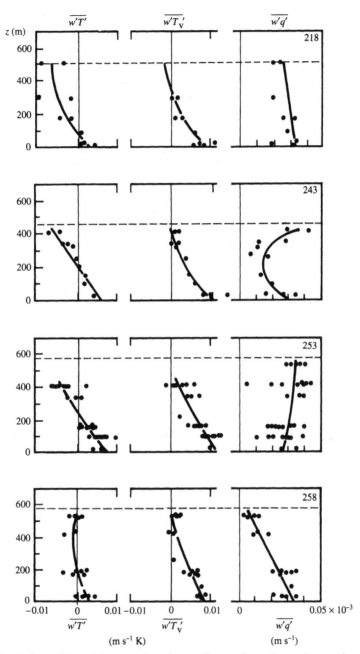

Fig. 6.23 Vertical profiles of heat and moisture fluxes throughout the marine ABL for four GATE days. The pecked straight lines define the cloud base which approximately defines the ABL top. Solid lines through the data are drawn by eye. Only day 258 has negligible cloud cover. From LeMone (1980), by kind permission of the author.

(sensible and virtual) become negligible (Nicholls, 1985). The latter depth scale appears to be that most appropriate to mixed-layer scaling for the clear-sky boundary layers, or the moderately unstable ABL associated with partial cloudiness (over the land or sea). Figure 6.23 shows GATE data for four partially cloudy days where cloud base and ABL top approximately coincide.

For the near-neutral or slightly unstable marine ABL ($h/L >$ about -5) where surface buoyancy is weak, the mixed-layer height h tends to be much less than the inversion level and, in the presence of cumulus convection, corresponds approximately with the cloud base (the local condensation level). In mid-latitudes it seems to obey neutral, barotropic Ekman scaling, with $h \approx 0.2u_{*0}/|f|$ and with the Ekman layer lying beneath both the cumulus and overlying stratocumulus layers (e.g. Nicholls, 1985). This stability regime is typical of open ocean sites. In some respects, however, the height scale h may not correspond with the ABL top defined in terms of turbulent motions associated with the clouds themselves, and the definition of the ABL becomes less exact. Only in moderate to strong instability, when buoyancy is strong (as with cold-air outbreaks when air–sea temperature contrasts are large), do we find that h corresponds to the inversion base (irrespective of cloud base), as is usual during the afternoon over land in moderately unstable conditions.

6.4 Mesoscale flow and IBL growth

We concern ourselves here with the coastal region in which the advection of air across the coastline results in the growth of a thermal internal boundary layer (TIBL) towards an equilibrium ABL some distance from the coastline. Characteristics of this equilibrium, possibly diurnally varying, ABL have been discussed in earlier chapters. The focus of this section is on mesoscale advection, in contrast with the small-scale advection discussed in Chapter 4. Fetches required to achieve growth of the ABL to its equilibrium state may extend to many hundreds of kilometres in the stably stratified case.

Most mesoscale studies focus on the structure and growth of the TIBL, and stem from the perceived relevance of the IBL to diffusion and pollution problems in the coastal region. Although the advection of air across the coastline relates to both surface roughness and temperature changes the primary consideration is often the response and growth of an IBL to a marked step-change in surface temperature. It is recognised that in real-world situations roughness changes also are present. The emphasis generally has been on the growth equation for the IBL depth in both convective and stable conditions, and the factors affecting the depth.

Definition of the thermal IBL is usually in terms of the marked changes in temperature and humidity gradient found near its top, and this seems to be most appropriate for the stable case. Observations of convective TIBL structure (usually for fetches less than about 50 km) during sea-breeze events and for general onshore flow conditions have found that the top defined from the θ profiles can be well below the level of minimum turbulent kinetic energy. Such observations of the mean and turbulence structure reveal the IBL as a fairly well-mixed layer, with large horizontal gradients in θ_v and turbulent kinetic energy for several tens of kilometres inland from the coast. In contrast, turbulence-closure-model studies of the TIBL structure and growth (e.g. Durand *et al.*, 1989) reveal that the TIBL top defined in terms of the θ_v inversion layer tends to coincide with the level of near-zero heat flux and turbulent kinetic energy, and with maximum temperature variance.

6.4.1 The convective thermal IBL

The convective IBL usually forms over land under clear skies during the daytime when there is an onshore flow of air. Simple models of convective IBL growth can be developed from a slab-model approach, based on mixed-layer dynamics. This can be used to derive a set of governing equations from which $h_b(x)$ can be determined, where the fetch is designated as the distance x from the coastline. Equations for mixed-layer wind components and temperature, the continuity equation, and an equation for h_b permit numerical solution, with appropriate boundary conditions. The h_b equation required for closure results from applying the TKE equation at the inversion level and utilizing a suitable entrainment assumption.

For practical application, it is possible to derive a suitable model, based on the steady state, pure advective case, with equations identical to the set 6.8, 6.11 and 6.12 for the growing CBL over a homogeneous surface (Fig. 6.24(*a*)). The transformation $\partial/\partial t = u_m \partial/\partial x$ is made, where u_m is the slab velocity, so that in direct analogy to the simple entrainment case (Eq. 6.18) we have

$$\gamma_\theta h_b u_m \partial h_b/\partial x = (1 + 2\beta)(\overline{w'\theta_v'})_0 \tag{6.84}$$

which can be readily integrated to give $h_b(x)$. If we assume that the lapse rate γ_θ is constant and that the surface heat flux is independent of x, and use the initial condition that $h_b = 0$ at $x = 0$, then

$$h_b^2 = 2(1 + 2\beta)(\overline{w'\theta_v'})_0 x/\gamma_\theta u_m \tag{6.85}$$

so revealing the $h_b \propto x^{1/2}$ behaviour (e.g. Venkatram, 1977). For practical application, the heat flux may be parameterized in terms of a bulk relation using the transfer relations developed in Section 3.4.

Most observational studies have been confined to inland fetches in the range of 5–50 km, since this is the typical distance required to approach the full depth of the ABL. Growth of the TIBL over this fetch range has generally supported the relation $h_b = ax^{1/2}$, with h_b and x both in metres, and with a in the approximate range 2–5. These numerical values can readily be shown to be consistent with that implied in Eq. 6.85. Let us take $\beta = 0.2$, and $\gamma_\theta = 0.01 \text{ K m}^{-1}$ (a rather large value), so that

$$h_b^2 \approx (2.8/0.01)(\overline{w'\theta_v'})_0 x/u_m.$$

Realistic values of $u_m = 5 \text{ m s}^{-1}$ and $H_{v0} = 100 \text{ W m}^{-2}$ give $a \approx 2$, whilst $u_m = 2 \text{ m s}^{-1}$ and $H_{v0} = 400 \text{ W m}^{-2}$ give $a \approx 7$. Decreasing γ_θ to the low value of 0.001 K m^{-1} increases these a values by a factor $10^{1/2}$. Thus, a values lying in the range 2–22 can be expected and are comparable with observed values (Hsu, 1986).

This $x^{1/2}$ behaviour is found because stationarity is assumed as well as

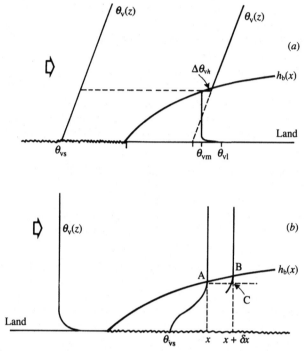

Fig. 6.24 (a) Schematic θ_v profiles over the land and over the sea under conditions of onshore flow and formation of a convective IBL of depth $h_b(x)$. The initial lapse rate is denoted by γ, which is also the lapse rate above the IBL; θ_{vs} and θ_{vl} are the sea and land surface temperatures respectively. (b) As in (a) but for offshore flow, and formation of a stable IBL. The θ_v profiles are shown over the land, and the sea at distances x and $x + \delta x$ offshore. Reference is made to points A, B and C in the text.

horizontal uniformity of the heat flux. The effect of a time-varying heat flux can be readily deduced if the heat flux is assumed to vary sinusoidally through the day (a good approximation). Let $H_{v0} = H_{v0}^{\max} \sin(2\pi t'/T)$, where $T = 24$ hours and t' is the time elapsed since sunrise and also the time at which air crosses the coastline. If t is the travel time from the coast to a point x inland, Eq. 6.18 can be integrated to give the TIBL height at time t:

$$h_b{}^2 = 2\gamma_\theta{}^{-1}(1 + 2\beta)\int_{t'}^{t+t'} (H_{v0}^{\max}/\rho c_p)\sin(2\pi t''/T)\,dt''$$

$$\approx 2\gamma_\theta{}^{-1}(1 + 2\beta)(H_{v0}^{\max}/\rho c_p)(T/2\pi)\{\cos(2\pi t'/T)$$

$$- \cos[2\pi(t + t')/T]\}. \tag{6.86}$$

The depth of the TIBL after a travel time $t(= x/u_m)$ depends on t' since this sets the average level of the heat flux during the passage of the air to a point x inland from the coastline. For $t' = 0$, the air crosses the coast at sunrise when the heat flux is zero over the land surface whilst if $t' = T/4$ the air crosses at midday when the heat flux is a maximum. Growth of the TIBL will be greater near midday than near sunrise. What typical travel times are appropriate? For the TIBL depth to approach the "equilibrium" CBL depth expected "far inland", t of order one hour suffices, noting that for $t' = 0$, the CBL will be shallow (≈ 100 m) and for $t' = T/4$, the CBL will be approaching its maximum daytime depth (≈ 1–2 km). Two solutions of Eq. 6.86 serve to illustrate the problem:

(i) with $t' = 0$ and $t = x/u_m (\ll T)$,

$$h_b{}^2 = 2\gamma_\theta{}^{-1}(1 + 2\beta)(H_{v0}^{max}/\rho c_p u_m)(2\pi x^2/u_m T); \qquad (6.87a)$$

(ii) with $t' = T/4$,

$$h_b{}^2 = 2\gamma_\theta{}^{-1}(1 + 2\beta)(H_{v0}^{max}/\rho c_p u_m)x. \qquad (6.87b)$$

Thus, after one hour's travel time inland from the coast ($x/u_m = 1$ hr) the midday IBL will be greater than the early morning IBL by a factor $(u_m T/2\pi x)^{1/2} \approx 2$. The analysis also shows that the stationarity assumption is likely to be much better around the time of maximum heat flux when, for an assumed sinusoidal variation, the heat flux undergoes its minimum rate of change with time.

At very large fetches, boundary-layer heights must tend towards an equilibrium value. That is, the square-root dependence must ultimately be invalid at large enough x where h_b tends to a constant value. In addition, the lapse rate γ_θ is likely to vary with x, particularly at fetches where the IBL top approaches the level of the subsidence inversion. In the context of the growth equation (6.84), a large increase in γ_θ at the inversion will reduce the rate of growth of the IBL considerably.

6.4.2 The stable thermal IBL

Much of the interest in the stable case is related to offshore flow in the coastal region from warm land to cool sea. Growth of the stable thermal IBL has mostly been studied by appeal to historical data and dimensional analysis (Mulhearn, 1981; Hsu, 1983), by use of numerical and simple physical models (Garratt, 1987), and by analysis of detailed aircraft data (Garratt and Ryan, 1989). Growth rates are found to be small, with fetches of several hundreds of kilometres required to develop an IBL several hundred metres deep. The nature of the θ_v profiles within the IBL over the sea is found to be quite different to that found in the stable boundary layer over land. Over the sea, the θ_v profiles are found to have large positive curvature with vertical gradients increasing with height, interpreted as reflecting the dominance of turbulent cooling within the layer. The behaviour contrasts with the behaviour in the nocturnal boundary layer over the land, where curvature is negative (vertical gradients decreasing with height) when radiative cooling is dominant. The observations shown in Fig. 6.17(*b*) can be represented by (cf. Eq. 6.71)

$$(\theta_v - \theta_s)/\Delta\theta_{vB} = (z/h_b)^2 \qquad (6.88)$$

where θ_s is the sea-surface temperature and $\Delta\theta_{vB}$ is the temperature difference between the IBL top and the sea surface. At and just above h_b, there exists a region of marked negative curvature so that $\partial\theta_v/\partial z$ tends to small values aloft.

Incidentally, these observations reveal no significant dependence upon fetch, suggesting an approximately self-preserving form of the temperature profile over the range of x experienced (x/h_b ranges between 300 and 1800).

Based on the above, a suitable model of IBL growth can be formulated. Its basic structure is shown in Fig. 6.24(b). We assume for simplicity that the advected continental mixed-layer air of virtual potential temperature θ_{vc} remains constant. Above the IBL we take $\gamma_\theta = 0$. The starting point is the two-dimensional steady-state θ_v equation integrated between the surface and h_b, with an assumed linear flux profile and assumed self-preserving forms for the profiles, i.e. $u/U = f_1(z/h_b)$, $(\theta_v - \theta_s)/\Delta\theta_{vB} = f_2(z/h_b)$. Here U is a large-scale, or geostrophic, wind at a small angle to the IBL mean wind direction along which the x-axis is defined. Thus, x is the actual fetch. Note that vertical advection is excluded in this analysis, since its impact is assumed to be small. We have

$$\int_0^{h_b} u\, \partial\theta_v/\partial x\, dz = \overline{(w'\theta_v')}_0 \qquad (6.89)$$

and we note that both $\partial\theta_v/\partial x$ and $\overline{(w'\theta_v')}_0$ will be negative. In order to solve for h_b from Eq. 6.89, we proceed as follows. Referring to Fig. 6.24(b), we consider an infinitesimal change from h_b to $h_b + \delta h$ over a fetch change from x to $x + \delta x$. Then, with θ_v assumed constant at $z = h_b(x)$ and equal to θ_{vc} (at A and B), the temperature $\theta_{vh}(x + \delta x)$ at C can be approximated as $\theta_{vc} - (\partial\theta_v/\partial z)\delta h$. It immediately follows that

$$\partial\theta_v/\partial x = [\theta_{vh}(x + \delta x) - \theta_{vh}(x)]/\delta x \approx -(\partial\theta_v/\partial z)(\delta h_b/\delta x).$$

Substituting into Eq. 6.89, if further follows that

$$\partial h_b/\partial x = -(A_0/U\Delta\theta_{vB})\overline{(w'\theta_v')}_0, \qquad (6.90)$$

where A_0 is a positive profile shape factor depending on f_1 and f_2, with a value O(1). Inspection of the right-hand side of Eq. 6.90 shows that $\partial h_b/\partial x \sim C_{gH}$, the boundary-layer heat transfer coefficient, which has a typical value of about 10^{-3}, and so illustrates the small growth rate of the stable IBL.

For practical application, we need to parameterize the heat flux in Eq. 6.90. One approach is to use a critical layer-averaged flux Richardson number $\overline{(Rf)}$ concept (refer to Eqs. 6.48–6.50) for a quasi-steady stable boundary layer. By assuming self-similar profile forms, $\overline{w'\theta_v'} = \overline{(w'\theta_v')}_0 f_3(z/h_b)$ and $u_*^2 = u_{*0}^2 f_4(z/h_b)$, $\overline{(w'\theta_v')}_0$ can be written as

$$\overline{(w'\theta_v')}_0 = -u_{*0}^2 U \overline{Rf}[(g/\theta_v)h_b f(z/h_b)]^{-1}, \qquad (6.91)$$

where $f(z/h_b)$ is a positive function of f_1, f_3 and f_4. Combining Eqs. 6.90 and 6.91 and integrating gives

$$h_b^2 = \alpha_1 U^2(g\Delta\theta_{vB}/\theta_v)^{-1}x \qquad (6.92)$$

where

$$\alpha_1 = 2A_0 f(z/h_b)\overline{Rf}\, C_g \qquad (6.93)$$

with C_g defined by $u_{*0}^2 = C_g U^2$ (Eq. 3.21). The $x^{1/2}$ prediction in Eq. 6.92, and the dependence on other parameters, is generally confirmed by observations

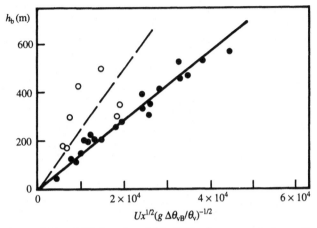

Fig. 6.25 Predictions of Eq. 6.92 for the stable IBL, for numerical results (solid circles) and observations (open circles). The straight lines represent least-squares linear fits to the data. It is probable that the differences between the two sets of results reflect differences in the angle between the geostrophic wind and the coastline normal. From Garratt (1990): reprinted by permission of Kluwer Academic Publishers.

and numerical simulations. Figure 6.25 shows numerical results and observational data that give $\alpha_1^{1/2} \approx 0.014$ and 0.024 respectively.

Notes and bibliography

Section 6.1

Power laws describing mean and turbulent quantities in the lower part of the CBL are extensively reviewed in

Kader, B. A. and A. M. Yaglom (1990), Mean fields and fluctuation moments in unstably stratified turbulent boundary layers, *J. Fluid Mech.* **212**, 637–62.

For an overview on scalar structure in the ABL (including the convective case) with many useful references, the reader should consult

Webb, E. K. (1984), Temperature and humidity structure in the lower atmosphere, in *Geodetic Refraction – Effects of Electromagnetic Wave Propagation Through the Atmosphere*, ed. F. K. Brunner, pp. 85–141. Springer, Berlin.

Two recent reviews on advances in the study of the CBL are

Wyngaard, J. C. (1988), Convective processes in the lower atmosphere, in *Flow and Transport in the Natural Environment: Advances and Applications*, eds W. L. Steffen and O. T. Denmead, pp. 240–269 (includes commentary by E. K. Webb). Springer-Verlag, London, and

Young, G. S. (1988), Convection in the ABL, *Earth-science Reviews* **25**, 179–98.

An advanced analysis on eddy structure in the CBL, with applications to diffusion problems, can be found in

Hunt, J. C. R., Kaimal, J. C. and J. E. Gaynor (1988), Eddy structure in the CBL – New measurements and new concepts, *Quart. J. Roy. Met. Soc.* **114**, 827–58.

The structure of the entrainment process as revealed in LES results is discussed in detail in

Schmidt, H. and U. Schumann (1989), Coherent structure of the convective boundary layer derived from large-eddy simulations, *J. Fluid Mech.* **200**, 511–62.

The following two papers describe a detailed model of dry convection from which the CBL growth equations are derived, and discussed at length:

Betts, A. K. (1973), Non-precipitating cumulus convection and its parameterization, *Quart. J. Roy. Met. Soc.* **99**, 178–96;

Carson, D. J. (1973), The development of a dry inversion-capped convectively unstable boundary layer, *Quart. J. Roy. Met. Soc.* **99**, 450–67.

The following paper reviews the parameterization of turbulent entrainment by the ABL in the context of a one-layer (slab) model:

Tennekes, H. and A. G. M. Driedonks (1981), Basic entrainment equations for the ABL, *Bound. Layer Meteor.* **20**, 515–31.

Two papers that consider mixed-layer growth and parameterization of the TKE equation in some detail are

Driedonks, A. G. M. (1982), Models and observations of the growth of the ABL, *Bound. Layer Meteor.* **23**, 283–306, and

Manins, P. C. (1982), The daytime planetary boundary layer: A new interpretation of Wangara data, *Quart. J. Roy. Met. Soc.* **108**, 689–705.

For further discussion on the range of application of Eq. 6.42, top-down and bottom-up diffusivities (Eqs. 6.43, 6.44) and modifications to Eq. 6.44 to ensure that K_b is well behaved, the reader should consult

Holtslag, A. A. M. and C.-H. Moeng (1991), Eddy diffusivity and countergradient transport in the convective atmospheric boundary layer, *J. Atmos. Sci.* **48**, 1690–8.

Section 6.2

The influence of longwave cooling on the structure of the NBL has been studied in the following two papers:

Garratt, J. R. and R. A. Brost (1981), Radiative cooling effects within and above the nocturnal boundary layer, *J. Atmos. Sci.* **38**, 2730–46;

André, J. C. and L. Mahrt (1982), The nocturnal surface inversion and influence of clear-air radiative cooling, *J. Atmos. Sci.* **39**, 864–78.

Section 6.3

There are many specific papers in the literature dealing with aspects of the marine ABL; the following paper gives an overview of the marine boundary layer, with emphasis on the tropics:

Lemone, M. A. (1980), The marine boundary layer, in *Workshop on the PBL*, ed. J. C. Wyngaard, pp. 182–231. American Meteorological Society, Boston.

Some relevant work related to the mid-latitude marine ABL can be found in

Royal Society (1983), *Results of the Royal Society Joint Air-Sea Interaction Project (JASIN)*, eds H. Charnock and R. T. Pollard. Royal Society, London. 449 pp.

The detailed structure of the marine ABL during a cold-air outbreak is described in

Atlas, D., Walter, B., Chou, S.-H. and P. J. Sheu (1986), The structure of the unstable marine boundary layer viewed by lidar and aircraft observations, *J. Atmos. Sci.* **43**, 1301–18.

Section 6.4

A review of the mesoscale IBL with many references on the subject can be found in

Garratt, J. R. (1990), The internal boundary layer – A review, *Bound. Layer Meteor.* **50**, 171–203.

7

The cloud-topped boundary layer

Much of the material to date has either ignored the presence of cloud because it has not been relevant to the problem in hand (e.g. formulation of the mean and turbulence equations; considerations of the surface energy balance and evaporation formulations) or has implicitly assumed clear skies. The presence of cloud has a large impact on boundary-layer structure and upon surface weather. Clouds that are limited in their vertical extent by the main capping or subsidence inversion are an intrinsic feature of the cloud-topped boundary layer (CTBL), and consist mainly of three cloud types: (i) shallow cumulus (Cu) in the form of fair-weather cumulus (either as random fields or cloud streets), (ii) stratocumulus (Sc) and (iii) stratus (St). In addition, fog may exist in the lower regions of the ABL in the form of radiation fog, frontal fog, advection fog and ice/snow fog.

The presence of clouds leads to considerable complications compared to a dry ABL because of the important role played by radiative fluxes and phase changes. In a dry ABL, the turbulent structure, the mean variables and their evolution in time are controlled by large-scale external conditions and by the surface fluxes. In a cloudy ABL, the surface fluxes may be important, but radiative fluxes produce local sources of heating or cooling within the interior of the ABL and therefore can greatly influence its turbulent structure and dynamics. The state of equilibrium of a CTBL is determined by competition between radiative cooling, entrainment of warm and dry air from above the cloud, large-scale divergence and turbulent buoyancy fluxes. The present chapter attempts to summarize the structure of the CTBL by appeal both to observations (Section 7.2) and numerical (mixed-layer) modelling (Section 7.5). In addition, we give a brief description of the role of radiative fluxes in the entrainment process and introduce the topic of entrainment instability leading to the breakdown of cloud layers (Sections 7.3 and 7.4).

7.1 General properties of the CTBL

7.1.1 Introduction

The cloud-topped boundary layer (CTBL) can be broadly identified with a turbulent region in which patterns and ensembles of stratus, stratocumulus and cumulus clouds reside beneath a capping inversion. It is a dominant feature of the weather of the lower atmosphere and of the climate conditions of many areas of the globe, particularly over the sea, and has been recognised in recent times as an important component of the climate system.

The reader should perhaps be reminded of the main features that distinguish a stratus from a stratocumulus layer; these features relate both to the visual appearance of the cloud and to the dynamical nature of the cloud and sub-cloud layers. In general, stratus clouds are found much closer to the surface than stratocumulus. Indeed, stratus may start life as low-level fog gradually developing into an elevated cloud layer of thickness anywhere between a few tens of metres to hundreds of metres. In contrast, stratocumulus is very much an elevated cloud layer often, though not exclusively, beneath the ABL top at a subsidence inversion. Its thickness may vary between a hundred to several hundreds of metres. When seen from above, by aircraft or satellite, stratocumulus, even in continuous sheets, has distinctive cellular patterns associated with convective motions within the cloud (Fig. 7.1). Stratus, however, is often (though not always) uniform and featureless with few undulations in cloud top. Dynamically, stratocumulus is driven by convection, particularly that due to buoyancy generation near the cloud top related to strong radiative cooling. Stratus, on the other hand, may be partly convection driven due to cloud-top radiative cooling or it may be associated with strong winds and large wind shear.

The presence of water in the ABL in the liquid and vapour phases adds complications to the analysis of a dry ABL. The phase change of water in a cloudy ABL introduces, in addition to θ and q, a third variable, in the form of the liquid water content q_1. The instantaneous virtual potential temperature θ_v (cf. Eq. 2.28) must now take into account the influence of the liquid-water mass upon the air density, whence

$$\theta_v \approx \theta(1 + 0.61q - q_1)$$
$$\approx \theta + \bar{\theta}(0.61q - q_1). \qquad (7.1a)$$

An analogous relation holds for mean θ_v, whilst for the fluctuations (cf. Eq. 2.30c),

$$\theta_v' \approx \theta' + \bar{\theta}(0.61q' - q_1'). \qquad (7.1b)$$

Note that Eqs. 7.1a, b reduce to Eqs. 2.28 and 2.30c in the absence of cloud ($q_1 = 0$).

In the presence of phase changes, an important thermodynamic variable is the equivalent potential temperature, θ_e, defined as the temperature attained by a parcel of air that ascends pseudo-adiabatically until all its water vapour has been condensed out and then descends dry-adiabatically to 1000 hPa. For ABL purposes, it is calculated with sufficient accuracy (e.g. Lilly, 1968; Betts, 1973) as

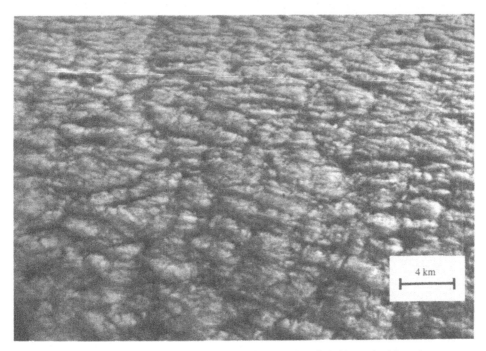

Fig. 7.1 Photograph of stratocumulus from an altitude of 5500 m, looking down at an angle of 45° to cloud top (which is at a height of approximately 1500 m). After Nicholls and Leighton (1986), *Quarterly Journal of the Royal Meteorological Society*.

$$\theta_e \approx \theta + (\lambda/c_p)q \tag{7.2a}$$

so that the fluctuations are given by

$$\theta_e' \approx \theta' + (\lambda/c_p)q'. \tag{7.2b}$$

Cloud base is determined approximately by the local condensation level (LCL) for rising unsaturated air parcels. If the LCL lies beneath the capping inversion, then a CTBL is assumed to exist, with cloud top usually coinciding with the inversion level or lower. Where there are even slightly unstable conditions, the CTBL often corresponds to a layer well-mixed with regard to θ_e and to total water specific humidity q_t or mixing ratio r_t (e.g. Deardorff, 1981; Driedonks and Duynkerke, 1989). A well-mixed layer in θ or θ_v may exist below cloud base, since neither of these temperatures is conserved within the cloud. Within the cloud layer, θ_v follows the moist adiabat.

Figure 7.2 shows in schematic form three examples of the CTBL where one or more cloud layers reside beneath the capping or subsidence inversion (see e.g. Turton and Nicholls, 1987). Type (a) is the classical CTBL, in which the cloud and sub-cloud layers are fully coupled due to turbulent mixing. Most of the discussion in this chapter is directed towards this type, where the ABL (depth h) extends from the surface to the cloud top at the subsidence inversion. In type (b), two or more cloud layers exist beneath the inversion though the ABL only extends to the top of the lowest cloud layer. In type (c), a radiatively driven

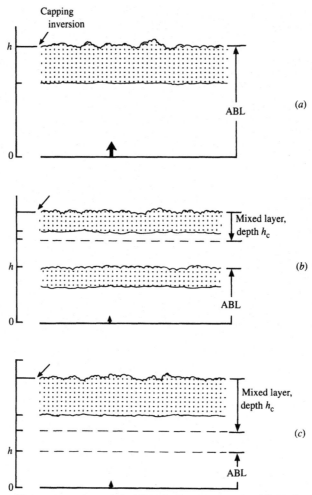

Fig. 7.2 Schematic representation of the CTBL for three cases. (*a*) Fully coupled system, with a well-mixed layer from surface to cloud top; the vertical arrow at the surface indicates significant surface fluxes. (*b*) Two cloud layers beneath the capping inversion (indicated by the upper slanted arrow), with a well-mixed ABL of height h and an elevated mixed layer of depth h_c. (*c*) Decoupled system and only one cloud layer, with weak surface fluxes and an elevated mixed layer of depth h_c.

elevated mixed layer (of depth h_c) containing the cloud layer is decoupled from the surface. Strictly, type (*c*) is not a cloud-topped boundary layer, though it is closely linked to type (*a*) since one may naturally evolve into the other. Types (*b*) and (*c*) are included in the later discussion, and considered as *de facto* ABL clouds.

In the presence of liquid water clouds within the ABL, equations for mean temperature and water variables must take into account the possibility of phase changes and must include a conservation equation for mean liquid water content (either as a specific humidity or mixing ratio). The equation for mean q can be written (cf. Eq. 2.37)

$$\partial \bar{q}/\partial t + \bar{u}_j \partial \bar{q}/\partial x_j = - \partial(\overline{u_j'q'})/\partial x_j + \bar{M}/\rho, \qquad (7.3)$$

where we have neglected the molecular term. Here, $\bar{M}$ is the mass of water vapour per unit volume per unit time being created by a phase change from liquid or solid. For liquid water the analogous equation for $\overline{q_l}$ can be written

$$\partial \overline{q_l}/\partial t + \bar{u}_j \partial \overline{q_l}/\partial x_j = - \partial(\overline{u_j'q_l'})/\partial x_j + W_q/\rho - \bar{M}/\rho \qquad (7.4)$$

where W_q is a net liquid water source/sink term (e.g. it is negative for net precipitation leaving the air parcel). The total water mean specific humidity $\overline{q_t}$ is defined by

$$\overline{q_t} = \bar{q} + \overline{q_l} \qquad (7.5)$$

and an equation for $\overline{q_t}$ can be simply derived by adding Eqs. 7.3 and 7.4 to give

$$\partial \overline{q_t}/\partial t + \bar{u}_j \partial \overline{q_t}/\partial x_j = - \partial(\overline{u_j'q_t'})/\partial x_j + W_q/\rho. \qquad (7.6)$$

In Eq. 7.6, $\overline{q_t}$ can be replaced by the total mean liquid water mixing ratio, $\overline{r_t}$, if the conservation equation for $\overline{r_t}$ is required.

In the absence of clouds, θ_v is conserved for dry adiabatic processes. With liquid water clouds, this is no longer the case and θ_v will generally follow the moist adiabat during parcel ascent through the cloud layer. However, the variables θ_e and q_t are conserved for both dry and moist adiabatic processes in the absence of precipitation (Betts, 1973). Both θ_e and q_t, together with the pressure of an air parcel, can be used to define the thermodynamic state and water content of the air. Note that, for example, radiative cooling affects θ_e, but not q_t; also, precipitation reduces q_t and affects θ_e.

Conservation equations for mean θ_v and θ_e are not strictly necessary, since these quantities can be evaluated from Eqs. 7.1a and 7.2a using solutions for $\bar{q}$ (Eq. 7.3), $\overline{q_l}$ (Eq. 7.4) and $\bar{\theta}$. When phase changes are taken into account, the θ equation becomes (cf. Eq. 2.34), with the molecular term neglected,

$$\partial \bar{\theta}/\partial t + \bar{u}_j \partial \bar{\theta}/\partial x_j = - \partial(\overline{u_j'\theta'})/\partial x_j + (\rho c_p)^{-1} \partial \bar{R}_j/\partial x_j - \lambda \bar{M}/\rho c_p. \qquad (7.7)$$

Because of the simple linear nature of Eq. 7.2a, an equation for mean θ_e, based on Eqs. 7.3 and 7.7, can be written as

$$\partial \overline{\theta_e}/\partial t + \bar{u}_j \partial \overline{\theta_e}/\partial x_j = - \partial(\overline{u_j'\theta_e'})/\partial x_j + (\rho c_p)^{-1} \partial \bar{R}_j/\partial x_j. \qquad (7.8)$$

It should also be noted that, based on Eq. 7.2b,

$$\overline{u_i'\theta_e'} = \overline{u_i'\theta'} + (\lambda/c_p)\overline{u_i'q'}. \qquad (7.9)$$

7.1.2 Boundary-layer regimes and cloud types

The incidence of boundary-layer clouds tends to be greater over the sea, mainly because they are very dependent on a low-level moisture source such as that provided by evaporation from the sea surface. Most observational studies have been made on the marine CTBL, with emphasis on tropical and sub-tropical cumulus, and upon sub-tropical and mid-latitude stratocumulus. Both theoretical and modelling work have tended to concentrate on the stratocumulus CTBL.

The cloudy ABL is found over relatively large areas under the following conditions:

(i) during cold-air outbreaks over the sea to the east of continents. A progression of cloud development occurs as distance from the coast increases, with clouds being initially cumuliform and developing into stratocumulus as they spread out beneath a capping inversion. Such a CTBL may exist for several days and is often associated with mesoscale cellular convection. Surface heat fluxes are generally large, perhaps as high as 200–300 $\mathrm{W\,m^{-2}}$;

(ii) in anticyclonic regions over the ocean in low temperate latitudes, to the west of continents in particular. This CTBL is chiefly stratocumulus, with partial to full cover, and may persist for several weeks at a time over any one area. Surface heat fluxes are usually small;

(iii) over land in mid-latitudes, where stratocumulus may form near the coast during onshore flow, or as a result of the spreading out of fair-weather shallow cumulus produced by daytime heating of the surface. In anticyclonic conditions, particularly during the cooler months, stratus clouds may persist for several days;

(iv) over polar regions, usually in the form of stratus, and particularly during summer. Multiple cloud layers may exist, some of which are decoupled from surface processes. The persistent layers of stratus over the Arctic Ocean are important modulators of the Earth's radiation budget and, in addition, exert a strong influence on the melting rate of the pack ice;

(v) in the trade-wind regions, usually in the eastern parts of the main oceans. Clouds are predominantly fair-weather cumulus and stratocumulus, and are found away from the disturbed region of the inter-tropical convergence zone.

7.1.3 Current knowledge of physical processes

Both observational and modelling studies suggest that several processes are important in determining the formation, maintenance and dissipation of clouds within the ABL. In general, the internal structure of the CTBL is determined by turbulent motions, entrainment, radiative transfer and cloud microphysical structure. The structure of the CTBL depends strongly on the dominant mechanism responsible for generating turbulence. This may be convective, resulting from cloud-top radiative cooling or surface heating, or shear driven, resulting from surface stress or shear at cloud top. In addition, radiative fluxes are affected by the cloud and produce local sources of heating and cooling that can influence the turbulent structure. It is generally considered that longwave radiative cooling extends only over the first 50 m or so beneath a stratocumulus cloud top, with solar heating extending deeper into the cloud layer (e.g. Fravalo et al., 1981). One main effect of the radiative cooling is the generation of an upward buoyancy flux across the ABL that in turn drives entrainment at cloud top. This entrainment brings warmer and drier air down into the ABL (and into the cloud), and subsequently promotes evaporation of cloud droplets and so cooling. This evaporative cooling may, under some circumstances, lead to an instability process in which the parcel cools further, becoming negatively buoyant and sinking through the cloud. This may lead to greater entrainment rates, and promote the breaking up of a solid cloud deck. Additional factors that are probably of relevance in the CTBL include the occurrence of drizzle, and possible sub-cloud evaporation of water droplets, with the related cooling affecting the vertical thermal stability of the ABL.

The importance of extended, quasi-permanent stratocumulus sheets for weather and climate lies primarily in their ability to influence the Earth's radiation budget, both in respect of the net heating or cooling of the global atmosphere and in terms of their effect on the heat and moisture fluxes in the ABL and on the radiation budget at the surface. The formation and maintenance of these extensive cloud layers are associated with several features, as follows (Driedonks and Duynkerke, 1989):

(i) unconditionally stable stratification in the mid-troposphere. This confines convection from the surface and cloud formation to a relatively shallow ABL under an intense inversion, with suppression of deep moist convection;

(ii) a supply of moisture from the surface with vertical mixing throughout the ABL. These are crucial requirements for the formation and maintenance of the cloud deck against factors that tend to dissipate it (i.e. that tend to dry out the ABL). These factors include cloud-top entrainment, large-scale subsidence, and heating of the ABL due to a surface heat flux or shortwave absorption. Sufficient moisture supply from the surface (via evaporation) to the cloud layer compensates for these dissipating processes and maintains the stratocumulus. As a direct consequence, stratocumulus layers are found mostly over the sea.

7.1.4 Models

Compared with the clear-air ABL, relatively few modelling studies have been made on the CTBL. One major task of the modelling approach is to provide a parameterization of the CTBL in GCMs and in climate models. Any realistic climate model has to produce, for example, the extensive stratocumulus sheets found over the oceans and the summertime stratus cover found over the polar regions, in order to incorporate and assess their influence on the radiation budgets and consequently on the climate.

Detailed models of the CTBL must determine the response of the boundary layer to external forcing, produce realistically the turbulent structure of the CTBL and the cloud distribution, and predict the time evolution of the relevant quantities. One major requirement is the incorporation of those processes affecting the formation, maintenance and dissipation of the cloud layer. Models of differing complexity can be, and have been, used in studies of the CTBL; they include large-eddy simulation (LES), explicit turbulence closure (higher-order and K-) and bulk models. Only the first two are able to resolve a number of sub-layers, including multiple cloud layers. All suffer, in practice, from the problems of adequately resolving the entrainment and radiatively cooled region near cloud top and of incorporating sufficiently detailed cloud microphysics, particularly in the context of the cloud–radiation interaction.

7.2 Observations

Knowledge of the structure of the CTBL, and of its response to changing boundary conditions, is very limited compared with what is known about the cloud-free, convective mixed layer driven by surface heating. Unfortunately, due to the diverse and complex nature of the measurements required and the relative inaccessability of the cloud to ground-based instrumentation, there have been relatively few comprehensive observational studies. Of these, most focus on the problem of the formation, maintenance and dissipation of stratocumulus. Much of the observational work has been concentrated in relatively few regions around the world, and mostly over the oceans where the semi-permanent cloud fields exist. In the subsections to follow, characteristic features of the CTBL are described by reference to profiles of mean variables and turbulence quantities. This will help to set the scene before the problem of modelling and the roles of radiative cooling and entrainment at cloud top in cloud maintenance are discussed.

7.2.1 Stratocumulus

Most observations of stratocumulus have been made over the ocean, of thin layers (thickness about 100 m), broken layers and thick or solid layers (thickness of several hundred metres). The data reveal the following broad categories of stratocumulus-topped ABL.

  (i) A fully mixed ABL, with a single elevated cloud layer *fully coupled* with the sub-cloud layer. This type of CTBL is usually associated with either significant surface heat fluxes or strong radiative cooling at cloud top.
 (ii) A single cloud layer *decoupled* from the surface, in which mixing is the result of cloud-top radiative cooling. The cloud forms part of a detached turbulent layer, whose base is separated from the true ABL (i.e. the mixed layer adjacent to the surface). Surface fluxes are usually quite small.
(iii) A fully mixed single layer, or layer in which two or more separated turbulent layers exist, which support *multiple* cloud layers (not necessarily all being stratocumulus). If more than one cloud layer exists, the intervening layer of clear air may serve as a barrier to vertical mixing.

Fully-coupled single layer

This layer may be predominantly surface driven due to heating or friction associated with strong winds, or it may be cloud-top driven due to strong radiative cooling. In the latter case, cloud-top radiative cooling may be sufficiently intense, in situations where surface fluxes are small, to support a convectively mixed layer extending from the cloud top to the surface.

Examples of mean profiles of thermodynamic variables and wind components are shown in Figs 7.3 and 7.4 for the three categories of thin, thick and broken stratocumulus. The ABL is fully mixed because of the combination of non-negligible surface fluxes in moderate to strong winds, and the existence of cloud-top radiative cooling. Through the cloud layer θ or θ_v increases according to moist adiabatic ascent, whilst θ_e is approximately constant. Of particular significance are the large jumps in properties across the ABL top, with changes

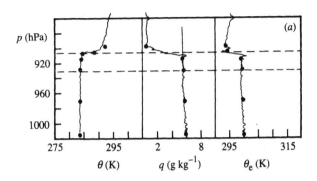

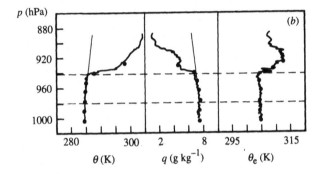

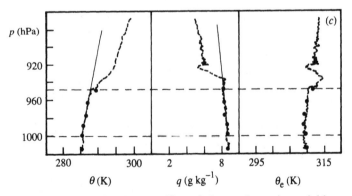

Fig. 7.3 Examples of observed mean profiles of thermodynamic variables made in the CTBL over the ocean offshore from California for (*a*) thin continuous Sc, (*b*) thick continuous Sc, (*c*) broken Sc. Pecked lines indicate the positions of cloud top (corresponding to ABL top) and base. Moist adiabatic distributions of θ and q are indicated in the cloud layer. The observed radiometric sea-surface temperatures were 286.9 K, 282.9 K and 287.2 K respectively. From Albrecht *et al.* (1985), *Journal of Atmospheric Sciences*, American Meteorological Society.

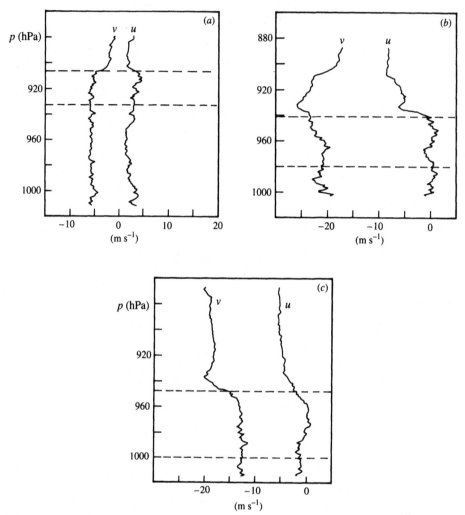

Fig. 7.4 Profiles of zonal (u) and meridional (v) wind components for the three cases shown in Fig. 7.3. From Albrecht *et al.* (1985), *Journal of Atmospheric Sciences*, American Meteorological Society.

in θ_e of both signs. Jumps are more moderate in the broken cloud case in Fig. 7.3. In all cases, small to moderate shear exists at cloud top, though within the ABL shear is small reflecting the well-mixed nature of the whole layer.

Decoupled single layer

An example of a decoupled stratocumulus layer is shown in Fig. 7.5. The cloud layer is barely 150 m thick and located beneath a sharp inversion. The dry adiabatic lapse rate and fluctuations in the wind components indicate that turbulent mixing extends beneath cloud base and into the sub-cloud layer. This whole mixed region has essentially the same values of θ_e and q_t, and is decoupled from the true ABL or surface-based Ekman layer, in which turbu-

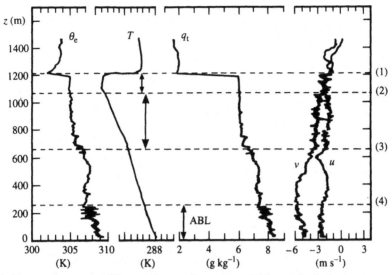

Fig. 7.5 Observed mean profiles of thermodynamic variables and wind components made in the CTBL over the ocean during JASIN, for a decoupled stratocumulus layer. The pecked horizontal lines delineate layer boundaries as follows: (1) cloud top; (2) cloud base; (3) bottom of subcloud layer; (4) top of the surface-related Ekman layer. After Nicholls and Leighton (1986), *Quarterly Journal of the Royal Meteorological Society*.

lence is maintained to a height of about 280 m by surface-related processes, as a result of an intervening stably stratified layer. Vertical wind shear is generally small from the surface to the top of the cloud layer.

It is unlikely that any single method of scaling data obtained in convective layers containing stratocumulus (and thus cloud-top driven) yields results that have universal applicability since the TKE balance is potentially much too complicated. However, the use of some form of mixed-layer scaling can be shown to be at least partly successful in this regard. The mixed layer of interest may comprise the whole layer between cloud top and the surface (surface coupled), or be an elevated layer extending from cloud top to a level between the surface and cloud base (decoupled). The mixed-layer depth h_c is defined with the cloud top as origin. In the case of negligible surface heat flux, the convective scales for velocity and temperature are defined in terms of the integrated buoyancy flux I across the mixed layer (Nicholls, 1989)

$$I = (g/\theta_v)\int_0^h \overline{w'\theta_v'}\,dz \qquad (7.10)$$

and

$$W_* = (2.5I)^{1/3} \qquad (7.11)$$

$$T_{V*} = W_*^2 T_v/gh_c. \qquad (7.12)$$

To estimate typical values of these scales, we take $h_c = 500$ m and a maximum in-cloud buoyancy flux of $0.02\ \mathrm{m\,s^{-1}\,K}$, suggesting $I \approx 0.15\ \mathrm{m^3\,s^{-3}}$. Thus, we expect $W_* \approx 0.75$ m s^{-1} and $T_{V*} \approx 0.03$ K, implying therefore similar values for downdraught velocities and temperature deficits.

The vertical profiles of scaled vertical velocity variance and heat flux shown in Fig. 7.6 for several runs reveal the relatively large impact of the cloud layer, and of cloud-top radiative cooling, on the generation of TKE and negative buoyancy. The curve in Fig. 7.6(a) (note that this uses the convective velocity scale w_*) is one which fits observations from the clear-sky CBL over heated land or water surfaces (see Section 3.5.1 and Fig. 3.21), and is of course plotted in an inverted form (z' is measured from cloud top downwards). This takes account of the buoyancy source due to cooling near cloud top, as opposed to the surface source in the clear CBL.

Multiple layers

Though not shown in diagrammatic form here, a good example of multiple layers can be found in Nicholls (1985). In this case (from the JASIN experiment in the north-east Atlantic), coupling between a stratocumulus layer of thickness about 350 m and the surface-based ABL of depth 400 m was generally weak and intermittent, mainly because of small surface heat fluxes related to near-neutral stability conditions. The LCL coincided with the top of the shallow ABL and, in the presence of conditional instability above ($\partial \theta_e / \partial z < 0$), a field of cumulus clouds extended into the overlying stratocumulus. Even though the cumulus clouds were observed to be intermittent and irregularly distributed in the horizontal, they appeared to be transporting mass (water vapour) from the ABL towards the stratocumulus layer.

7.2.2 Stratus

Observations of stratus cloud layers have tended to concentrate on the Arctic region in summertime, with very few comprehensive data from elsewhere. Nevertheless, observations to date seem to suggest at least two main categories of stratus-topped ABL.

(i) A fully mixed ABL, with a single elevated cloud layer fully coupled with the sub-cloud layer. This type of CTBL is associated with moderate to strong winds and negligible buoyancy effects. An excellent example is shown in Fig. 7.7 for marine stratus, with the cloud layer occupying most of the 950 m deep ABL in which θ_e is constant through most of the sub-cloud and cloud layers. The wind profiles suggest that shear-generation of TKE dominates. The vertical velocity variance typically shows little enhancement near cloud top so that buoyancy effects are probably not significant (Nicholls and Leighton, 1986).

(ii) An ABL above which several low-level cloud layers exist.

We can summarize important observational aspects of the stratus- and stratocumulus-topped ABL as follows.

(i) As with the clear ABL, strong jumps in θ_v and q occur across cloud top (defined as the ABL top), with θ_e often showing both positive and negative jumps. The change in θ_e across the cloud top, $\Delta^c \theta_e$, is an

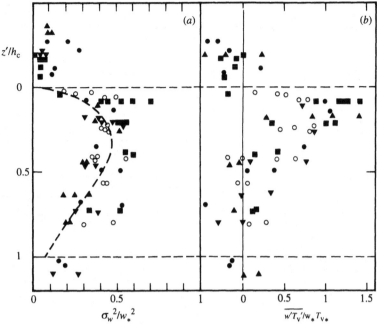

Fig. 7.6 Observed normalized profiles of (*a*) vertical velocity variance and (*b*) buoyancy flux for a stratocumulus-topped mixed layer driven by radiative cooling at cloud top. Data are for five separate aircraft flights, involving one coupled cloud layer (open circles) and four decoupled cloud layers, with the mixed-layer depth h_c measured from the cloud top downwards. The pecked curve in (*a*) is from Lenschow *et al.* (1980), i.e.

$$\sigma_w^2/w_*^2 = 1.8(z'/h)^{2/3}(1 - 0.8z'/h)^2$$

and is given very closely by Eq. 3.109. Note that the curve use the scale w_*, whilst the data use the scale W_*. The data are for flights over sea areas around the United Kingdom. After Nicholls (1989), *Quarterly Journal of the Royal Meteorological Society*.

important parameter characterizing entrainment, and is closely associated with strong cloud-top radiative cooling (here, the jump operator Δ^c is defined as the above-cloud value minus the in-cloud value).

(ii) In the presence of strong winds, large surface fluxes or large entrainment effects (related to cloud-top radiative cooling), the CTBL is well mixed, with the cloud layer coupled to the sub-cloud layer and to the surface. The top of the ABL usually corresponds to cloud top.

(iii) In light to moderate winds, with negligible surface fluxes of sensible heat and momentum, the cloud layer is decoupled from the surface. In this case, the CTBL is not well mixed.

(iv) The CTBL may consist of more than one cloud layer.

(v) Precipitation in the form of drizzle may occur, which tends to produce stable thermal stratification of the CTBL. The stability is affected because the cloud layer is associated with the latent heat of condensation, whilst in the sub-cloud layer evaporative cooling occurs.

(vi) In some instances, the cloudy stratiform ABL is convective in nature and

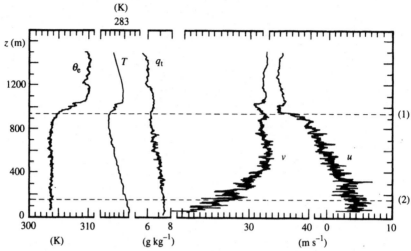

Fig. 7.7 Observed mean profiles of thermodynamic variables and wind components made in the CTBL over the ocean during JASIN, for a thick stratus layer. The pecked horizontal lines delineate layer boundaries as follows: (1) St top; (2) St base. After Nicholls and Leighton (1986), *Quarterly Journal of the Royal Meteorological Society*.

has a number of similarities with the clear CBL heated from below. Generally though, the main source of buoyancy for this CTBL is at the entraining boundary and there are additional sources and sinks of TKE associated with radiative effects and phase changes. It is unlikely that any single method of scaling data obtained in convective layers containing stratocumulus will yield results that have universal applicability since the TKE balance is potentially much too complicated. At other times, particularly under strong wind conditions, the CTBL is indistinguishable in a dynamical sense from the neutral ABL without clouds.

Many of these features are relevant to the problem of modelling the CTBL, and to parameterizing its impact on the atmosphere in numerical models.

7.2.3 Cumulus

Boundary-layer cumuli are cumulus clouds whose vertical extent is limited by the main subsidence or capping inversion. Such clouds are generally non-precipitating, and are usually referred to as *fair-weather cumulus*, *cumulus humilis*, or *trade-wind cumulus*. Boundary-layer cumuli are often composed of an ensemble of *forced*, *active* or *passive* clouds (Stull, 1985) with the deepest of these, the active clouds, ascending no more than a few kilometres above the main inversion.

Forced cumulus clouds form within the top of mixed-layer thermals while the thermals are overshooting into the overlying stable layer. These thermals remain negatively buoyant during the overshoot, even though there is warming from condensation. Thus, the visible thermal rises above the LCL, but fails to reach its level of free convection (where a parcel will have positive buoyancy relative to the environment). Active cumulus clouds ascend above the level of free convection and thus become positively buoyant. Because of this they tend to ascend to greater heights than forced clouds and develop circulations that are independent of the original triggering thermals. In the forced case, little or no transport (venting) of ABL air into the free troposphere occurs whilst, for the active case, transport does take place. Two mechanisms are possible for moving ABL air into an active cloud base. One is the continued updraught forced by the triggering thermal, whilst the other is the cloud-induced updraught associated with the negative pressure perturbation often found at the bottom of a positive buoyancy region. Finally, passive clouds are the decaying remnants of formerly active clouds and, in general, have no interaction with the ABL, except that they shade the ground (as do other non-ABL clouds).

How are the turbulence structure and the vertical fluxes affected by the presence of cumulus clouds? According to GATE observations on days of scattered cumulus, the normalized profiles of buoyancy flux are little affected by the presence or absence of scattered cloud (see Chapter 6, Fig. 6.23). In contrast, the moisture flux profiles appear to be directly related to cloudiness. Profiles of vertical velocity variance and vertical fluxes for the trade-wind cloud layer are shown in Fig. 7.8, for both observations and three-dimensional LES simulation results. Generally speaking, fluxes and variances tend to have maxima near or below cloud base, with no significant increase in the cloud layer.

7.2.4 Diurnal variations

It is a well-known fact that stratocumulus and cumulus fields are closely related, in their activity over land, to the time of day. Marine boundary-layer clouds also undergo diurnal variations, and only recently have detailed modelling studies revealed the underlying physical processes that govern this behaviour (e.g. Bougeault, 1985; Turton and Nicholls, 1987; Duynkerke, 1989).

The diurnal variation in the thickness of Californian coastal stratus is shown in Fig. 7.9 based on mean soundings made during an international field experiment. The data show a rise in cloud base during the day and a fall at night. Cloud top changes hardly at all so the cloud-layer thickness changes from about 240 m at night to only 30–40 m in the afternoon. The behaviour is strongly linked to solar heating of the cloud layer, resulting in daytime warming, stabilization of the sub-cloud layer and decoupling of the cloud layer from the surface.

Model studies show that the decoupling process is closely related to the strong diurnal variation in cloud thickness. The tendency for the layer to become decoupled from surface processes is promoted by both solar warming in the cloud and by small surface buoyancy fluxes. Following decoupling, and the creation of an intermediate stable sub-cloud layer, water vapour fluxes into cloud base are reduced, as is the liquid water content of the cloud.

7.3 Radiation fluxes and cloud-top radiative cooling

At the top of a stratocumulus layer, longwave radiative losses are primarily due to radiation from the cloud water droplets themselves, so the rate of cooling

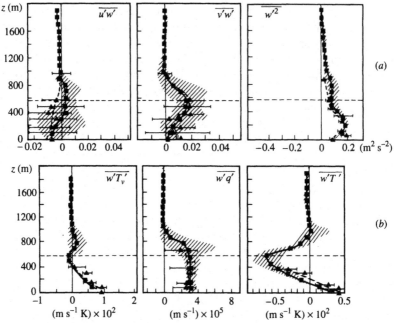

Fig. 7.8 Comparison of profiles of turbulence quantities for a fair-weather marine boundary layer topped by cumulus clouds. Turbulence quantities were generated in a three-dimensional numerical model (LES) and observed from airborne instrumentation during one day of GATE. The horizontal pecked line identifies cloud base. (*a*) Comparison of momentum fluxes and vertical velocity variance from the model (squares and solid curves) and aircraft observations (triangles and pecked curves). The range of observed values is indicated by bars; the shaded areas denote the range of modelled values. Curves are fitted by eye. (*b*) As in (*a*), for heat and moisture fluxes. Note that, at cloud base, the virtual heat flux is about -0.2 times the surface value (see also Fig. 6.23), a behaviour that seems independent of cloud activity. After Nicholls *et al.* (1982), *Quarterly Journal of the Royal Meteorological Society*.

depends strongly on the liquid water content. In the presence of a sharp transition between clear and cloudy air, large changes in net longwave flux are expected near cloud top. In the absence of shortwave fluxes, this leads to significant cooling rates over a relatively thin layer. In such conditions, the primary effect of radiative transfer is to destabilize the cloud layer, generating turbulence through the action of buoyancy. When the sun shines on the cloud top, the amount by which the infrared cooling rates are compensated by solar absorption in the cloud will depend on the rate at which shortwave is reduced as it penetrates the cloud. Longwave radiative cooling is a primary forcing mechanism for a stratocumulus-topped ABL, whilst solar absorption will produce heating in the cloud and act to dissipate the cloud layer.

7.3.1 Observations

Shortwave fluxes
Observations show that a significant fraction of the incident shortwave energy is reflected from the typical stratocumulus layer, with albedos being around

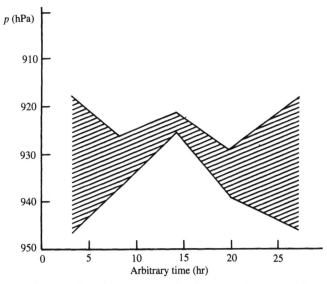

Fig. 7.9 Diurnal variation of the pressure at cloud-base and at cloud-top based on the analysis of soundings taken through coastal stratocumulus off California during FIRE (First ISCCP Field Experiment) 1987. From Betts (1990), by kind permission.

0.6–0.7. Of the remainder, about 25 per cent is absorbed in the cloud layer (about $60 \, \text{W m}^{-2}$ in the mid-latitude summer over the sea) with the rest reaching the surface. Profiles of observed shortwave fluxes related to marine strato-cumulus are shown in Fig. 7.10, together with theoretical curves to illustrate their excellent simulation of the flux profiles. The decrease in the shortwave flux evidently occurs across the whole of the layer. The upwards and downwards shortwave fluxes above the cloud imply a cloud albedo of about 0.7, whilst the fluxes between the surface and cloud base imply a sea-surface albedo of about 0.1.

Longwave fluxes
Observations show that the greatest changes in net longwave flux occur within a very thin layer at, and just beneath, cloud top. The radiative extinction length, over which a fraction $1 - e^{-1}$ (equal to 0.632) of the change occurs, is typically less than 50 m. Figure 7.11 gives some observed longwave flux profiles for both marine and continental nocturnal stratocumulus, and also theoretical curves, which together illustrate the sharp transition at cloud top. Net cloud absorption is typically about $-60 \, \text{W m}^{-2}$, i.e. a net emission. The corresponding cooling rates can be calculated from the radiative part of Eq. 2.46, i.e.

$$(\partial T / \partial t)_{\text{rad}} = (\rho c_{\text{p}})^{-1} \, \partial R_{\text{N}}^{\text{lw}} / \partial z \qquad (7.13)$$

where R_{N}^{lw} is the net longwave flux. Values of the cooling rate between -5 and $-10 \, \text{K hr}^{-1}$, corresponding to differences in R_{N}^{lw} of 50–$100 \, \text{W m}^{-2}$ over 30 m, are not unusual within a thin layer, and emphasize the problem of vertical resolution of such a process in a cloud model (see later).

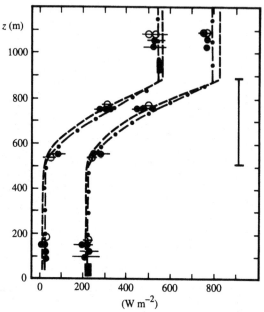

Fig. 7.10 Profiles of upward and downward shortwave fluxes for a Sc-topped ABL, based on aircraft observations (open and solid circles) and theoretical calculations (pecked curves). Measurements have been corrected to local noon; the position of the cloud layer is also indicated. The theoretical curves are based on radiation schemes of Slingo and Schrecker (1982) and Schmetz *et al.* (1981), as discussed in the source reference. After Slingo *et al.* (1982a), *Quarterly Journal of the Royal Meteorological Society*.

The distribution of radiative cooling near cloud top influences the vertical distribution of buoyancy fluxes, which is important in determining the entrainment rate. This has been an important issue (e.g. Deardorff, 1981), particularly regarding the assumptions used in mixed-layer models of the CTBL, and not until the late seventies onwards were observations available to help resolve the problem. The observations described briefly above, and others, tend to show that longwave cooling occurs almost entirely in the cloud layer, well below the position of the inversion.

7.3.2 Modelling

The shortwave and longwave radiative fluxes basically depend on the height of the cloud, the temperature and humidity profiles, the surface temperature and the cloud microphysics (particularly liquid water content and drop size distribution). Theoretical values based on various radiative transfer schemes are shown in Fig. 7.10 for shortwave fluxes, and in Fig. 7.11 for longwave fluxes, and are obviously in close agreement with the observations. Such theoretical calculations allow the distribution of cooling and heating through the cloud to be determined as a function of cloud thickness, i.e. of liquid water path through the cloud. Figure 7.12(*a*) gives an example of shortwave calculations through a model

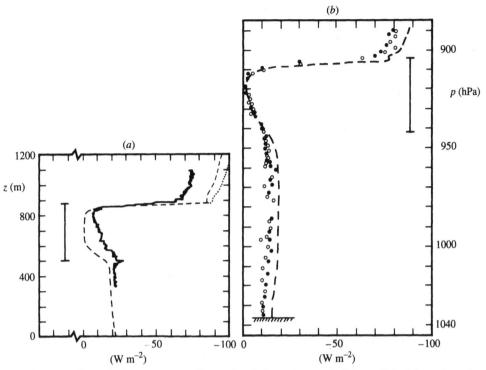

Fig. 7.11 Profiles of the longwave fluxes for (a) marine and (b) continental nocturnal stratocumulus, based on aircraft observations (continuous trace in (a) and circles in (b)) and theoretical calculations (pecked curves). The position of the cloud layer is indicated in each case.

In (a), the theoretical fluxes are based on the radiation scheme of Schmetz and Raschke (1981), and in (b) upon the scheme of Roach and Slingo (1979). After Slingo *et al.* (1982a) (a) and Slingo *et al.* (1982b), (b), *Quarterly Journal of the Royal Meteorological Society*.

stratocumulus cloud layer for a range of liquid water values. These show that shortwave heating (related to the local vertical gradient of shortwave flux) is always more or less uniformly distributed through the cloud.

In contrast, the longwave calculations shown in Fig. 7.12(b) reveal a strong dependence on the amount of liquid water in the cloud. For small values ("thin" cloud), the infrared cooling is small and is distributed throughout the cloud layer. With increasing liquid water path, the infrared cooling grows and tends to become more localized near the cloud top. For the largest values ("thick" cloud) the cooling is similar to that usually assumed in many mixed-layer models (see later), with only small cooling occurring above the cloud top. In the real world, at night time, no shortwave effect is present, but during the day shortwave heating tends partly to compensate the longwave cooling in the upper part of the cloud, and reinforce longwave warming (at large enough liquid water content) near cloud base. This implied variation in heating and cooling through the cloud impacts on the buoyancy fluxes, as mentioned previously.

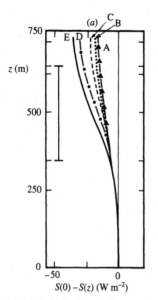 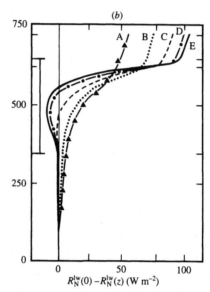

Fig. 7.12 (*a*) Calculated shortwave fluxes through a model stratocumulus cloud layer as a function of cloud liquid water path: A, $0.004\,\mathrm{kg\,m^{-2}}$; B, $0.008\,\mathrm{kg\,m^{-2}}$; C, $0.016\,\mathrm{kg\,m^{-2}}$; D, $0.032\,\mathrm{kg\,m^{-2}}$; E, $0.064\,\mathrm{kg\,m^{-2}}$. The abscissa is the difference between the surface shortwave flux and the flux at height z. (*b*) Calculated net longwave fluxes through a model stratocumulus cloud layer as a function of cloud liquid water path; values as in (*a*). The abscissa is the difference between the surface longwave flux and the flux at height z. In both cases, the position of the cloud layer is identified. After Fravalo *et al.* (1981), *Journal of Atmospheric Sciences*, American Meteorological Society.

7.4 Entrainment and entrainment instability

7.4.1 Entrainment

The problem of entrainment into the clear CBL was discussed in Chapter 6. In the absence of cloud, the main source of TKE is located at the heated lower boundary with energy supplied to the entrainment interface by turbulent transport through a mixed layer. In the cloudy case, latent heat release and cloud-top radiative cooling may both provide additional buoyancy giving rise to significant internal sources of TKE. Thus, entrainment in the cloudy case may also be maintained by sources actually adjacent to the interface between cloud top and clear air.

Observations at the top of stratocumulus clouds show that mixing (entrainment) between cloud and inversion air occurs in a relatively shallow layer estimated from observations to be only a few tens of metres deep (Nicholls and Turton, 1986). The region is referred to as the *entrainment interface layer*. In considering this mixing region, it is important to distinguish between a local interface layer and an averaged interface layer. Most observations are of a local, line measurement of cloud-top structure while modelling results usually represent a horizontally averaged structure. Local thicknesses of the interface layer may be of the order of 10 m or less, but horizontally averaged thicknesses, as computed in models for example, may be nearer 50–100 m.

In the entrainment interface layer, the air is often sub-saturated, with the mixing taking place as radiatively cooled parcels detach themselves from the base of this layer and draw in filaments of warmer air from within it. These become organized downdraughts at slightly lower levels and from there are mixed down into the rest of the cloud. The presence of the cloudy layer means that both radiative and evaporative cooling, processes not relevant in the clear ABL case, can have a significant influence on the density fluctuations that result from mixing within this entrainment region near the interface. So far as their relative effectiveness for producing temperature reductions within the interface layer is concerned, either effect may dominate (e.g. Nicholls and Turton, 1986). Radiative cooling is more important when little mixing occurs, otherwise evaporative cooling dominates. The generation of negatively buoyant parcels (with a deficit in θ_v compared to the cloudy air) is dominated by the radiative cooling of air that has undergone little mixing with the inversion, so that cloud-layer downdraughts tend to be formed preferentially from such air.

7.4.2 The structure of eddies at cloud top

The primary convective elements in the upper part of the cloud layer are negatively buoyant downdraughts forced by the strong radiative cooling near cloud top. Properties of these downdraughts have been investigated primarily by aircraft observations made through horizontally uniform, unbroken marine stratocumulus (e.g. Nicholls, 1989).

Even a casual visual inspection of in-cloud time series data reveals that coherent downdraughts are the most striking feature, as illustrated in Fig. 7.13(*a*). They tend to occupy a maximum fractional area of between 30 and 40 per cent just below cloud top, and are found in relatively narrow regions ($\sim 0.1h_c$–$0.15h_c$ wide) around the periphery of larger-area updraughts (diameter $\sim 0.5h_c$–$0.75h_c$). Here h_c is the depth of the elevated mixed layer, whose boundaries are located at a level within the sub-cloud layer and at cloud-top height. On average, the downdraughts are both cooler and drier than their surroundings in the upper part of the cloud layer, as shown in Fig. 7.13(*b*) for conditionally sampled downdraughts. The contrasts are greatest near cloud top, decreasing to around zero near $z' = 0.5h_c$, as downdraughts mix with surrounding cloud (z' is measured downwards from the cloud top). At lower levels, the downdraughts have slightly positive buoyancy.

Downdraughts contain, on average, only a small proportion of air originating from above cloud top, and the incorporation of radiatively cooled, cloudy air into the downdraughts is the primary mechanism by which an overall negative buoyancy is produced. The downdraughts are found to contribute over a half of the total fluxes of heat, water vapour and liquid water in the upper part of the cloud layer despite occupying only an average 35 per cent of the horizontal area.

The mean buoyancy deficit of the downdraughts is more than sufficient to explain their downward acceleration, but the poor correlation between w' and T_v' within identified downdraught events suggests that the well-defined vertical velocity signature results from the horizontal convergence of circulations constrained by the overlying inversion rather than from direct buoyant instability at

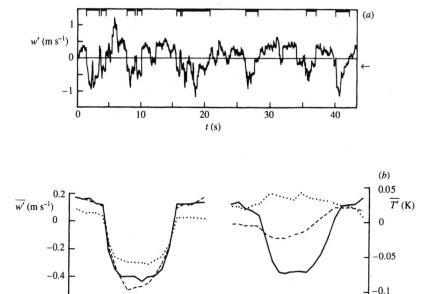

Fig. 7.13 (*a*) An example of downdraught events detected in an aircraft time series of vertical velocity, during a horizontal traverse near the top of continuous marine stratocumulus. The horizontal arrow indicates the selected threshold ($-0.5\,W_*$) for conditional sampling of the events. (*b*) Composite conditionally sampled data of vertical velocity and temperature fluctuations within downdraughts from three levels during a single flight. Solid curve, $Z'/h_c = 0.11$; broken curve, $Z'/h_c = 0.49$; dotted curve, $Z'/h_c = 1.04$; $Z'/h_c = 0$ corresponds to cloud top. After Nicholls (1989), *Quarterly Journal of the Royal Meteorological Society*.

cloud top. In this interpretation (Nicholls, 1989), horizontal circulations formed by updraughts spreading out under the inversion scour the cloud-top interface, incorporating air that has been radiatively cooled and/or partially mixed with inversion air before being forced down in well-defined, narrow convergence zones. These downdraughts subsequently accelerate because of their net negative buoyancy, mixing internally and with the surrounding cloud. The downward mass flux causes compensating updraughts to rise until they near the inversion, where they are forced to spread out.

7.4.3 Entrainment instability

The air above the stratocumulus cloud top is often observed to have a much higher potential temperature and a much lower specific humidity than the cloudy air. Conceptually, one might then be led to expect cloud-top entrainment to tend to destroy a cloud layer. Further considerations suggest a possible stability criterion as follows (Lilly, 1968): if a parcel of the upper air is mixed into the cloudy air by the entrainment process, evaporation of cloud droplets into the initially dry parcel will reduce its temperature. If the mixed parcel reaches saturation at a lower temperature than that of the cloud top it will be negatively buoyant and will sink freely into the cloud. In such a case, the evaporation and penetration process will occur spontaneously and continue until the cloud is evaporated. For the cloud top to become unstable to entrainment then requires the simple *conditional instability criterion*

$$\Delta^c \theta_e < 0 \tag{7.14}$$

where Δ^c refers to a difference between non-cloudy inversion air and cloudy air. Unfortunately, Eq. 7.14 does not take into account the effects of water vapour and liquid water on buoyancy (i.e. the use of θ_v rather than θ). Thus, convection will occur in the cloud when $\delta\theta_v < 0$, where $\delta\theta_v$ is the difference in θ_v between its value in the mixed parcel within the cloud, θ_{vp}, and its value in the cloud itself, θ_{vb}. To investigate the problem of cloud-top evaporative instability, and the possible range of density fluctuations which can be generated by mixing, we consider the following (see e.g. Nicholls, 1989). If a mass χ of warm, dry air from above the cloud in the inversion layer (parcel labelled 'a') is mixed isobarically with a mass $1 - \chi$ of cool, moist air from within the cloud just below the inversion layer (parcel labelled 'b'), then the resulting mixed parcel (labelled 'p') has variables

$$\theta_{ep} = \chi\theta_{ea} + (1 - \chi)\theta_{eb} \tag{7.15}$$

$$q_{tp} = \chi q_{ta} + (1 - \chi)q_{tb}. \tag{7.16}$$

Here, χ is comparable to a mass mixing ratio. Note that Eq. 7.15 can be written as $\theta_{ep} - \theta_{eb} = \chi\Delta^c\theta_e$, since $\Delta^c\theta_e = \theta_{ea} - \theta_{eb}$. Since $\delta\theta_v = \theta_{vp} - \theta_{vb}$, use of Eqs. 7.1a and 7.2a, together with Eqs. 7.15 and 7.16, yields, after considerable manipulation (e.g. Nicholls and Turton, 1986),

$$\delta\theta_v \approx \chi(\Delta^c\theta_e - \theta_{eb}\Delta^c q_t) - (q_p - q_b)(\lambda/c_p - 1.61\theta_{eb}). \tag{7.17}$$

In the derivation of Eq. 7.17, a number of minor terms have been neglected. For given thermodynamic conditions, the mixture is saturated if χ is less than some critical value χ_{cr} where the liquid water content of the mixed parcel goes to zero. For $\chi < \chi_{cr}$, we need to find the condition that makes the right-hand side of Eq. 7.17 negative. In this case, q_p and q_b are both saturation values. The reader should note that when the air samples are mixed togther, all possible ratios of mixtures may be present. However, only a fraction of these ratios are actually negatively buoyant. Furthermore, the maximum amount of negative buoyancy actually available is typically a few tenths of a degree (in virtual temperature), or less.

For small temperature differences $\theta_p - \theta_b$ we can write

$$s = \partial q^*/\partial T \approx (q_p - q_b)/(\theta_p - \theta_b)$$

and replacing θ by θ_e using Eq. 7.2a yields

$$q_p - q_b = s\gamma(\theta_{ep} - \theta_{eb})/(s + \gamma) \tag{7.18}$$

where $\gamma = c_p/\lambda$ is the psychrometric constant. Using Eq. 7.15, and substituting back into Eq. 7.17, gives the buoyancy excess of a parcel of air that has mixed with sufficient inversion air so as to remain just saturated:

$$\delta\theta_v = \chi\theta_{eb}(\gamma\Delta^c\theta_e/a_0 - \Delta^c q_t) \tag{7.19}$$

where

$$a_0 = (s + \gamma)\theta_{eb}(1 + 1.61s\theta_{eb})^{-1}. \tag{7.20}$$

For values of θ_b equal to 274, 283, 293 and 303 K the coefficient a_0 has values of 0.18, 0.22, 0.29 and 0.37 respectively. If the term within the parentheses of Eq. 7.19 is negative, all mixtures with $0 < \chi < \chi_{cr}$ will be negatively buoyant. Thus, a simple theory of evaporative instability gives a modified criterion for cloud-top entrainment instability (Randall, 1980b):

$$\Delta^c\theta_e < (a_0/\gamma)\Delta^c q_t. \tag{7.21}$$

It should be noted that, since $\Delta^c q_t < 0$ for typical stratocumulus, the criterion in Eq. 7.21 is not as strict as that in Eq. 7.14 in the sense that Eq. 7.21 predicts cloud top stability even when $\Delta^c\theta_e$ is slightly negative. Thus, taking $\Delta^c q_t = -0.002$ and $a_0 = 0.2$ gives -1 K as the cut-off value for $\Delta^c\theta_e$. The buoyancy excess $\delta\theta_v$ given by Eq. 7.19 is typically -0.06χ, with $\theta_{eb} = 300$ K and the term in parentheses equal to -0.002. Taking $\chi = 0.1$ gives $\delta\theta_v \approx -0.06$ K, thus emphasizing the small values referred to earlier.

It is not clear that the onset of this instability suffices to break apart a solid stratocumulus deck. For example, the sinking of negatively buoyant parcels into the cloud top and their generation of TKE does not necessarily imply that the parcels will sink entirely through the cloud. The observations of persistent stratocumulus decks and trade cumulus fields shown in Fig. 7.14 show a surprisingly large number that violate the stability condition given by Eq. 7.21 (data to the left of the curve representing the equation). In other words, other mechanisms must exist that affect the break-up of a solid cloud layer into a broken stratocumulus layer or scattered cumulus. For any of the points to the left of the curve in Fig. 7.14, evaporative instability is not sufficient to guarantee cloud break-up. Surface evaporation and upward transport processes can apparently provide enough water vapour to compensate for the evaporative effects associated with the weak evaporative instability.

Indeed, it can be argued (Randall, 1984) that under a fairly wide range of realistic conditions cloud-top entrainment may tend to deepen an existing cloud layer. This will occur when the net effect of entrainment is to cause the cloud top to rise more quickly than the cloud base.

7.5 Numerical modelling of the CTBL

Both LES and higher-order closure ensemble-averaged models of the CTBL exist, and modelling studies have been reported in the literature (see references in Table 8.1). One main advantage of these models, assuming detailed cloud microphysics and radiation schemes are included, lies in their ability to resolve the thin region near the stratocumulus top, so allowing the radiative cooling to be calculated realistically. To close this chapter we will concentrate, however, on mixed-layer modelling since such models are quite widespread and are at the

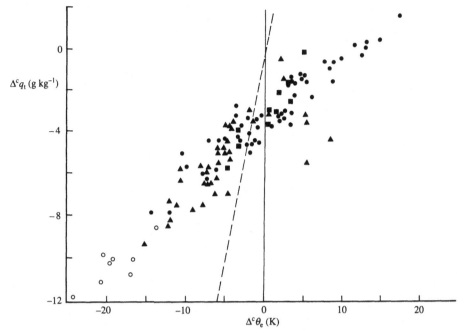

Fig. 7.14 Observational data from many sources plotted as $\Delta^c q_t$ against $\Delta^c \theta_e$, for both stratocumulus (solid symbols) and trade cumulus (open symbols). Stratocumulus observations: solid squares, general mid-latitude; solid triangles, general sub-tropical; solid circles, FIRE sub-tropical. The critical thermodynamic instability curve representing Eq. 7.21, with $a_0 = 0.2$, is shown as the pecked curve. Note that the majority of stratocumulus data are to the left of the critical curve, and hence violate the predictions of the thermodynamic theory of evaporative instability. Based on Kuo and Schubert (1988).

basis of CTBL parameterization in large-scale numerical models of the atmosphere (and will be in the forseeable future).

Mixed-layer modelling is directly analogous to the slab modelling discussed in Chapter 6 regarding the dry, clear CBL and, in relation to the stratocumulus-topped ABL, was pioneered by Lilly (1968). The mixed-layer (jump) model analogous to that for the dry case is sketched in Fig. 7.15, with the interface region magnified to illustrate the partitioning of radiative flux divergence between the cloud layer and the inversion layer, and the associated turbulent flux variations. The major point of interest concerns the fraction r of radiative cooling occurring in the inversion layer compared with the fraction occurring in the cloud. If $r = 1$, the effect of radiation on entrainment is direct and radiative cooling is involved directly in an increase in the ABL height. On the other hand, if $r = 0$ the influence of radiation on entrainment is indirect. That is, radiative cooling takes place exclusively within the cloud layer, where it produces TKE. Observations tend to show that most of the cooling occurs in a thin cloudy region near cloud top of depth about 10–50 m, although model results suggest a dependence of the vertical distribution of radiative cooling upon liquid water content, for example.

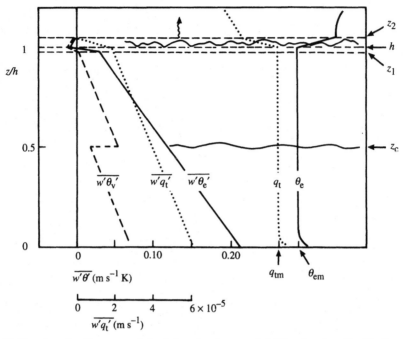

Fig. 7.15 Simple mixed-layer model of the cloud-topped ABL, showing idealized profiles of q_t and θ_e and their jump across the interface region (suitably magnified), and profiles of fluxes $\overline{w'\theta_v'}$, $\overline{w'q_t'}$ and $\overline{w'\theta_e'}$. Cloud base is at $z = z_c$ and cloud top is between $z = h$ and $z = z_2$. The jump in $\overline{w'\theta_e'}$ between z_1 and h compensates for the lower portion of the cloud-top longwave radiative flux divergence. Radiative cooling, directly related to the change $\Delta^c R_N$, occurs between z_1 and z_2, with a fraction r occurring above cloud top and within the inversion layer ($h < z < z_2$). For the given $\overline{w'q_t'}$ and $\overline{w'\theta_e'}$ profiles (with $r = 0.5$, $\Delta^c\theta_e = 5\,\text{K}$, $\Delta^c q_t = -0.002$, $z_c/h = 0.5$, $\theta_m = 284\,\text{K}$, $q_{tm} = 0.006$ and $\Delta^c R_N = 75\,\text{W m}^{-2}$), the $\overline{w'\theta_v'}$ profile is calculated from Eqs. 7.29 (sub-cloud) and 7.31b (cloud). After Deardorff (1976), *Quarterly Journal of the Royal Meteorological Society*.

In the dry ABL, the simplest model has equations involving four unknowns, θ_{vm}, $\Delta\theta_{vh}$, h and $(\overline{w'\theta_v'})_h$, assuming the surface flux is given, and the set is closed by including a suitable entrainment assumption. In the moist case (stratocumulus-topped ABL), the model is necessarily more detailed because of the need to include radiative cooling effects near cloud top and the effects of condensation. The model's purpose is to enable ABL growth to be evaluated in the presence of a cloud layer and also to enable the evaluation of entrainment fluxes at cloud top, diagnostics related to the cloud-top entrainment instability criterion, and possibly changes in the height of cloud base (thus, changes in the thickness of the cloud). The moist model incorporates equations for the conservative variables θ_e and q_t, together with h, entrainment fluxes $(\overline{w'\theta_e'})_h$, $(\overline{w'q_t'})_h$ and suitable entrainment assumptions. In a mixed-layer model, a stratocumulus cloud layer is assumed to exist in the upper part of the ABL if the diagnosed or calculated ABL top is higher than the LCL.

Let property changes across the ABL top be denoted $\Delta^c\theta_e$ and $\Delta^c q_t$ (between h and z_2), and $\Delta^c R_N$ (between z_1 and z_2). The fraction of $\Delta^c R_N$

occurring above cloud top and within the inversion layer ($h < z < z_2$) is designated r. As in the dry case, the conserved variables averaged through the mixed layer can be derived by vertical integration of Eqs. 7.6 and 7.8 (W_q is set to zero) simplified to horizontally homogeneous conditions:

$$\partial \theta_{em}/\partial t = [(\overline{w'\theta_e'})_0 - (\overline{w'\theta_e'})_h]/h + (1 - r)\Delta^c R_N/\rho c_p h \qquad (7.22)$$

$$\partial q_{tm}/\partial t = [(\overline{w'q_t'})_0 - (\overline{w'q_t'})_h]/h. \qquad (7.23)$$

In Eq. 7.22 the radiative flux divergence between level z_1 and the surface has been neglected compared with that between z_1 and h. This is appropriate for small r, but in general the term can be readily incorporated only at the expense of increased detail. The fluxes vary linearly with height across the mixed layer to a height $z = z_1$, then jump to new values at $z = h$ because of the non-zero radiative flux divergence. At $z = z_2$ the fluxes go to zero (Fig. 7.15). Application of the conservation equations in an infinitesimally thin layer at cloud top yields

$$\Delta^c \theta_e (\partial h/\partial t - w_h) = - (\overline{w'\theta_e'})_h + r\Delta^c R_N/\rho c_p \qquad (7.24)$$

$$\Delta^c q_t (\partial h/\partial t - w_h) = - (\overline{w'q_t'})_h. \qquad (7.25)$$

Equations 7.24 and 7.25 can be regarded as predictive equations for h, and in order that they predict h in a consistent manner we require that

$$\Delta^c q_t(- (\overline{w'\theta_e'})_h + r\Delta^c R_N/\rho c_p) = - \Delta^c \theta_e (\overline{w'q_t'})_h \qquad (7.26)$$

should be satisfied. The presence of the radiation terms in Equations 7.22 and 7.24 represents the impact of radiation on the mixed-layer turbulence and entrainment respectively. The reader can readily demonstrate that Eqs. 7.24 and 7.25 reduce to Eqs. 6.8 and 6.9 appropriate to the cloud-free case, by setting $q_1 = 0$ and $\Delta^c R_N = 0$.

Equations for the differences $\Delta^c \theta_e$ and $\Delta^c q_t$ can be either diagnostic or prognostic. In the former case, their definitions

$$\Delta^c \theta_e = \theta_e^+ - \theta_{em} \qquad (7.27)$$

$$\Delta^c q_t = q_t^+ - q_{tm} \qquad (7.28)$$

require the values at $z_2(\theta_e^+, q_t^+)$ to be specified, since the mixed-layer variables are obtained from Eqs. 7.22 and 7.23. Prognostically, rate equations for the two differences can be written by analogy with that for $\Delta\theta_{vh}$ in the dry case (Eq. 6.12).

In our simple model, there are six equations (Eqs. 7.22–7.25 and 7.27, 7.28) and ten unknowns (θ_{em}, q_{tm}, h, two entrainment fluxes, $\Delta^c \theta_e$ (or θ_e^+) and $\Delta^c q_t$ (or q_t^+), two surface fluxes and $\Delta^c R_N$). It is assumed that r and w_h are specified. The fluxes at the surface involve the surface quantities $(\overline{w'\theta'})_0$ and $(\overline{w'q'})_0$, which are assumed known from the standard flux formulations involving transfer coefficients, surface and mixed-layer properties. In addition, the radiative fluxes will be either specified or, more properly, calculated from a radiation sub-model. Even so, there remain more unknowns than equations, so $\partial h/\partial t$ must be obtained by way of an entrainment assumption.

The assumption for the entrainment rate requires knowledge of the vertical flux of θ_v, because the buoyancy term $(g/\theta_v)\,\overline{w'\theta_v'}$ is responsible for most of the TKE generation in the mixed layer, and because the turbulence so generated causes the entrainment. It follows from Eqs. 7.1b and 7.9 that *in the sub-cloud layer* $(z < z_c)$ where $q_1 = 0$,

$$\overline{w'\theta_v'} \approx \overline{w'\theta_e'} - e_1\overline{w'q'} \qquad (7.29)$$

with

$$e_1 = (\lambda/c_p) - 0.61\theta_m. \qquad (7.30)$$

In the cloud layer $(z_c < z < h)$, which is assumed to be wholly saturated $(q = q^*)$, we find

$$\overline{w'\theta_v'} \approx \overline{w'\theta_e'} - (e_1 - \theta_m)\overline{w'q^{*\prime}} - \theta_m\overline{w'q_t'} \qquad (7.31a)$$

where q has been replaced by its saturated value, and is thus a unique function of temperature. Following Deardorff (1976), this allows $\overline{w'q^{*\prime}}$ to be rewritten as $e_2\overline{w'\theta_e'}$ based on Eq. 7.9 and Eq. A18; here $e_2 = e_3(1 + e_3/\gamma)^{-1}$ and $e_3 = 0.622\lambda q^*(\theta_m)/(R_v\theta_m^2)$. Hence, Eq. 7.31a can be simplified to

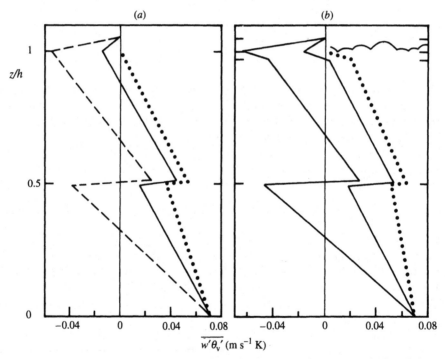

Fig. 7.16 Theoretical profiles of the virtual heat flux based on numerical solutions of Eqs. 7.22–7.34, for values of the parameters used in Fig. 7.15, $(\overline{w'\theta'})_0 = 0.06\ \mathrm{m\,s^{-1}\,K}$ and $(\overline{w'q'})_0 = 6 \times 10^{-5}\ \mathrm{m\,s^{-1}}$. In (a), $\Delta^c R_N = 0$, $\partial h/\partial t = 0.0042\ \mathrm{m\,s^{-1}}$; in (b), $\Delta^c R_N/\rho c_p = (\overline{w'\theta'})_0$, $\partial h/\partial t = 0.01\ \mathrm{m\,s^{-1}}$. The solid profiles are for calculations using Eq. 7.34 (with $\beta_1 = 0.2$) for the entrainment hypothesis; the dotted profiles use Eq. 7.33 with $\beta_1 = 0$ (zero buoyancy flux); the pecked profiles use Eq. 7.33 with $\beta_1 = 1$ (minimum buoyancy flux, or maximum entrainment). After Deardorff (1976), *Quarterly Journal of the Royal Meteorological Society*.

$$\overline{w'\theta_v'} \approx e_4\overline{w'\theta_e'} - \theta_m\overline{w'q_t'}, \tag{7.31b}$$

where $e_4 \approx 1 - e_1e_2$. For θ_m equal to 275 K and 290 K, it is readily shown that e_4 is about 0.7 and 0.5 respectively. Assuming linear profiles in $\overline{w'\theta_e'}$ and $\overline{w'q_t'}$, profiles for $\overline{w'\theta_v'}$ can be readily evaluated for both sub-cloud and cloud layers, noting that there is a discontinuity in $\overline{w'\theta_v'}$ at $z = z_c$ because θ_v is not conserved for a vapour–liquid phase change (as is θ_e and q_t). The buoyancy flux at h can be written, after combining Eqs. 7.24, 7.25 and 7.31b,

$$(\overline{w'\theta_v'})_h \approx e_4r\Delta^c R_N/\rho c_p - (e_4\Delta^c\theta_e - \theta_m\Delta^c q_t)(\partial h/\partial t - w_h). \tag{7.32}$$

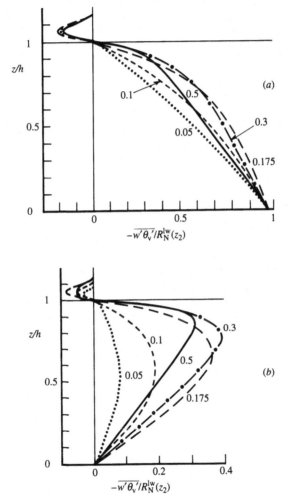

Fig. 7.17 Illustration of the sensitivity of the buoyancy flux profiles through a strato-cumulus-topped ABL to the radiative cooling fraction r. In (a), $R_N^{lw}(z_2)$, the above-cloud net longwave radiative flux, is taken to be equal in magnitude to $\rho c_p(\overline{w'\theta_v'})_0$; in (b), $(\overline{w'\theta_v'})_0 = 0$. Values of r are indicated against each curve, and are based on Fig. 3 of Deardorff (1981) for $\Delta h/h = 0.1$, where Δh is the thickness of the entrainment region. Profiles are based on the numerical calculations of Deardorff (1981).

It should be noted that, for the case of no cloud ($q_1 = 0$), Eq. 7.32 reverts to Eq. 6.8 (where note must be made of the assumptions leading from Eq. 7.31a to Eq. 7.31b) and Eq. 7.31a reverts to Eq. 7.29.

The entrainment hypothesis usually chosen is similar to that discussed for the dry case in Section 6.1.5, with the TKE balance setting maximum and minimum bounds on the entrainment (Lilly, 1968). In this approach, it is usual to use the ratio of maximum and minimum buoyancy fluxes, with

$$\overline{(w'\theta_v')}_{\min}/\overline{(w'\theta_v')}_{\max} = -\beta_1. \tag{7.33}$$

Maximum entrainment occurs when all buoyant production is used in the entrainment process, whilst the minimum occurs when all energy dissipation balances the buoyant production within the mixed layer. A physically realistic interpolation sets the flux $\overline{(w'\theta_v')}_h$ as a certain fraction of the average buoyancy flux within the mixed layer, i.e.

$$\overline{(w'\theta_v')}_h = -h^{-1}[2\beta_1/(1-\beta_1)]\int_0^h \overline{w'\theta_v'}\,\mathrm{d}z \tag{7.34}$$

with the fraction equal to $1/2$ if $\beta_1 = 0.2$ (Deardorff, 1976).

In order to illustrate the sensitivity of buoyancy flux profiles to several important quantities and parameters, we show theoretical profiles of $\overline{w'\theta_v'}$ in Fig. 7.16 where $\Delta^c R_N$ and β_1 are varied for given external and surface flux conditions. The discontinuity at cloud base is very evident, as is also the double-jump structure in the interface region associated with the longwave radiative flux divergence with $r = 0.5$. This is a case with a moderate surface heat flux ($\overline{(w'\theta')}_0 = 0.06\,\mathrm{m\,s^{-1}\,K}$) and evaporation (6.6 mm day^{-1}). The associated profiles of $\overline{w'\theta_e'}$ and $\overline{w'q_t'}$ are shown in Fig. 7.15.

The sensitivity of the buoyancy flux profiles to the value of r is illustrated in Fig. 7.17 for a "dry" cloud only (latent heat effects are removed), based on results of Deardorff (1981). In Fig. 7.17(a), the most relevant feature relates to the value of r which maximizes the vertically integrated buoyancy flux within the mixed layer – this value is near 0.2. In Fig. 7.17(b), where the surface flux is zero, the dependence of the profiles upon r is more substantial, and the effect of the cloud-top radiative cooling in maximizing $\overline{w'\theta_v'}$ in the upper regions of the mixed layer is clearly evident. These profiles show strong similarities to the stratocumulus observations described earlier (see Fig. 7.6b).

Notes and bibliography

Section 7.1

A collection of articles dealing with the status of research on the CTBL as of 1985 can be found in

World Meteorological Organization (1985), *Report of the JSC/CAS Workshop on Modelling of Cloud Topped Boundary Layer*, WMO World Climate Programme, Report No. WCP-106. World Meteorological Organization, Geneva, 29 pp. plus extensive appendices.

A recent review of the turbulence structure of the ABL capped by a cloud deck, based on observations and model studies, is

Driedonks, A. G. M. and P. G. Duynkerke (1989), Current problems in the stra-
    tocumulus-topped ABL, *Bound. Layer Meteor.* **46**, 275–303.
This paper contains a number of relevant references to papers on both observational and
modelling studies.
The classic work on the modelling of the CTBL and the role of entrainment is
Lilly, D. K. (1968), Models of cloud-topped mixed layers under a strong inversion,
    Quart. J. Roy. Met. Soc. **94**, 292–309.
This paper also discusses and derives the stability criterion for entrainment.

Section 7.3

Theoretical studies on the distribution of radiative cooling near cloud top include
Deardorff, J. W. (1981), On the distribution of mean radiative cooling at the top of a
    stratocumulus-capped mixed layer, *Quart. J. Roy. Met. Soc.* **107**, 191–202, and
Nieuwstadt, F. T. M. and J. A. Businger (1984), Radiative cooling near the top of a
    cloudy mixed layer, *Quart. J. Roy. Met. Soc.* **110**, 1073–8.

Section 7.4

The discussion, in Section 7.4.2, of the eddy structure near cloud top and the role of
downdraughts in the entrainment process leans heavily upon
Nicholls, S. (1989), The structure of radiatively driven convection in stratocumulus,
    Quart. J. Roy. Met. Soc. **115**, 487–511.
In this paper, data from a series of aircraft flights in marine stratocumulus are described
and interpreted.
The entrainment rate into a stratocumulus deck, and its dependence upon the vertical
distribution of radiative cooling, is discussed in
Deardorff, J. W. (1976), On the entrainment rate of a stratocumulus-topped mixed layer,
    Quart. J. Roy. Met. Soc. **102**, 563–82, and in
Randall, D. A. (1980), Entrainment into a stratocumulus layer with distributed radiative
    cooling, *J. Atmos. Sci.* **37**, 148–59.
The criterion for instability of dry air mixing into a cloud top is discussed in
Deardorff, J. W. (1980), Cloud top entrainment instability, *J. Atmos. Sci.* **37**, 131–47,
    and in
Randall, D. A. (1980), Conditional instability of the first kind upside-down, *J. Atmos.
    Sci* **37**, 125–30.
Hanson, H. P. (1984), Stratocumulus instability reconsidered: A search for physical
    mechanisms, *Tellus* **36A**, 355–68.

Section 7.5

Observations of daytime marine stratocumulus and results from a one-dimensional
mixed-layer model are compared in
Nicholls, S. (1984), The dynamics of stratocumulus: Aircraft observations and compari-
    sons with a mixed layer model, *Quart. J. Roy. Met. Soc.* **110**, 783–820.

8

Atmospheric boundary-layer modelling and parameterization schemes

It is my intention in this chapter to summarize, and comment upon, the parameterization methods used in ABL and larger-scale numerical models of the atmosphere, which allow vertical fluxes to be evaluated. The chapter draws on much of the material presented earlier in this book, particularly as this impacts on turbulence closure schemes and the calculation of surface fluxes over the land. The latter, in turn, demand that consideration be given to the nature of the surface and the calculation of surface temperature and surface humidity.

8.1 Introduction

Numerical solutions of the averaged conservation equations require both a suitable closure scheme to account for the turbulent flux terms and, usually, a set of physical parameterizations of the Earth's surface, clouds and radiation. Both one-dimensional ABL models, and two- and three-dimensional atmospheric models with horizontal grid scales much greater than the typical ABL length scale (~ 1 km), utilize ensemble-averaged equations. In contrast, in large-eddy simulation (LES) models (which potentially represent the best approach to calculating the three-dimensional time-dependent structure of the ABL), true volume averaging is incorporated. In LES models, sub-grid turbulence parameterization in the limit of small grid scale (relative to the turbulence length scale) is likely to be far simpler, and universally more applicable, than in the ensemble-averaged case. This is the case because large eddies are explicitly resolved and small eddies are likely to take on a more universal character. Unfortunately, in many applications, LES is impracticable – in mesoscale and large-scale models, for example – and emphasis here will be given to the details of closure schemes used with the ensemble-averaged equations.

ABL modelling has as its central theme the representation of turbulent mixing and evaluation of the second moments appearing in the mean equations (Eqs. 2.34, 2.37 and 2.41). ABL models are structured in integral (slab) form, or in a high (vertical) resolution form utilizing either first-order or higher-order closure.

8.1.1 Integral (slab) models

The integral approach predicts the vertically averaged properties of the ABL, so that details of the vertical profile structure of any property are unavailable. The approach is particularly suited to cases where vertical gradients are small throughout much of the ABL, or where vertically averaged quantities are specifically required. In the former case, the daytime entraining CBL is the best example (see Chapter 6) whilst GCM schemes that have limited vertical resolution are examples of the latter. We have already discussed a typical slab or integral model in the context of the CBL in Chapter 6 and the cloud-topped boundary layer in Chapter 7. Closure of the slab equations for both the clear and cloudy ABL requires knowledge of surface and entrainment fluxes and the ABL depth.

8.1.2 High-resolution models

"High-resolution" as used here implies multiple levels in the vertical, so allowing the internal structure of the ABL to be evaluated. To solve for the mean fields, the Reynolds flux-divergence term must be approximated at each level. There are two main categories of this type of model: (i) those utilizing the ensemble-averaged equations, or volume-averaged equations that approximate ensemble averages because the averaging scale is much greater than the ABL scale; and (ii) LES models with volume averaging and explicit representation of the large-eddy structure. The ensemble-averaged models may be structured either as a specific ABL model, or as an interactive component of a mesoscale or general circulation model.

8.1.3 Large-eddy simulation models

In these, the averaging volume is sufficiently small that the largest energy-containing eddies are resolved explicity, at least well away from the lower boundary and the overlying inversion layer. This is of paramount importance, since turbulent flows tend to differ from one another mainly in their large-eddy structure whereas the small scales in all turbulent flows tend to be statistically similar. The large eddies are very sensitive to the environment (geometry and stratification), and in particular to the buoyancy forcing. Thus, only the less sensitive small scales need to be parameterized. LES models allow the use of relatively simple sub-grid closure schemes, including the first- and second-order schemes used in ensemble-averaged models.

Solution of the volume-averaged equations provides variables that are partly random. Thus, to compute mean (in the ensemble sense) vertical profiles of any variable, there is a need to average over a series of horizontal planes and/or over a number of time steps, and/or several runs or simulations. For area-averaged turbulent statistics the total value of the property will be the sum of the resolved and unresolved (parameterized sub-grid) parts. These components are shown for the CBL in Fig. 8.1 as a function of height for both normalized TKE and heat flux, based on typical LES results with a horizontal grid scale of only 50 m. At this resolution, sub-grid scales above the surface layer contribute less than 12 per cent of the TKE and negligible heat flux. Note how, as the

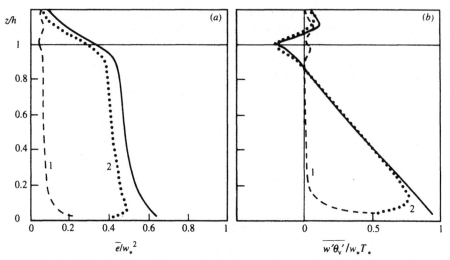

Fig. 8.1 Vertical distributions of (a) normalized TKE and (b) normalized buoyancy flux in the CBL, based on results from a LES model. Curves 1 and 2 represent the sub-grid and resolvable scales respectively, with their sum represented by the solid curves. After Moeng and Wyngaard (1989), *Journal of Atmospheric Sciences*, American Meteorological Society.

lower boundary is approached, the parameterized part becomes a significant fraction of the surface value.

The major strength of the LES approach is its ability to resolve many of the energetic and flux-carrying eddies in the ABL. Some of the more relevant LES studies undertaken in the recent past are summarized in Table 8.1.

Before one can determine the surface fluxes, and fluxes throughout the ABL, in any model it is often necessary to evaluate a number of surface characteristics. Some properties, including roughness length and albedo, are most likely specified, but this is not generally the case with surface temperature and humidity. If we are dealing with the sea surface, the surface temperature may well be prescribed and the surface humidity taken as the saturated value at this temperature.

8.2 Surface temperature

Numerical models of the atmosphere usually compute the ground surface temperature T_0 from either Eq. 5.1. (a diagnostic form of the surface energy balance (SEB) equation) or Eq. 5.4 (a prognostic form of the SEB equation, i.e. a rate equation for T_0).

8.2.1 Diagnostic solutions

In the diagnostic case, the ground heat flux may be parameterized very crudely, e.g. as a constant fraction of R_N or by assuming the heat capacity of the ground is zero and setting $G_0 = 0$. Alternatively, it may be computed using a full treatment of soil heat diffusion in a multi-level soil model (see Section 5.1.2 for sample solutions). All the terms in Eq. 5.1 are either computed separately or

Table 8.1. *A summary of the main LES studies of the ABL*

Topic	Reference
Neutral and unstable ABL	Deardorff (1972)
Depth and mean structure of the CBL	Deardorff (1974)
Cumulus-topped (trade wind) ABL	Sommeria (1976)
Stratocumulus-topped ABL	Deardorff (1980)
CBL turbulence	Moeng (1984)
Diffusion in the CBL (bottom-up, top-down)	Wyngaard and Brost (1984)
Sub-grid eddy diffusivities (channel flow)	Mason and Callen (1986)
Stratus-topped ABL	Moeng (1986)
Neutral ABL	Mason and Thompson (1987)
Plume dispersion in the ABL	Nieuwstadt and de Valk (1987)
Passive and buoyant plumes in the CBL	Haren and Nieuwstadt (1989)
Non-local mixing in the CBL	Ebert *et al.* (1989)
CBL flow structure	Mason (1989)
Coherent structures in the CBL	Schmidt and Schumann (1989)
The CBL – diffusion and chemistry	Schumann (1989)
Turbulent sheared convection	Sykes and Henn (1989)
CBL over a heated slope	Schumann (1990)

parameterized in terms of T_0, and the equation is solved iteratively using the Newton–Raphson method (e.g. Jacobs & Brown, 1973; Pielke, 1984, Chapter 11).

8.2.2 Prognostic solutions

In the prognostic case, the soil heat flux may be crudely parameterized or even set equal to zero. One method that is widely used is based on Eq. 5.4, with G_1 determined from a two-layer soil model (Fig. 8.2). In this, the surface temperature is approximated by the temperature of the thin upper layer, of temperature T_s and thickness $\Delta z'$. The deeper soil layer is assumed to be of sufficient thickness (d_0) that over time scales of interest the flux of heat at the bottom is zero. In order to solve Eq. 5.4, the soil heat flux between the upper and lower soil layers is represented as

$$G_1 = \mu(T_s - \bar{T}). \tag{8.1}$$

Here $\bar{T}$ is the temperature of the deep soil, which acts as a heat reservoir. Combining Eqs. 5.4 and 8.1 gives, with $C_g = C_s \Delta z'$ (C_s is the volumetric heat capacity),

$$C_g \partial T_s / \partial t = (R_N - H_0 - \lambda E_0) - \mu(T_s - \bar{T}). \tag{8.2}$$

This predictive equation for surface soil temperature is often referred to as the *force-restore method* because the forcing by $(R_N - H_0 - \lambda E_0)$ is modified by a restoring term containing $\bar{T}$ (e.g. Deardorff, 1978; Dickinson, 1988). This term restores T_s exponentially towards $\bar{T}$ if the forcing is removed. An analogous equation for $\bar{T}$ can be derived from Eq. 5.2 in approximate form as

$$(C_s d_0) \partial \bar{T} / \partial t = (R_N - H_0 - \lambda E_0). \tag{8.3}$$

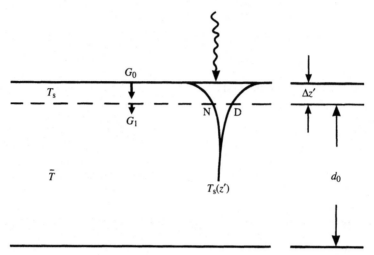

Fig. 8.2 Schematic two-layer soil model for calculation of soil heat flux and "surface" temperature. The thin surface layer, of thickness $\Delta z'$, has temperature T_s, whilst the thicker deep-soil or reservoir layer, of thickness d_0, has a temperature of $\bar{T}$. The thicker vertical arrows indicate the direction of the soil heat fluxes corresponding to temperature profile D (daytime); profile N is typical of night-time over land under clear skies.

Equation 8.3 is used with d_0 taken as the e-folding (damping) depth of the annual wave, i.e. $d_0 = 365^{1/2} D/2$, where D is the damping depth of the diurnal wave given by Eq. 5.9. For time periods of less than a day or so $\bar{T}$ is approximately constant.

For practical use, the coefficients C_g (hence $\Delta z'$) and μ must be determined. The outcome of the analysis of Bhumralkar (1975) and Blackadar (1979), for example, showed that

(i) C_g and μ can be chosen so that the amplitude and phase of the upper soil layer temperature are identical to the surface temperature of a real soil layer of uniform thermal conductivity k_s and volumetric heat capacity C_s (see Section 5.1.2);

(ii) C_g and μ can be calculated from Ω, the angular velocity of the Earth, which determines the diurnal period, k_s and C_s.

For a sinusoidal forcing, e.g. as given by Eq. 5.10 for the soil heat flux at the surface G_0, Eq. 5.8 for T_s is a solution of Eq. 5.7 giving, at the surface,

$$T_s = \bar{T} + A_0 \sin \Omega t. \tag{8.4}$$

With G_0 given by Eq. 5.10 it is required that T_s given by Eq. 8.4 be a solution of Eq. 8.2 for T_s. Careful consideration shows that this is so if we take

$$C_g = C_s(\kappa_s/2\Omega)^{1/2} \tag{8.5a}$$

and

$$\mu = \Omega C_g. \tag{8.5b}$$

It should be noted that Eq. 8.5a implies $\Delta z' = (\kappa_s/2\Omega)^{1/2} = D/2$. Referring to

Table A7 for moist sand (with moisture content $\eta = 0.2$), $D \approx 0.15$ m, $C_g = C_s D/2 \approx 1.7 \times 10^5$ J m^{-2} K^{-1} and $\mu \approx 12.5$ W m^{-2} K^{-1}.

Since the surface forcing is not really sinusoidal in nature, the values of C_g and μ given by Eq. 8.5 need to be modified to account for higher harmonics. This was done by Blackadar, for example, using a Fourier series representation and calibrated by reference to solar insolation under clear skies at the time of the equinoxes, normalized by the midday value. The procedure gives

$$C_g = 0.95 C_s (\kappa_s / 2\Omega)^{1/2} \tag{8.6a}$$

and

$$\mu = 1.18 \Omega C_g \tag{8.6b}$$

which are commonly used in numerical models.

8.3 Surface humidity (soil moisture)

In determining the surface humidity for a bare soil surface, two quite different approaches, interactive and non-interactive, need to be recognized. The non-interactive approach means that the surface humidity or soil wetness does not respond to atmospheric forcing in a realistic way, if at all.

8.3.1 Non-interactive (diagnostic) schemes

The evaporation from, and hence the surface temperature of, a bare soil surface is highly dependent on the surface humidity, and thus upon the near-surface soil moisture content. This is readily apparent in Figs. 5.9(a) and 5.9(b). The problem of computing q_0 (or the surface relative humidity r_h) for a bare soil has been bypassed in many numerical models of the atmosphere by specifying some measure of the surface wetness. For some purposes this may be acceptable, but it is at best a very crude representaton of the evaporation process and, in particular, fails to include properly the feedback between the atmosphere and surface moisture status. Several examples of non-interactive schemes are (Carson, 1982) as follows.

(i) The Bowen ratio B is specified, which implies (Eqs. 3.46b, 3.51)

$$q_0 = \gamma B^{-1}(\theta_0 - \theta) + q \tag{8.7}$$

where γ is the psychrometric constant c_p/λ.

(ii) The actual evaporation is set to a constant fraction x of the potential evaporation E_P:

$$E_0 = x E_P. \tag{8.8}$$

This gives (Eqs. 3.63 and 5.23),

$$q_0 = x q^*(T_0) + (1 - x)q, \tag{8.9}$$

where T_0 is the temperature of the drying soil surface. (It should be noted that a more practicable temperature is T_{0P}, the temperature that the surface would have if it were actually wet, and q and T were held constant. However, the hypothetical temperature is unknown and the use of T_0 in Eq. 8.9 results in unrealistically high values of E_P (Eq. 5.23) when the surface is dry. The problem can be avoided by using the potential evaporation written as a combination-type equation (Eq. 5.26)).

(iii) The surface relative humidity is set to a constant value, i.e. $r_h = q_0/q^*(T_0)$ = constant, with the evaporation given by Eq. 5.40.

In (i) and (ii) in particular, an undesirable feature from a conceptual viewpoint is the dependence of q_0 on q, the specific humidity at the first atmospheric level, which gives rise to an implicit dependence on the choice of reference level. For many purposes, it is necessary to have an interactive scheme with a realistic dependence of the evaporation upon soil moisture, but one which is less detailed than the multi-level soil model approach based on solution of the one-dimensional moisture diffusion equations described in Section 5.3.

8.3.2 Interactive schemes

In Equation 8.8, x is made to depend on soil moisture status according to two approaches involving the moisture content of either a thick upper layer of soil acting as a bucket of water (Method 1) or a thin near-surface layer of soil (Method 2).

Method 1 ("bucket" method)
In this approach (Fig. 8.3), x is given by

$$x = \min(1, \eta_{bb}/\eta_k) \qquad (8.10)$$

where η_{bb} is the volumetric moisture content of the soil layer of thickness d_1 and η_k is a critical value; for $\eta_{bb} > \eta_k$ the surface behaves as if it were saturated. This is only one of many expressions for x. In fact, experiments have never given a consistent form for x, dealing, as they do, with real soils. This is the major limitation of Eqs. 8.8 and 8.10. The main virtue of this approach is its simplicity, giving values of evaporation constrained between wet and dry limits.

At the base of the layer, the moisture flux is taken to be zero, so applying Eq. 5.22 across the layer yields

$$\partial\eta_{bb}/\partial t = (P - E_0/\rho_w)/d_1 \qquad (8.11)$$

with P in units of $m\,s^{-1}$. Here, $0 \leqslant \eta_{bb} \leqslant \eta_s$, where η_s is the maximum (saturation) value above which runoff is assumed to occur. For time scales of a few days, d_1 in the range 0.5–1 m is typical of the values used in numerical models, with $\eta_k \approx 0.75\eta_s$. For longer time scales, drainage out through the bottom of the layer may become significant though it is often ignored or assumed to be part of the general runoff.

The main shortcoming of the bucket approach is that the evaporation does not respond to short-period occurrences of precipitation, which in fact change η_{bb}, and hence the evaporation, only gradually (Deardorff, 1978). For example, assuming $d_1 = 0.5$ m, rainfall of 1 cm in three hours would increase η_{bb} from 0.08 to 0.10 relative to a saturation value of 0.3, while E_0 (as calculated from Eq. 8.8) would only increase by 10 per cent of its value towards E_P (for a wet surface). The thick layer is analogous to a bucket that holds, say, 15 cm of water

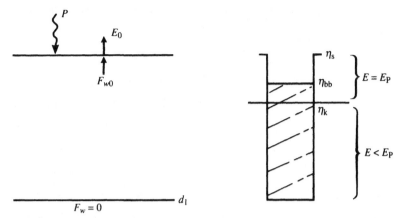

Fig. 8.3 Schematic single-layer, or bucket, model of soil moisture. At depth d_1, the water flux is assumed to be zero. The soil moisture content of this thick upper layer is η_{bb}, whose time evolution is governed by Eq. 8.11. The critical value, η_k, determines whether the evaporation rate is equal to or smaller than the potential rate. If η_{bb} equals the saturated value, η_s, and there is precipitation, runoff is assumed to occur (the bucket overflows).

at saturation (it overflows if more water is added from rainfall). It empties or fills at a rate governed by the time scale implicit in Eq. 8.11 (e.g. at a rainfall rate of 1 cm in three hours, with a bucket water depth of 15 cm, the time scale is 1–2 days). In reality, the actual surface becomes saturated after only a short period of rainfall, and dries after only a short period of high evaporation.

Method 2 ("force restore")
Near-surface soil moisture can be treated in an anologous way to surface temperature using a two-layer soil model (see Section 8.2.2). As with temperature, the model must represent the rapid response of the near-surface moisture to forcing by precipitation or evaporation and must also include a source of moisture from the deep soil to the surface when there is no precipitation (e.g. Deardorff, 1977)

The evaporation is evaluated as a fraction of E_P (Eq. 8.8), with E_P defined by either Eq. 5.23 or 5.26. The fraction x is now dependent on the near-surface soil moisture content η_g. To calculate this quantity, we consider a region of soil comprising two layers, each with uniformly distributed moisture and other properties (see Fig. 8.4). Applying a moisture budget to the thin surface layer, of depth d_2 and of moisture content η_g, gives (Eq. 5.22)

$$\rho_w \partial \eta_g / \partial t = (\rho_w P - E_0)/d_2 - F_{w1}/d_2. \tag{8.12}$$

The flux at depth $z' = d_2$ needs to be parameterized; its form can be deduced from Eq. 5.44, applied at $z' = d_2$ in finite difference form, i.e.

$$F_{w1}/\rho_w = -D_\eta(\eta_g - \eta_b)/\Delta z'' - K_\eta$$
$$= -(D_\eta/\Delta z'')(\eta_g - \eta_b + K_\eta \Delta z''/D_\eta)$$

suggesting that F_{w1} is given by

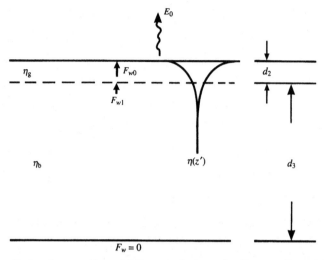

Fig. 8.4 Schematic two-layer soil model for calculation of soil water flux and near-surface moisture content. The thin surface layer, of thickness d_2, has a moisture content of η_g, whilst the thicker deep-soil or reservoir layer, of thickness d_3, has a moisture content of η_b. The vertical arrows indicate the magnitude and direction of the soil water fluxes corresponding to the left-hand η profile.

$$F_{w1} = \rho_w d_2 b_0 (\eta_g - \eta_{eq}) \tag{8.13}$$

with $\eta_{eq} \approx \eta_b - K_\eta \Delta z''/D_\eta$ where gravity effects are represented through the hydraulic conductivity K_η. The variable η_{eq} represents an equilibrium moisture content when the force of gravity balances capillarity forces. It is possible for the equilibrium values to be significantly lower than the mean moisture, particularly for coarse-grained soils such as sand. This is illustrated in Fig. 8.5. Thus, a sand sample with an initial mean moisture content of about 0.4 will lose water by gravitational drainage until the equilibrium value of about 0.2 is reached.

An equation for η_b can be derived readily by applying Eq. 5.22 to the thick soil layer, or thickness d_3. If the flux from the base of the layer is assumed zero,

$$\partial \eta_b / \partial t = F_{w1}/\rho_w d_3 \tag{8.14}$$

with F_{w1} given by Eq. 8.13. For real soils, and to take account of any sink of deep moisture due to vegetation (see Section 8.4), Eqs. 8.12 and 8.14 are rewritten in the form

$$\partial \eta_g / \partial t = a_0 (P - E_0/\rho_w) - b_0 (\eta_g - \eta_{eq}) \tag{8.15}$$

$$\partial \eta_b / \partial t = c_0 (\eta_g - \eta_{eq}) - E_{tr}/\rho_w d_3. \tag{8.16}$$

For future reference, a term representing the canopy transpiration rate E_{tr} is included in Eq. 8.16; this is set to zero for a bare soil surface. The coefficients a_0, b_0 and c_0 need to be derived for each soil type over a wide range of η_g and η_b values. This can be done by forcing the two-layer model with evaporation rates generated in a fully interactive, one-dimensional numerical model of the ABL and soil layer, based on solutions to the full moisture equations given in Section 5.3 (Eqs. 5.42–5.49) (e.g. Noilhan and Planton, 1989). The moisture

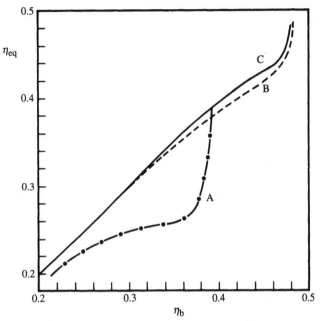

Fig. 8.5 Approximate relation between the equilibrium moisture content η_{eq} and the actual bulk moisture content η_b for three soil types: A, sand; B, silt loam; C, clay. From Noilhan and Planton (1989), *Monthly Weather Review*, American Meteorological Society.

values η_g and η_b are matched to the solutions for soil moisture content as a function of depth in the soil z', $\eta(z')$, and optimum values of the coefficients deduced. These are illustrated in Fig. 8.6 for three soil types (see Table A9). In general, the behaviour is consistent with three drying stages of an evaporating soil surface. In stage 1, where η_g is relatively large, a_0 is small and evaporation is at the potential rate and controlled by atmospheric factors. In stage 2, a_0 takes on intermediate values and the evaporation falls below potential rates. Here, evaporation is limited by soil moisture transport to the surface, and soil moisture diffusivity is a strong function of moisture content (Eq. 5.48 and Table A9). Finally, in stage 3, a_0 is large and evaporation is small, and is governed essentially by vapour transfer to the surface and by adsorption.

With η_g determined, the evaporation can be found from Eq. 8.8 with the factor x given as a function of η_g, or from Eq. 5.40, with r_h given as a function of η_g. In the former, this might be a relation analogous to Eq. 8.10, but more realistic functional forms can be found based on the calculations discussed in the previous paragraph. Since E_P, the potential evaporation defined as in Eq. 5.23, becomes a problematic concept for dry and drying soil conditions (when the daytime surface temperature is high), it is better to use the potential evaporation written in the form of Eq. 5.26 (i.e. E_L in Eq. 5.38) and to deduce the evaporation from

$$E_0/E_P = f_{st}(\eta_g/\eta_s). \tag{8.17}$$

The function f_{st} can be readily found from numerical simulations, which

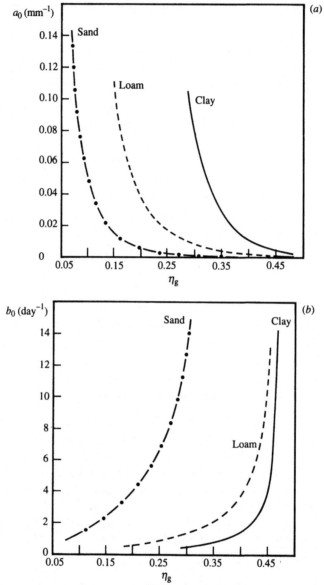

Fig. 8.6 (a) Dependence of the "force-restore" coefficient a_0 upon η_g for the three soil types in Fig. 8.5. (b) As in (a), but for the coefficient b_0. After Noilhan and Planton (1989), *Monthly Weather Review*, American Meteorological Society.

generate the curves shown in Fig. 5.9(b), for example. These are based on the averaged properties of real homogeneous soils. The application of Eq. 8.17, and similar relations for evaporation (e.g. Eq. 8.8), to model grid areas where average soil properties are unlikely to be known should be viewed with some caution. However, it does have the advantage of allowing evaluation in model simulations of the comparative behaviour of the atmosphere in the presence of quite dissimilar soils such as sand and clay.

8.4 Canopy parameterization

The presence of vegetation over an area of ground modulates the evaporation from the soil, and contributes further to the vertical flux of water vapour into the ABL through transpiration. A realistic canopy formulation must ultimately represent the effects of vegetation (averaged over the grid square in a three-dimensional numerical model) upon evaporation, energy partitioning, rainfall interception and soil moisture, as well as albedo and aerodynamic roughness. Inclusion of canopy effects allows the deep-soil moisture (in the root zone) to act as a source for evapotranspiration. Except when completely wet, the canopy foliage exerts some degree of physiological control upon the evaporation rate, and the surface humidity becomes indeterminate. Under these conditions, a canopy or surface resistance (conductance) is introduced into the evaporation formulation, and the resistance (conductance) concept is at the heart of most canopy models.

Single-level canopy formulations are the most appropriate for use in mesoscale and general circulation models, and these will be emphasized here. GCMs, for example, have the option of full canopy or bare soil grid coverage. For partial canopy cover, either as a sparse uniformly distributed cover or as full cover occupying only a fraction of the grid area, more complexity is involved. Detailed descriptions of multi-level canopy models, where the variation of fluxes through the canopy is evaluated, are beyond the scope of this book (see Finnigan and Raupach, 1987; Raupach, 1988).

For a complete vegetation cover, the simplest canopy model uses a constant r_s in Eqs. 5.31 and 5.37 for evaporation, with r_s values consistent with known bulk stomatal resistances, together with specified values of albedo and z_0. In contrast, a complex single-level canopy model contains many parameters with which to evaluate fluxes from the soil beneath the canopy, from open areas between the canopy elements as well as from the foliage itself (e.g. the SiB model of Sellers *et al.*, 1986 and the BATS model of Dickinson *et al.*, 1986). In addition the component fluxes are averaged over the grid area in some realistic way. With this approach, quite sophisticated treatments for the surface resistance can be used.

Let us consider the major requirements for simple, but realistic, canopy models by describing full-canopy formulations, for both isothermal and non-isothermal surfaces. In the isothermal case, both canopy and surface-soil layers are assumed to have the same temperature, but in the non-isothermal approach, the canopy and soil temperatures are allowed to differ. These models are essentially one-dimensional formulations applied to a grid area that might comprise either uniform cover (soil or vegetation) or an assumed distribution of patches of bare soil and full canopy.

8.4.1 Isothermal canopy/soil model

This model represents a combined overstorey and soil layer, and assumes that the soil surface and the foliage are at the same temperature T_{0f} (Noilhan and Planton, 1989). Figure 8.7 illustrates the main elements of this approach. The main task is to compute the turbulent fluxes of heat and water vapour, H_0 and

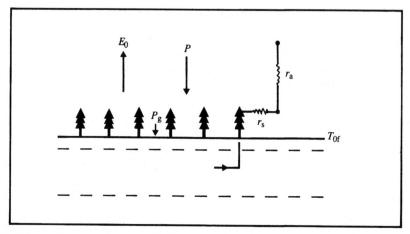

Fig. 8.7 Schematic representation of the main elements of an isothermal dense canopy model. Linked to both the atmosphere (via resistances r_s and r_a) and the deep soil (via evapotranspiration), the canopy and upper soil layer are assumed to be at a temperature T_{0f}. P_g is the precipitation reaching the soil surface.

λE_0 respectively, from the canopy to the air. The heat flux is given by the surface-layer relation Eq. 3.62, with θ_{0f} the canopy (surface) potential temperature:

$$H_0 = \rho c_p (\theta_{0f} - \theta)/r_{aH}. \tag{8.18}$$

For evaporation, a distinction is made between dry and wet canopies. Thus, for a wet canopy, evaporation is at the potential rate, given by Eq. 5.23:

$$E_{wc} = E_P = \rho[q^*(T_{0f}) - q]/r_{aV}. \tag{8.19}$$

For a dry canopy, with evapotranspiration under physiological control, evaporation is given by Eq. 5.31:

$$E_{dc} = E_{tr} = \rho[q^*(T_{0f}) - q]/(r_{aV} + r_s) \tag{8.20}$$

where r_s is defined for unit area of land surface. Use of Eq. 8.19 or 8.20 depends on the amount of liquid water residing on the foliage, due either to rainfall or dewfall. In addition, the soil moisture depends on precipitation reaching the ground, and so the effects of *canopy interception of rainfall* should be included. This is done by carrying an equation for m, the depth of water residing on the foliage. It follows that

$$\partial m/\partial t = P - E_{wc}/\rho_w - P_g \tag{8.21}$$

where P is precipitation at the top of the canopy (units of velocity, e.g. mm hr^{-1}) and P_g is the runoff from the interception reservoir, dependent upon the canopy interception of rain. We define m^+ as the maximum depth of water that can reside on the foliage per unit surface area; then, if $m > m^+$, rain is no longer intercepted but reaches the ground as throughfall. Values of P_g are set according to the following constraints:

$$\text{if } m < m^+ \quad \text{then} \quad P_g = 0,$$

$$\text{if } m \geq m^+ \quad \text{then} \quad P_g = P.$$

Now if $m = 0$, the canopy is assumed dry and the surface evaporation E_0 will be given by

$$E_0 = E_{dc} \tag{8.22a}$$

whilst if the canopy is completely wet, with $m \geq m^+$ (evaporation), or $m > 0$ (condensation),

$$E_0 = E_{wc}. \tag{8.22b}$$

Equation 8.22b can represent evaporation of water from the canopy (m decreasing) or condensation (m increasing) through dewfall ($E_{wc} < 0$).

As an aside, it is worth noting that in general circulation models, for example, where grid scales may be hundreds of kilometres, some account of spatial variability in rainfall interception is made. Thus, when $m < m^+$, it is assumed that only a fraction m/m^+ of the canopy is wetted. Evaporation from the canopy, averaged over the whole grid area, is then given by

$$E_0 = (m/m^+)E_{wc} + (1 - m/m^+)E_{dc} \tag{8.22c}$$

whilst, if condensation is occurring through dewfall with $E_{wc} < 0$, Eq. 8.22b applies, i.e. the whole canopy is assumed moistened.

In order to solve for H_0 and E_0, the turbulent fluxes at the top of the canopy, the canopy–soil surface temperature must be evaluated. This is done through a surface energy balance equation, with the soil–canopy system treated as a two-layer structure, and canopy temperature (T_{0f}) calculated from the force-restore method. Thus, in analogy with Eq. 8.2 for soil,

$$C_v \partial T_{0f}/\partial t = (R_N - H_0 - \lambda E_0) - \mu_v(T_{0f} - \bar{T}) \tag{8.23}$$

where C_v and μ_v are directly analogous to C_g and μ defined by Eq. 8.6 for bare soils. In fact, the heat capacity of a stand of vegetation is typically much smaller than that for soil and is in the range 0.1–1 mm of water per unit leaf area index (C_v lies between, say, 2000 and $2 \times 10^4 \, \mathrm{J\,m^{-2}\,K^{-1}}$ for LAI = 5, representing crops and forests) compared with 15–60 mm of water for soils (C_g lies between, say, 7×10^4 and $2.5 \times 10^5 \, \mathrm{J\,m^{-2}\,K^{-1}}$ for dry sand and wet clay).

In the above equations, specification of r_s is crucial to the evaluation of E_{dc}, and in addition m^+ needs to be set. For full canopy cover, both can be expected to depend on foliage density and hence upon leaf area index, L_A. In the case of the maximum canopy interception, a simple relation that covers a range of grasses, crops and tree species is (Dickinson, 1984)

$$m^+ \approx 0.2 L_A \tag{8.24}$$

with m^+ in mm of water. Typical values are 1 mm for a thick green canopy, suggesting that it would take approximately two hours for the foliage to become fully "saturated" from a dry state, at a rainfall rate of 0.5 mm per hour with no evaporation from the leaves.

In the case of r_s, identity with a bulk stomatal resistance is assumed. Numerous formulations exist that purport to represent the dependence of the stomatal resistance upon a number of factors, including soil moisture availability, solar radiation, absolute temperature, atmospheric water vapour deficit and carbon-dioxide concentration. The stomatal resistance for an individual leaf, r_{sti}, can be written in general functional forms as

$$r_{sti} = r_{sti}^+ F_1 F_2 F_3 F_4 F_5 \qquad (8.25a)$$

whilst the resistance for the whole canopy is assumed to have the form

$$r_{st} = r_s^+ F_1 F_2 F_3 F_4 F_5 \qquad (8.25b)$$

implying $r_s^+ = r_{sti}^+/L_A$. In the above, F_1 gives the radiation dependence, F_2 gives the soil moisture dependence (water stress), F_3 gives the vapour pressure deficit dependence, F_4 gives the temperature dependence and F_5 gives the CO_2 dependence. Detailed expressions for the functional forms for each of F_1 to F_5 are beyond the scope of the book (but see Jarvis, 1976; Sellers *et al.*, 1986). For many purposes only F_1 and F_2 are important, although for some species and in moderately extreme climates, the effects of high temperature and large vapour deficits can cause significant reduction in evapotranspiration (usually after local noon). Values of the unconstrained surface resistance r_s^+ are given in Table 5.1 and are typically $50\ \mathrm{s\,m^{-1}}$. They are generaly smaller than the minimum daytime values of r_{st} because of the influence of the F functions, whose product is always greater than unity even under optimal conditions.

How well does the isothermal canopy model perform? Figure 8.8 shows model simulations and related observations of the surface energy fluxes above a moderately dense forest in south-west France. The comparisons are remarkably good, but the parameterizations have yet to be fully tested under widely different surface and atmospheric conditions – including partial canopies, rainy periods and longer time scales.

8.4.2 Non-isothermal canopy/soil model

The main purpose of deriving the equations in this section is to obtain modified, more realistic, expressions for the vertical fluxes from the ground foliage system to the atmosphere. The major features of this modified model are shown schematically in Fig. 8.9. In this, the soil surface temperature T_g is allowed to

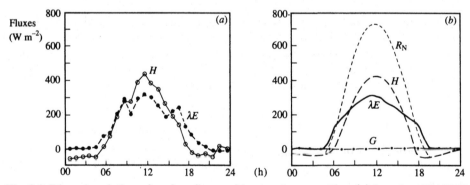

Fig. 8.8 Diurnal variation of surface fluxes of heat and evaporation (*a*) based on HAPEX observations over a forest with sandy soil in south-west France and (*b*) numerical simulations using an interactive atmosphere–isothermal-canopy model. The net radiation and soil heat flux are also shown in (*b*). After Noilhan and Planton (1989), *Monthly Weather Review*, American Meteorological Society.

differ from the foliage temperature T_f, resulting in fluxes from both underlying ground (E_g, H_g) and foliage (E_f, H_f) to the atmosphere (Deardorff, 1978). The total fluxes, H_0 and E_0, are given by

$$H_0 = H_f + H_g = \rho c_p (\theta_0 - \theta)/r_{aH} \tag{8.26}$$

$$E_0 = E_f + E_g = \rho (q_0 - q)/r_{aV}. \tag{8.27}$$

In the above, θ and q are properties evaluated at some level in the air above the canopy, and θ_0, q_0 are effective surface values that are related to an unknown canopy air-space temperature and humidity. In order to solve for the component fluxes, the temperature T_0 must be evaluated from a solution of the SEB equations applied at the canopy top (for T_f) and at ground level (for T_g). The individual fluxes can be described as

$$H_f = \rho c_p (\theta_f - \theta_0)/r_b \tag{8.28}$$

$$H_g = \rho c_p (\theta_g - \theta_0)/r_d \tag{8.29}$$

$$E_f = \rho (q_f - q_0)/r_b \tag{8.30}$$

$$E_g = \rho (q_g - q_0)/r_d \tag{8.31}$$

where the resistances r_b and r_d for heat and vapour transfer are assumed equal. Note that the component fluxes are written in terms of property differences between canopy or ground and canopy air space. Written formally in this way, Eqs. 8.26 and 8.27 imply

$$T_0 = \alpha_1 T_f + \alpha_2 T_g + \alpha_3 T \tag{8.32a}$$

$$q_0 = \alpha_1 q_f + \alpha_2 q_g + \alpha_3 q. \tag{8.32b}$$

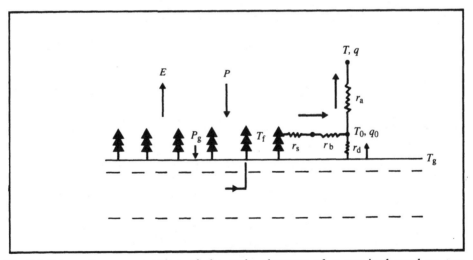

Fig. 8.9 Schematic representation of the main elements of a non-isothermal or two-component canopy model. Linked to the atmosphere (via resistances r_s, r_b and r_a), to the soil or undergrowth (via resistance r_d) and the deep soil (via evapotranspiration), the canopy and upper soil layer are at temperatures T_f and T_g. P_g is the precipitation reaching the soil surface.

These equations are required for determination of the effective surface quantities T_0 and q_0. The coefficients are given by $\alpha_1 = r_t/r_b$, $\alpha_2 = r_t/r_d$ and $\alpha_3 = r_t/r_{aH}$, with the resistances-to-transfer acting in parallel, and

$$r_t^{-1} = r_b^{-1} + r_d^{-1} + r_{aH}^{-1}. \qquad (8.33)$$

In practice, Eqs. 8.30 and 8.31 are not used because of the unknown humidities q_f and q_g. For E_f, an equation analogous to Eq. 8.22 is more appropriate, with

$$E_f = (m/m^+)E_{wc}^f + (1 - m/m^+)E_{dc}^f \qquad (8.34a)$$

if $E_{wc}^f \geq 0$, and, for condensation, with $E_{wc}^f < 0$,

$$E_f = E_{wc}^f. \qquad (8.34b)$$

The wet and dry canopy fluxes are given by

$$E_{wc}^f = \rho[q^*(T_f) - q_0]/r_b \qquad (8.35)$$

$$E_{dc}^f = \rho[q^*(T_f) - q_0]/(r_b + r_s). \qquad (8.36)$$

For E_g, an equation analogous to Eq. 8.17 is used, requiring soil moisture to be calculated as previously described.

The foliage temperature is evaluated from a solution of the diagnostic form of an SEB equation (see Section 8.2.1) appropriate to the foliage layer. This can be written, with canopy storage neglected,

$$(1 - \alpha_f)R_{s0} + \varepsilon_f R_{L0}^d + \varepsilon_g \sigma T_g^{\,4} - 2\varepsilon_f \sigma T_f^{\,4} \approx H_f + \lambda E_f \qquad (8.37)$$

where subscripts f and g refer to foliage and ground surfaces respectively. R_{s0} and R_{L0}^d are the shortwave and downwards longwave fluxes (Section 5.2) at canopy top. The two terms in T^4 reflect the expectation that the foliage layer emits longwave radiation both upwards and downwards (a loss) and receives longwave radiation from the underlying ground (a gain).

For the ground temperature, solution of a second SEB equation is required, either in diagnostic or prognostic form. In diagnostic form, the equation can be written

$$(1 - \alpha_g)R_{sg} + \epsilon_f \sigma T_g^{\,4} - G_0 \approx H_g + \lambda E_g \qquad (8.38)$$

where the shortwave radiation at the ground beneath the canopy, R_{sg}, needs to be parameterized in terms of R_{s0} and the decrease in flux through the canopy. If G_0 is calculated by the force-restore method, Eq. 8.38 must be recast in prognostic form as described in Section 8.2.2.

For the equations to be solved, and heat fluxes evaluated, knowledge of the aerodynamic resistances r_{aH}, r_{aV}, r_b and r_d is required. The first two are simply the surface-layer resistances described in earlier sections; r_b is like a boundary-layer resistance between the leaves and foliage air space and must be formulated accordingly, and, finally, r_d is a resistance to transfer between the ground and canopy air space. Both r_b and r_d can be expected to be on the order of, or greater than, r_a (see e.g. Deardorff, 1978).

8.4.3 Models of partial canopy cover

The canopy formulations discussed above are best applied to vertical exchange from an area of uniform, dense vegetation cover. Their extension to sparse or partial canopy cover is not straightforward, although progress has been made in recent times (Shuttleworth and Wallace, 1985; Shuttleworth and Gurney, 1990). In the real world, partial canopy cover is structured in an infinite number of ways, but two extreme structures suffice to illustrate the problem. In the first, a low-density uniform cover exists across the grid area; in the second, the area is comprised of patches of dense canopy and bare soil. From a grid area perspective, the fractional canopy cover σ_f may be the same for each. However, there is obviously a great deal of difference between E_f and E_g when they are the foliage and ground evaporation in a one-dimensional layered system (uniform sparse cover), and when they represent spatially separate areas. Yet in many schemes used in large-scale models, no allowance is made for such differences in the sub-grid distribution of vegetation.

Irrespective of the distribution, the simplest schemes evaluate the area-averaged turbulent fluxes as contributions from both canopy and bare soil. One approach assumes a linear combination of these component fluxes, whilst also incorporating suitably modified equations for surface temperature (the SEB), intercepted water and individual fluxes. The total fluxes away from the surface are then given by relations such as

$$H_0 = \sigma_f H_{fc} + (1 - \sigma_f) H_{g0} \tag{8.39}$$

$$E_0 = \sigma_f E_{fc} + (1 - \sigma_f) E_{g0} \tag{8.40}$$

if the isothermal canopy model is used. Here, H_{fc} and E_{fc} are hypothetical fluxes from a full canopy occupying a fraction σ_f of the grid area; H_{g0} and E_{g0} are fluxes from bare ground. The above relations ensure the correct asymptotic approach as σ_f tends to zero and unity.

Highly detailed, single-level, non-isothermal canopy models have been developed in the last few years for incorporation into GCMs. Their performance should exceed that of their isothermal counterparts, particularly where temperature differences between foliage and ground are likely to be significant, that is, in partial canopies where shortwave radiation penetrates to the ground. To date, detailed comparisons between model simulations and observations have been few. Those shown in Fig. 8.10 illustrate the improvement found when replacing a simple bucket hydrology scheme (Fig. 8.10(*b*)) with a non-isothermal canopy model (Fig. 8.10(*a*)), though we cannot say whether a simpler canopy scheme would have produced as great an improvement.

8.4.4 Comments on canopy models

It is not yet clear what level of detail is required in a canopy scheme for use in a numerical model of the atmosphere (whether boundary-layer, mesoscale or general circulation model). Given the range of uncertainties implicit in any model, including the many simplifying assumptions, a single-layer description of the canopy and a two-layer description of the soil probably represent the *minimum* level of detail required for many purposes. In such an approach, bulk stomatal (surface) resistance, roughness length, albedo and canopy interception storage capacity need to be specified. The surface resistance should include dependence on solar radiation and bulk soil moisture, so that transpiration

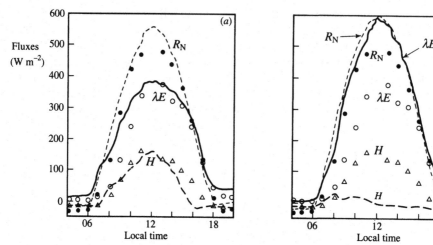

Fig. 8.10 Observed (data points) and simulated (curves) 30-day mean surface energy balance components for a tropical forest. Simulations were performed (*a*) in a GCM with a non-isothermal canopy model and (*b*) in a GCM with no canopy, and "bucket" hydrology only. After Sato *et al.* (1989), *Journal of Atmospheric Sciences*, American Meteorological Society.

becomes small when soil moisture decreases below some "wilting" value. For the soil, several moisture-dependent coefficients are required, together with a functional relationship between evaporation and soil moisture. One cannot yet say whether the SiB or BATS models on the one hand, or a simple isothermal canopy model on the other, are suitable as single-layer canopy schemes. Both approaches must take into account the effects of horizontal heterogeneity on grid-area averaging and the impact of complex terrain on ABL flow (Chapter 9).

What about the relative merits of single- and multi-layer canopy models? Broadly speaking, single-layer models are appropriate where one is concerned with vegetation as a permeable lower boundary to the atmosphere in systems with a length scale much larger than that of the vegetation itself. For example, they would be appropriate in studies of the ABL and the atmospheric general circulation, or in hydrological modelling of medium-scale to large-scale catchments. Such models allow evaporation to be determined as if the plant canopy were no more than a partly wet plane at the lower boundary of the atmosphere. This notional plane – often referred to as a "big leaf" – is ascribed a physiological and aerodynamic resistance to water vapour transfer. In contrast, multi-layer models are appropriate when it is necessary to resolve detail within the canopy, either because the detail is important in its own right or because the height scale of the canopy is comparable to that of the system under study. For example, they may be used to describe the interaction between microclimate and physiology, or the hydrology of small forested catchments. These models aim to describe not only the evaporation from the entire canopy, but also the partitioning of the evaporation between various parts of the canopy – soil, understorey and crown – together with other aspects of the canopy microclimate.

In the above three sections, a number of schemes have been described that allow surface properties, in particular surface temperature and humidity, to be evaluated. We now turn to the problem of computing the turbulent surface fluxes. Most model schemes are based on Monin–Obukhov theory for the surface layer, and thus on formulations described in Chapter 3.

8.5 Surface fluxes

Basic formulations for the surface fluxes can be found in Section 3.3 (surface-layer theory), Section 3.4 (generalized boundary-layer theory) and for evaporation specifically, in Section 5.3. These, together with the surface energy balance equation (Section 5.1), form the basis of the flux evaluations in most numerical models of the atmosphere. Methods for evaluating the surface values of temperature and humidity upon which the fluxes depend have been discussed above.

8.5.1 Iteration procedures for all fluxes

The turbulent fluxes at the surface are characterized by the turbulent scales u_{*0}, θ_{*0} and q_{*0}. In many model computational procedures, these scales are computed iteratively from a "first guess", with the solutions depending on specified forms of the Ψ stability functions and on solutions for the surface temperature and humidity. Computationally, this double iterative method is time consuming and an alternative analytical approach is available for the stability-dependent functions.

8.5.2 Analytical form of the transfer coefficients

The bulk transfer relations for surface fluxes can be written as follows:

$$u_{*0}^2 = C_{DN} F_M(z/z_0, Ri_B)u^2 \tag{8.41}$$

$$u_{*0}\theta_{v*0} = C_{HN} F_H(z/z_0, z/z_T, Ri_B)u(\theta_v - \theta_0) \tag{8.42}$$

$$u_{*0}q_{*0} = C_{EN} F_E(z/z_0, z/z_q, Ri_B)u(q - q_0) \tag{8.43}$$

with Ri_B defined by Eq. 3.45 and C_{DN}, C_{HN}, C_{EN} given by Eqs. 3.43 and 3.48 (we take $z_T = z_q$ and $C_{HN} = C_{EN}$). In the above, surface-layer wind coordinates are used, so that u is the surface-layer (total) wind. The functions F represent the ratios C_D/C_{DN}, C_H/C_{HN} and C_E/C_{EN}. When written in analytical form the fluxes can be calculated without the need for double iteration of the solutions. In addition, allowance for the correct asymptotic behaviour can be incorporated; for example, use of the Monin–Obukhov surface-layer similarity theory does not give the correct free-convection behaviour since the fluxes are u_* dependent (as $u \to 0$, the fluxes become indeterminate). In essence, the curves shown in Fig. 3.7 need to be represented analytically. Simple analytical formulae that approximate the known C/C_N behaviour at finite wind speeds and have the desired asymptotic form (requiring $F \propto Ri_B^{1/2}$ as $u \to 0$ to give a finite heat flux) are

$$Ri_B < 0 \qquad F_{M,H} = 1 - a\,Ri_B/(1 + b_{M,H}|Ri_B|^{1/2}) \tag{8.44}$$

$$Ri_B > 0 \qquad F_{M,H} = (1 + c\,Ri_b)^{-2}. \tag{8.45}$$

For an assumed set of Ψ functions (based on the results of Businger *et al.*, 1971), and $z_0 = z_T$, Louis (1979) derived the following values for the coefficients: $a = 2c = 9.4$; $b_M = 7.4 C_{DN} a(z/z_0)^{1/2}$ and $b_{H,E} = 0.72 b_M$. For the assumed set of Ψ functions used in the evaluations of the curves given in Fig. 3.7, numerical values of the constants are slightly different. In addition, allowance must be made for differences between z_0 and z_T. Thus, we take

$$b_M = b_M^* C_{DN} a(z/z_0)^{1/2} \text{ and } b_H = b_H^* C_{HN} a(z/z_T)^{1/2}$$

with $a = 2c = 10$. Least-squares fitting in the range $-5 < \zeta < 0$ gives, for $z_0 = z_T$, $b_M^* = 4.9$ and $b_H^* = 2.6$, and for $z_0/z_T = 7.4$, $b_M^* = -0.34 C_{DN}^{-1/2} + 13.7$ and $b_H^* = -0.18 C_{DN}^{-1/2} + 6.3$. With these values, the analytical expressions given by Eqs. 8.44 and 8.45 provide the best fit to the curves in Fig. 3.7 giving F_M and F_H to within 5 to 10 per cent of their true values in the range $-5 < \zeta < 1$.

Calculation of the surface fluxes is the most important component of a boundary-layer parameterization scheme in a large-scale model. In addition, it may be necessary to evaluate

(i) the depth h of the ABL, in order that momentum, heat and water vapour can be distributed vertically;
(ii) the values of the turbulent fluxes and other turbulent properties within the ABL, in order that the vertical flux divergence can be calculated.

8.6 Rate equation for ABL depth

For the clear CBL, a rate equation for the ABL depth was discussed in Chapter 6 (Section 6.1.4) for simple encroachment and entrainment cases, based purely on thermodynamical considerations. For simple entrainment, Eq. 6.18 is a suitable relation, but in more complex situations where shear and finite interfacial layer thickness may be important, or where clouds are present, this is unsuitable. A more elaborate form was developed by Deardorff (1974) for use in atmospheric models. In full form, and with the assumption of horizontal homogeneity relaxed, this can be written

$$\partial h/\partial t - w_h = w_{e0} X/Y - u_h \partial h/\partial x - v_h \partial h/\partial y \qquad (8.46)$$

which allows for horizontal variations in h, with u_h and v_h the mean horizontal wind components at h. In Eq. 8.46, w_{e0} is the growth rate due to entrainment by convective penetration given by Eq. 6.18, i.e.

$$w_{e0} = \partial h/\partial t = (1 + 2\beta)\overline{(w'\theta_v')}_0/\gamma_\theta h \qquad (8.47)$$

where β is the entrainment flux ratio, and

$$X = 1 + 1.1(u_{*0}^3/w_*^3)(1 - 3|f|h/u_{*0}) \qquad (8.48)$$

$$Y = 1 + 9w_*^2[\gamma_\theta(g/\theta_v)h^2]^{-1}(1 + 0.8u_{*0}^2/w_*^2). \qquad (8.49)$$

The reader should note that in Eq. 8.47 the factor $(1 + 2\beta)$ replaces a numerical value of 1.8 to be found in Deardorff's paper. The rate equation (8.46) serves as a useful interpolation formula, combining the following asymptotic cases (we take $w_h = 0$ and ignore advection).

(i) For $u_{*0} \to 0$, we have entrainment by convective penetration only. Thus, $X \to 1$ and $Y \to 1 + 9w_*^2[\gamma_\theta(g/\theta_v)h^2]^{-1} \approx 1$ for entrainment at a strong overlying inversion, giving $\partial h/\partial t \approx w_{e0}$. With $\gamma_\theta = 5\,\mathrm{K\,km^{-1}}$, $h = 1000\,\mathrm{m}$ and $(\overline{w'\theta_v'})_0 = 0.2\,\mathrm{m\,s^{-1}\,K}$, then $\partial h/\partial t \approx 0.056\,\mathrm{m\,s^{-1}}$ or $200\,\mathrm{m\,hr^{-1}}$ (see Section 6.1.4).

For the case where growth is through a near-adiabatic layer ($\gamma_\theta \approx 0$), and provided u_{*0}^2/w_*^2 is small, $X \to 1$ and $Y \to 9w_*^2[\gamma_\theta(g/\theta_v)h^2]^{-1}$ whence $\partial h/\partial t \to 0.11(1 + 2\beta)w_* \approx 0.15w_*$ (about $0.28\,\mathrm{m\,s^{-1}}$ using the above parameter values).

(ii) In the absence of buoyant energy, growth due to shear through a neutral layer, with both $(\overline{w'\theta_v'})_0$ and $\gamma_\theta \to 0$, is given by

$$\partial h/\partial t \approx 0.2u_{*0}(1 - 3|f|h/u_{*0}) \tag{8.50}$$

for the horizontally homogeneous case. Equation 8.50 is thus appropriate for the neutral ABL, and as $h \to 0.33u_{*0}/|f|$, the depth of the equilibrium neutral ABL is recovered with $\partial h/\partial t \to 0$ as required (the value of the numerical constant is discussed in Appendix 3). In the absence of rotation effects ($|f| \to 0$) $\partial h/\partial t \to 0.2u_{*0}$, close to values found in neutral laboratory flows (e.g. Lundgren and Wang, 1973).

For the stable or nocturnal situation, Eq. 6.70 represents the appropriate rate equation in the horizontally homogeneous case, and in equilibrium, h is given by Eq. 6.55.

For the cloudy ABL, Eq. 7.32 represents a rate equation suitable for a slab model, though it neglects shear contributions to the entrainment.

8.7 Turbulence closure schemes

We consider here the closure schemes used in association with both the ensemble-averaged equations, and the volume-averaged equations used in LES modelling. Their principle aim is to allow calculation of vertical fluxes throughout the turbulent ABL, covering the whole range of stabilities.

8.7.1 First-order closure (ensemble models)

First-order closure is often referred to as K-closure and has already been discussed in Chapter 2 (Section 2.5). It uses a flux-gradient relation such as that expressed by Eqs. 2.56–2.59, and transfers the problem of the unknown covariances to that of specifying, in a physically realistic manner, the eddy diffusivities. An example of this approach can be found in the Ekman solution of the momentum equations, described in Section 3.1.2. In this, K was simply set to a constant value.

For the surface layer, Prandtl's mixing length approach (Section 3.1.1) is an evident improvement on a constant K, though the Monin–Obukhov similarity theory allows K for any property to be specified in terms of a range of flow properties. Such relations are fully covered in Chapter 3. Thus

$$K_s = ku_{*0}z/\Phi_s(\zeta) \tag{8.51}$$

for any property s, where the Φ functions and Monin–Okukhov stability parameter ζ are defined in Section 3.3.1. Above the surface layer, specification of K is less rigorous, but usually follows one of three approaches: (*a*) prescribing K values, (*b*) prescribing K-profile shapes, (*c*) prescribing K dynamics. The first of these is relatively trivial – the derivation of the Ekman spiral, where K is set constant, is a good example (Section 3.1.2). The third approach is discussed in the next section, where K is related to turbulent kinetic energy. The second approach is very widely used, and is associated with numerous assumptions and approximations. Thus, the relation

$$K_s = l^2|\partial\mathbf{V}/\partial z|f_s(Ri) \tag{8.52}$$

characterizes an implicit approach where K is allowed to depend on flow structure and stability, represented here by the gradient Richardson number Ri. This form for K_s arises from consideration of the TKE balance between local dissipation and production (shear and buoyancy), with suitable scaling of the terms. In Eq. 8.52, both l, the mixing length, and the stability function $f_s(Ri)$ must be specified. Asymptotically, $l \to kz$ in the surface layer, and $f_s(Ri) \to 1$ in neutral conditions.

There are numerous semi-empirical forms for these two quantities. In unstable conditions, a typical formulation is

$$f_s(Ri) = 1 - a\, Ri\, (1 + b\, Ri^{1/2})^{-1} \tag{8.53a}$$

and in stable conditions,

$$f_s(Ri) = (1 + c\, Ri)^{-2}. \tag{8.53b}$$

In the above, a, b and c are empirical constants that may differ for momentum, heat and mass transfer. The mixing length in both stable and unstable conditions is written as (e.g. Blackadar, 1962)

$$l = kz(1 + kz/\lambda)^{-1}, \tag{8.54}$$

where the asymptotic mixing length λ is an adjustable parameter. An alternative empirical formulation for the stability functions sets

$$f_s(Ri) = (1 - 18\, Ri)^{1/2} \tag{8.55a}$$

and

$$f_s(Ri) = 1.1(1 - Ri/Ri_c) \tag{8.55b}$$

for unstable and stable conditions respectively, where Ri_c is the critical Richardson number. In this example, the mixing length has a slightly different form to that given by Eq. 8.54, e.g. $l = kz$ for $z < 200$ m; $l = 80$ m for $z > 200$ m. The assumption in the above formulations is that the turbulence can be characterized by a mixing length that varies linearly with height in the surface layer and is limited by a constant value λ aloft.

Rather than utilize a specific mixing length in stable conditions, numerical simulations of the NBL have provided a similarity-type relation which, in contrast to the above formulations, requires knowledge of the ABL depth. The diffusivity is written in the form of a non-dimensional K, as a function of z/L and z/h (Brost and Wyngaard, 1978), i.e.

$$K/ku_{*0}h = f(z/h)/(1 + 5z/L) \tag{8.56}$$

with

$$f(z/h) = (z/h)(1 - z/h)^{3/2} \tag{8.57}$$

based solely on curve fitting. If h is known, Eq. 8.56 serves as an alternative formulation to Eq. 8.55b: it is worth noting that Eqs. 8.56 and 8.57 give $K = 0$ for $z = 0$ and $z = h$, and $K \rightarrow ku_{*0}z$ when $z/h \rightarrow 0$ and $z/L \rightarrow 0$.

An alternative scheme to that represented by Eq. 8.52 uses an explicit representation for K in unstable conditions, which asymptotically approaches the surface-layer form at small z and becomes zero at the top of the ABL. This scheme is popular in mesoscale models of the atmosphere, and can be written (O'Brien, 1970)

$$K(z) = K(h) + A_1[K(h_s) - K(h) + (z - h_s)A_2] \tag{8.58}$$

where

$$A_1 = (h - z)^2/(h - h_s)^2 \tag{8.59a}$$

$$A_2 = (\partial K/\partial z)_{z=h_s} + 2[K(h_s) - K(h)]/(h - h_s). \tag{8.59b}$$

Here h_s is the height of the surface layer. Equation 8.58 is valid for $z > h_s$, so that $K(h_s)$ must satisfy Eq. 8.51 when $z = h_s$. The disadvantage of this approach lies in the lack of any major dependence of K upon flow properties, except through the influence of $K(h_s)$ in Eq. 8.58.

Figures 8.11 and 8.12 give sample K_M profiles based on Eqs. 8.56, 8.57 for stable conditions and Eqs. 8.58, 8.59 for unstable and neutral conditions. They are physically realistic in the sense that maximum values occur in mid-regions of the ABL, decreasing towards small values at top and bottom. In general, maximum values of K_M (of magnitude several times that at the top of the surface layer) occur at a height of about $h/3$, of magnitude in neutral conditions $\approx 0.06u_{*0}h$ ($\approx 18 \text{ m}^2 \text{ s}^{-1}$ with $u_{*0} = 0.4 \text{ m s}^{-1}$ and $h = 750 \text{ m}$). In moderately stable conditions, the maximum value reduces to $0.03u_{*0}h$ ($\approx 1.5 \text{ m}^2 \text{ s}^{-1}$ with $u_{*0} = 0.25 \text{ m s}^{-1}$ and $h = 200 \text{ m}$). In the CBL, maximum values of K_M in mid-levels are about $0.05w_*h$ ($\approx 150 \text{ m}^2 \text{ s}^{-1}$ with $w_* = 2 \text{ m s}^{-1}$ and $h = 1500 \text{ m}$), much greater than the neutral values, whilst maximum K_H values are likely to be several times greater still.

As discussed in Chapter 6, K-theory is likely to fail in strongly buoyant conditions (the CBL) where length scales of the energy-containing eddies are comparable with the depth of the layer over which transfer is occurring. In terms of a flux-gradient relation for heat transfer, some schemes used in large-scale models in particular have used a modified form to allow for countergradient fluxes and to avoid the use of negative K. Thus, for use in the

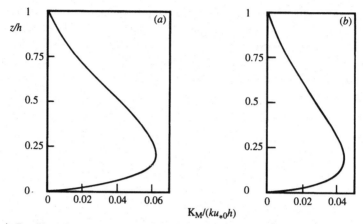

Fig. 8.11 (a) Profile of the dimensionless eddy viscosity in the stable ABL based on Eq. 8.56, with $f(z/h) = (z/h)(1 - z/h)^{1.5}$ and $h/L = 1.4$. (b) As in (a), for $h/L = 2.3$. Based on Brost and Wyngaard (1978).

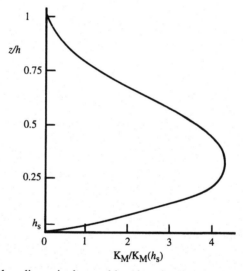

Fig. 8.12 Profile of the dimensionless eddy viscosity in the neutral ABL based on Eqs. 8.58 and 8.59, with $h_s = 0.04h$. From Pielke (1984), by kind permission of Academic Press and the author.

CBL above the surface layer, the heat flux is calculated from

$$\overline{w'\theta_v'} = - K_H(\partial\theta_v/\partial z - \gamma_\theta') \tag{8.60}$$

with the lapse-rate correction γ_θ' usually taken as $\approx 0.0007\,\mathrm{K\,m^{-1}}$ and K_H calculated from a relation such as Eq. 8.58. It is of interest to note that Eq. 8.60 is consistent with Eq. 2.65b, derived from the simplified prognostic equation for $\overline{w'\theta'}$. In this case, we see that identity requires

$$\gamma_\theta' = (g/\theta_v)\overline{\theta'\theta_v'}/\overline{w'^2}. \tag{8.61}$$

For a CBL of depth 1000 m, and a surface buoyancy flux of 250 W m^{-2}, values of the variances can be estimated from the normalized profiles given in Chapter 3 (Figs. 3.21, 3.22). Typically, $\sigma_\theta \approx 0.15$ K and $\sigma_w \approx 0.85$ m s^{-1}, whence $\gamma_\theta' \approx 0.001$ K m^{-1}, close to that specified above. If a closure scheme uses Eq. 8.60 then, for consistency, similar formulae for other scalar fluxes should be used, with

$$\overline{w'c'} = -K_c(\partial c/\partial z - \gamma_c'). \tag{8.62}$$

For flow simulations in the CBL in particular, K-dynamics based on local higher-order closure (usually to second order in the atmosphere) on the one hand (e.g. Wyngaard, 1982), and an approach using non-local closure on the other (e.g. Fiedler, 1984), are definite alternatives to first-order local closure (K-theory).

8.7.2 One-and-a-half-order closure (ensemble models)

Equations 2.62b and 2.65b suggest that the diffusivity, at least above the surface layer, might be better determined if it is related directly to the turbulent intensity. In fact, it is physically realistic to expect that K will be closely related to the TKE ($\bar{e}$), and the usual assumption is then

$$K = \Lambda \bar{e}^{1/2}. \tag{8.63}$$

This approach requires a set of empirical length scales Λ to be specified for each property; note that the TKE $\bar{e}$ is defined by Eq. 2.71. In order to solve the mean equations, use of Eq. 8.63 requires solution of the TKE equation (for the horizontally homogeneous case, as given by Eq. 2.74a) and hence approximations for the unknown pressure covariance, third-order and dissipation terms.

In the simplified form of the TKE equation (Eq. 2.74b), where we are concerned only with the combined pressure correlation and third-order term $\overline{w'(p' + e)}$ and the dissipation rate ε, it is usual to make a downgradient-diffusion assumption, and set

$$\overline{w'(p' + e)} = -K_e \partial \bar{e}/\partial z \tag{8.64}$$

with K_e given by Eq. 8.63. For the dissipation rate ε, the approximation is based on scaling arguments consistent with balanced conditions in the neutral surface layer, and it is usual to set (Mellor and Yamada, 1974)

$$\varepsilon = \bar{e}^{3/2}/\Lambda_1. \tag{8.65}$$

The basis of Eq. 8.65 is that ε is determined by the rate of energy transfer across the spectrum from the energy-containing eddies (the energy cascade), even though viscous dissipation occurs at the smallest scales. Thus, the dissipation rate can be expressed in terms of large-eddy properties, in particular the length scale Λ_1.

The additional equation to be carried can be written as (cf. Eq. 2.74a)

$$\partial \bar{e}/\partial t = K_M[(\partial u/\partial z)^2 + (\partial v/\partial z)^2] - (g/\theta_v)K_H \partial \theta_v/\partial z$$
$$+ \partial/\partial z(K_e \partial \bar{e}/\partial z) - \bar{e}^{3/2}/\Lambda_1 \tag{8.66}$$

with the diffusivities given by Eq. 8.63: $K_M = \Lambda_m \bar{e}^{1/2}$, $K_H = \Lambda_h \bar{e}^{1/2}$, $K_e = \Lambda_e \bar{e}^{1/2}$. It is fundamental to higher-order closure schemes that all length scales are everywhere proportional to each other; thus, it is usual to set

$$(\Lambda_1, \Lambda_m, \Lambda_h, \Lambda_e) = (a_1, S_m, S_h, S_e)\, l \qquad (8.67)$$

where l is a master turbulent length scale (Mellor and Yamada, 1982). In the above, the various constants and functions can be determined from experimental data, usually in flow situations where turbulent production and dissipation approximately balance. It is found that $a_1 \approx 5$, S_e is usually taken as equal to S_m, and both S_m and S_h can be specified as functions of stability (becoming zero at the critical Richardson number).

The length scale l needs to coincide with Prandtl's mixing length close to the surface, to give $l = kz$. The main weakness with this approach lies in the use of l for all flow simulations. Two quite different schemes that allow l to be determined can be identified. In the first, l is represented algebraically by Eq. 8.54 with the asymptotic mixing length λ specified accordingly (e.g. Yamada and Mellor, 1975). In the second scheme, l is determined with the aid of a prognostic equation for l; this is based on the integral of a two-point correlation function (Rotta, 1951), and requires some quite complicated closure assumptions. The equation is usually cast in terms of the variable $l\bar{e}$ (Mellor and Herring, 1973). It can be seen as an alternative to yet another approach in which one tries to avoid the problem of specifying numerous length scales implied by Eq. 8.67. This approach – referred to as $\bar{e}-\varepsilon$ closure – involves an additional prognostic equation for ε itself, usually in highly parameterized form (Lumley, 1979; Zeman, 1981). Obtaining ε in this way does allow the length scales required in Eq. 8.63 to be determined through equations like Eq. 8.65 that involve prognostic variables.

8.7.3 Second-order closure (ensemble models)

Second-order closure has been used in turbulence calculations for at least two decades, initially for shear flows and then progressively for ABL modelling in general (see Zeman, 1981). Part of the motivation for any higher-order closure is the hope that "if a crude assumption for nth moments predicts $(n-1)$th moments adequately, perhaps a similar assumption for $(n+1)$th moments will predict nth moments just as well" (Lumley and Khajeh-Nouri, 1974).

The second-order approach avoids the need to parameterize the fluxes using K coefficients in flux-gradient relations. Rather, the covariances and variances are evaluated by solving their respective rate equations, Eq. 2.60 for $\overline{u_i'u_k'}$, Eq. 2.63 for $\overline{u_i'\theta'}$ and Eq. 2.67 for the temperature variance. Note that, with $k = i$, Eq. 2.60 reverts to the TKE equation (Eq. 2.74a in simplified form) discussed above in Section 8.7.2. As with one-and-a-half-order closure, the complete set of equations must be closed by parameterization of the pressure covariance, third-order transport and molecular dissipation terms. Usually, second-order closure involves downgradient-diffusion approximations for the third moments, and length-scale assumptions for the dissipation rates. We here present some of the simplified closure schemes, and refer the reader to Zeman (1981), Mellor and Yamada (1982) and Moeng and Wyngaard (1989) for a more detailed discussion.

Molecular dissipation

These terms are important only in the TKE and temperature variance equations. For TKE (Eq. 2.74a), the rate of viscous dissipation ε is given by Eq. 8.65 above, while χ in Eq. 2.69a can be similarly approximated as

$$\chi = \bar{e}^{1/2}\overline{\theta'^2}/\Lambda_2 \tag{8.68}$$

where $\Lambda_2 = a_2 l$.

Third-order moments

(i) The quantity $\partial(\overline{u_i' u_j' u_k'})/\partial x_j$ occurs in the equation (2.60) for the covariance $\overline{u_i' u_j'}$, and in the simplified $\overline{u'w'}$ equation (Eq. 2.61a) as $\partial(\overline{u'w'^2})/\partial z$. These transport terms tend to be negligible in the neutral and stable ABL, but significant in convective conditions. Consideration of the rate equation for the third-order moments reveals that, if buoyancy influences are neglected, a relatively simple form can be used, i.e.

$$\overline{u_i' u_j' u_k'} = \Lambda_3 \bar{e}^{1/2} (\partial\overline{u_i' u_j'}/\partial x_k + \partial\overline{u_i' u_k'}/\partial x_j + \partial\overline{u_j' u_k'}/\partial x_i) \tag{8.69a}$$

which is the anticipated flux-gradient form, with $\Lambda_3 \propto \Lambda_e$. For use in the simplified $\overline{u'w'}$ equation, Eq. 8.69a becomes

$$\overline{u'w'^2} = 2\Lambda_3 \bar{e}^{1/2}\, \partial\overline{u'w'}/\partial z \tag{8.69b}$$

(ii) With $k = i$, the quantity $\partial(\overline{u_i'^2 u_j'})/\partial x_j$ occurs in Eq. 2.72 for the TKE, and in the simplified form (Eq. 2.74a) as $\partial\overline{w'e}/\partial z$. The latter term, for example, is approximated in gradient-diffusion form in Eq. 8.64 above.

(iii) The quantity $\partial\overline{u_i' u_j' \theta'}/\partial x_j$ occurs in Eq. 2.63 for the covariance $\overline{u_i' \theta'}$ and in the simplified $\overline{w'\theta'}$ equation (Eq. 2.64a) as $\partial(\overline{w'^2\theta'})/\partial z$. In analogy with the velocity triple correlation, a relatively simple form for closure is

$$\overline{u_i' u_j' \theta'} = \Lambda_4 \bar{e}^{1/2}(\partial\overline{u_i' \theta'}/\partial x_j + \partial\overline{u_j' \theta'}/\partial x_i) \tag{8.70a}$$

with

$$\overline{w'^2\theta'} = 2\Lambda_4 \bar{e}^{1/2}\, \partial\overline{w'\theta'}/\partial z. \tag{8.70b}$$

(iv) The quantity $\partial(\overline{\theta'^2 u_j'})/\partial x_j$ occurs in Eq. 2.67 for the temperature variance, and in the simpler form (Eq. 2.69a) as $\partial(\overline{w'\theta'^2})/\partial z$. A suitable closure consistent with the above relations is

$$\overline{u_j' \theta'^2} = \Lambda_5 \bar{e}^{1/2}\, \partial\overline{\theta'^2}/\partial x_j \tag{8.71a}$$

with

$$\overline{w'\theta'^2} = \Lambda_5 \bar{e}^{1/2}\, \partial\overline{\theta'^2}/\partial z. \tag{8.71b}$$

Here, $\Lambda_4 = lS_{u\theta}$ and $\Lambda_5 = lS_\theta$, with $S_{u\theta}$ and S_θ specified functions of stability (see Moeng and Wyngaard, 1989).

Pressure covariance

Quite extensive analysis is involved in deriving suitable approximate forms for the pressure covariance terms appearing in the velocity and temperature covariance equations. Simplified versions of these closure approximations are acceptable for many applications and utilize, in particular, the energy redistribution hypothesis of Rotta and "return-to-isotropy" arguments (Mellor and Yamada, 1982). This is the case in the following relation for the pressure term appearing in Eq. 2.60 for the covariance $\overline{u_i'u_k'}$, i.e.

$$\rho^{-1}(\overline{u_k'\partial p'/\partial x_i} + \overline{u_i'\partial p'/\partial x_k}) = (\bar{e}^{1/2}/\Lambda_6)[\overline{u_i'u_k'} - (2/3)\delta_{ik}\bar{e}] \quad (8.72)$$

the simplified one-dimensional case of which is represented by Eq. 2.62a, with the time scale τ_1 identical with $\Lambda_6/\bar{e}^{1/2}$. In the TKE equation (Eq. 2.72), the term $\partial\overline{p'u_i'}/\partial x_i$ is interpreted as pressure diffusion and is assumed to behave as velocity diffusion, i.e. it can be approximated in a similar fashion to the transport term. This approach is incorporated into Eq. 8.64 for use in the simplified TKE equation (Eq. 2.74b).

In the equation for the covariance $\overline{u_i'\theta'}$ (Eq. 2.63), the quantity $\overline{\theta'\partial p'/\partial x_i}$ can be represented by

$$\overline{\theta'\partial p'/\partial x_i} = (\bar{e}^{1/2}/\Lambda_7)\overline{u_i'\theta'} \quad (8.73)$$

which reduces to Eq. 2.65a in the one-dimensional case, with the time scale τ_2 identical with $\Lambda_7/\bar{e}^{1/2}$. In analogy with the specification of previous length scales, we set $\Lambda_6 = a_6 l$ and $\Lambda_7 = a_7 l$, with $a_6 \approx 2$ and $a_7 \approx 2$. Table 8.2 summarizes values of the closure constants taken from the literature.

8.7.4 Non-local closure (ensemble models)

The concept of a first-order non-local closure for turbulence becomes attractive in the light of the failure of K-theory in the CBL. It is non-local in the sense of relating the flux of a quantity at a given level to properties at levels throughout the ABL. Formulation of non-local flux parameterization has been described in the literature in terms of transilient turbulence and spectral diffusivity theories, and involves an integral closure model that can physically account for the non-local effects causing upgradient diffusion and a source-dependent diffusivity in the CBL (e.g. Stull, 1988, Chapter 6).

The basis of this approach rests on the concept of advection by large eddies, such that the concentration of a passive tracer at level i is affected by air mixing in from many adjacent levels. Although the non-local approach has a number of attractions, for many purposes local higher-order closure is seen as the alternative to K-theory, particularly with increased use of LES models.

8.7.5 Closures in LES modelling

It is as well to note several features of LES models that affect the nature and accuracy of the solution, in addition to the turbulent closure assumption that is adopted. Firstly, the pressure field must be evaluated non-hydrostatically (because of the small grid scales involved) and, secondly, the prescription of boundary conditions at all boundaries has a significant influence on the solution.

Table 8.2. *Turbulence closure constants, based on Mellor and Yamada* (1982) (*MY*82) *and Wichmann and Schaller* (1986)

The reader should be aware of a potential source of confusion regarding the values of closure constants, since in the above two papers equations are developed in terms of a quantity $q^2 = \overline{u_i'^2}$, and in this book (and other texts) in terms of $\bar{e} = \overline{u_i'^2}/2 = q^2/2$.

Table 2 in Wichmann and Schaller (1986) is recommended for more information on the values of constants. Their equations are written in terms of constants k (not the von Karman constant) and k_1–k_4, whilst in Mellor and Yamada (1982) equations contain analogous constants A_1, A_2, B_1, B_2 and functions S_m, S_h, S_q, $S_{u\theta}$ and S_θ. The relationships of these to our constants and functions are indicated.

Equation	Relevant constant	Value	Comments
8.65, 8.67	a_1	5.75	$B_1 = k^{-1} = 2^{3/2}a_1$
8.68	a_2	7.44	$B_2 = (kk_4)^{-1} = 2^{1/2}a_2$
8.72	a_6	2.08	$3A_1 = (kk_1)^{-1} = 2^{1/2}a_6$
8.73	a_7	2.14	$3A_2 = (kk_3)^{-1} = 2^{1/2}a_7$
8.66, 8.67	S_m S_h S_e		Values of S_m and S_h from Fig. 4 in MY82, noting that $S_{m,h} = 2^{1/2}S_{m,h}^{MY}$.
8.70, 8.71	$S_{u\theta}$ S_θ		$S_e = S_{u\theta} = S_\theta = 0.2$, noting that $S_e = 2^{1/2}S_q$.

Surface values of turbulent fluxes are usually determined from standard surface-layer similarity theory.

Values of the sub-grid-scale fluxes appearing in the volume-averaged equations (which have essentially the same form as the ensemble-averaged equations) are generally based on K-theory in association with the TKE equation (one-and-a-half-order closure), or second-order closure (rarely). Because of the relative lack of sensitivity of LES simulations to the closure scheme, the TKE approach is seen as an acceptable approach using basically the same formulations and paramaterizations as those described in Section 8.7.2. In general, choice of the master length scale is more readily made and is simpler in form than in the ensemble-averaged case. This is the case because, in most situations, the scale of the turbulence (being sub-grid) is set by the grid scale itself.

First order

The sub-grid fluxes τ_{ij} and H_i are parameterized as (Mason and Thomson, 1987; Mason, 1989)

$$\tau_{ij} = K_M(\partial u_i/\partial x_j + \partial u_j/\partial x_i) \tag{8.74}$$

$$H_i = -K_H\partial\theta_v/\partial x_i \tag{8.75}$$

where $K_H = K_M/P_t$ and P_t is the turbulent Prandtl number (which can be assumed to be unity for small-scale turbulence). The eddy diffusivity K_M is specified by Eq. 8.52, with Eq. 8.53 replaced by

$$Ri \leqslant 0 \qquad f_s(Ri) = (1 - Ri)^{1/2} \tag{8.76a}$$

$$Ri > 0 \qquad f_s(Ri) = (1 - \beta\, Ri)^{1/2} \tag{8.76b}$$

with $\beta = 5$. Equation 8.76a, combined with Eq 8.52, can be deduced from consideration of local TKE balance (Eq. 2.74b, with the transport and pressure terms set to zero), with $\varepsilon \sim u_{*0}^3/l \sim K^3/l^4$. Equation 8.76b ensures $f_s(Ri) \to 0$ as Ri approaches the critical value of β^{-1} (0.2). The length scale l is given by Eq. 8.54 with $\lambda = c_s\Delta$; here Δ is the grid length and $c_s \approx 0.23$.

One-and-a-half order
The eddy diffusivities are described by Eq. 8.63, with each Λ given by Eq. 8.67 (typically, with $S_m \approx 0.1$ and $S_h > S_m$ in some studies) and l defined in a number of ways, but typically as

$$Ri \leqslant 0 \qquad l = \Delta = (\Delta x\, \Delta y\, \Delta z)^{1/3} \qquad (8.77a)$$

$$Ri > 0 \qquad l = \min(c_1 \bar{e}^{1/2}/N, \Delta) \qquad (8.77b)$$

where $c_1 \approx 0.5$ and $N^2 = (g/\theta)\partial\theta/\partial z$. The TKE is evaluated from the rate equation Eq. 2.74a, with pressure covariance, transport and dissipation terms parameterized as described in Section 8.7.2. In LES studies, $K_e \approx K_M$ and $\Lambda_1 = c_2 l$, with $c_2 \approx 0.2$–4 and l given by Eq. 8.77.

Second order
Equations for $\overline{u_i'u_k'}$, $\overline{u_i'\theta'}$ and variances are carried, with the pressure covariance, transport and dissipation terms parameterized as described in Section 8.7.3. Apart from a suitable choice for the closure constants, the length scale l is given by Eq. 8.77a, or an equivalent form. For example, in Deardorff's modelling studies, he set l according to Eq. 8.77a and used values of the constants introduced in Section 8.7.3 as follows: $a_1 = 1.4$, $S_e = 0.2 = S_{u\theta} = S_\theta$, $a_6 = a_7 = 0.25$.

Irrespective of the closure scheme, surface fluxes are evaluated from surface-layer similarity theory using the same relations as the ensemble-average approach.

8.7.6 Comments on turbulence closure
Consideration of the vertically integrated equations of mean properties reduces the problem of closure to determination of the boundary fluxes and the depth of the ABL. The equations are solved for average properties only, so no information on internal structure is available. Slab models, however, can often give better predictions of growth of a daytime CBL than first-order models. For many situations, vertical structure within the ABL is needed and some form of turbulence closure must be sought.

The application of first-order closure (K-theory) has been widespread, and an enormous number of K formulations can be found in the literature (see Wipperman's (1973) monograph for work up to the early seventies, for example). To some extent, this reflects the difficulty of expressing K analytically, or measuring it, above the surface layer. The eddy diffusivity, being a flow property, cannot be accurately prescribed from the outset because it depends on the flow structure to be determined. The principal problem in first-order closure is finding a rational basis for parameterizing the diffusivity. Nevertheless, most eddy-diffusivity models appear to be quite satisfactory in the neutral and stable ABL. They can reproduce many features of the mean structure, but of course can give no information on the turbulence statistics. The K approach fails in well-mixed layers where mean gradients are close to zero, and usually therefore in the buoyancy-dominated convective boundary layer. This has been well demonstrated in the case of bottom-up and top-down diffusion and its impact on mean scalar gradients (Sawford and Guest, 1987). In the CBL, turbulence may become entirely decoupled from the mean gradients and the concept of eddy diffusivity then becomes questionable. This breakdown in the K approach seems to be related to the large time scales (the large eddies have a "memory") and the vertical inhomogeneity of the turbulence (Weil, 1990).

So far as the ABL is concerned, second-order closure is, in principle, less restrictive than integral modelling and it provides information on the vertical distribution of turbulence statistics. Further, second-order models remedy some of the shortcomings of the K approach (in the CBL, for example) by shifting the problem of closure to higher-order moments. The second-moment equations can also give useful insight into K behaviour, by suitable scaling of the higher-order terms and recasting the steady-state covariance equations into flux-gradient form; see Eqs. 2.62b, 2.65b and Holtslag and Moeng (1991).

In many respects, the second-order closure problem reduces to finding approximations for the molecular destruction, turbulent transport, and pressure covariance terms in the second-moment equations. The pressure transport term in the TKE equation is typically either neglected or simply absorbed into the turbulent transport term. Closure for the pressure covariance in the flux equations is usually an extended version of Rotta's return-to-isotropy hypothesis.

The turbulent transport (triple correlation) terms are not significant in the neutral and stable ABL, but in the CBL they may dominate the flow dynamics, and the entrainment at the ABL top. The most commonly used closure for the transport terms is the downgradient diffusion model, involving a suitable turbulent length scale (Moeng and Wyngaard, 1989). Finally, most second-order closure models parameterize molecular dissipation rates in terms of velocity and length scales based on standard scaling arguments, thus avoiding the need to carry a rate equation for the dissipation itself.

Closures summarized above were orginally developed to model turbulent neutral shear flows, and the associated adjustable constants were mostly obtained from laboratory data. Although transport modelling contains some fundamental problems, these are nowhere as critical as the problems associated with modelling the pressure and dissipation terms. In the case of the transport closure, the downgradient-diffusion assumption is clearly inadequate in the convective ABL where K-theory fails. For example, observations reveal positive values of both $\overline{w'^3}$ (e.g. Wyngaard, 1988) and $\partial\overline{w'^2}/\partial z$ (Fig. 3.21) in the convective surface layer, implying that the vertical turbulent flux of $\overline{w'^2}$ is upgradient there. LES results have verified the poor performance of the downgradient assumption in the CBL, particularly in the prediction of some turbulent statistics.

Another major deficiency lies in the use of length scales, and the fact that all process scales are ultimately related back to one master turbulent scale.

Determination of this length scale and the associated constants relating it to the subsidiary length scales (e.g. Eq. 8.67) is at the heart of this closure problem. Although length-scale prescriptions cause fewer problems in modelling, they often oversimplify the physics by constraining the modelled turbulence to an expected state. The problem can be partly overcome by carrying an equation for ε, which has the advantage of improving the representation of the physics.

Second-order modelling of the ABL is still evolving, though the increasing use of volume-averaged models that resolve the large-eddy structure in the ABL (LES models) suggests that the refinements to second-order schemes will not be so crucial in the future. The approach has been called "turbulence engineering" (Wyngaard, 1982), and a wide variety of models can be found in the literature, some based entirely on shear-flow closures and some tuned specifically for buoyancy-dominated turbulence.

For completeness, we close this chapter on ABL parameterization schemes with a brief commentary on the problem of ABL clouds and how their effects are, or should be, represented in large-scale models.

8.8 ABL cloud parameterization

In many ABL and large-scale models, ABL layered cloud is assumed to exist if the ABL top lies above the local condensation level, i.e. if the relative humidity at $z = h$ is 100 per cent. In many GCMs, low-level stratiform clouds are assumed to be present when the grid-point relative humidity at a specified level in the model exceeds some critical value (about 85–95 per cent). Between this critical value and saturation, fractional cloudiness is allowed to increase according to some empirical relation (e.g. a quadratic relation), becoming full cover at 100 per cent relative humidity. Some schemes then introduce a dependence of cloudiness upon other variables, e.g. (i) low-level stability and (ii) the cloud-top entrainment instability, as described by Eqs. 7.14 or 7.21.

In both slab and high-resolution ABL models, incorporation of additional equations for water variables and modification of existing thermodynamic equations allows the cloud presence to interact with both the radiation and turbulence fields, and to affect both the surface energy balance, for example, and the depth of the ABL.

In many GCMs in use today, the depth of the ABL is not governed by a suitable rate equation and may not even be computed (although it may be set equal to a fixed value). Hence, the presence of stratiform cloud is unaffected by boundary-layer dynamics. Such a non-interactive system can never aspire to simulate the real-world behaviour and distribution of boundary-layer cloud.

Because the ABL controls the evaporation and turbulent redistribution of water substance into the atmosphere, it strongly determines the global distribution of both cumuliform and stratiform clouds. A comprehensive parameterization of ABL processes for a numerical model of large-scale atmospheric circulations must take into account the interaction of the ABL with clouds. This aspect of the ABL parameterization problem has an importance comparable to that of determining the turbulent fluxes.

Notes and bibliography

Section 8.1

A very useful document outlining the application of LES modelling in the atmospheric context is

Wyngaard, J. C., ed. (1984), *Large-eddy Simulation – Guidelines for Its Application to PBL Research*. Final Report, US Army Research Office, Contract No. 0804, Boulder, CO, 122 pp.

Section 8.4

The following essays are strongly recommended to the reader for a source of critical comments on the value of canopy models, and their use in GCMs for climate simulation. They are quoted extensively in Section 8.4.4.

McNaughton, K. G. (1987), Comments on "Modeling effects of vegetation on climate", in *The Geophysiology of Amazonia*, ed. R. E. Dickinson, pp. 339–42, John Wiley and Sons, New York.

Raupach, M. R. and J. J. Finnigan (1988), Single-layer models of evaporation from plant canopies are incorrect but useful, whereas multilayer models are correct but useless: Discuss, *Aust. J. Plant Physiol.* **15**, 705–16.

Section 8.7

The use of Eq. 8.60 for heat transport and Eq. 8.62 for scalar transport has been proposed by

Holtslag, A. A. M. and C.-H. Moeng (1991), Eddy diffusivity and countergradient transport in the convective atmospheric boundary layer, *J. Atmos. Sci.* **48**, 1690–8.

They propose expressions for the eddy diffusivities K_H, K_c and countergradient parameters γ_θ', γ_c' utilizing the turbulent heat-flux equation (2.64a), the top-down and bottom-up decomposition and results from large-eddy simulations.

The review paper by Mellor and Yamada (1982), in particular, is well worth reading, since it traces the development of a turbulence closure model for application to a wide range of geophysical fluid problems. These include free convection, pollutant dispersion, diurnal ABL structure, two- and three-dimensional flow with orography, cloud simulation and inclusion in global atmospheric and oceanographic simulations.

For more detailed discussions on non-local closure and transilient turbulence theory, the reader is referred to Stull's book (1988, Chapter 6) and references contained therein.

9

The atmospheric boundary layer, climate and climate modelling

In this final chapter, I wish to explore the relationship between the ABL and climate from a conceptual perspective, and from the modelling perspective to make use of the results from some recent general circulation model studies. The chapter concludes with some comments on priority research areas of particular relevance to the climate modelling problem.

9.1 Introduction

Climate refers to the long-term mean state and variability of the atmosphere over a range of space scales, so that a distinction is normally made between local, regional and global climate problems. At the outset, there are two main questions that need to be considered as central to the theme of the ABL and climate. Firstly, does the ABL have an effect on climate, and secondly, do climate changes influence the ABL?

9.1.1 The effect of the ABL on climate

In answer to the first question, it is necessary to pay attention to two aspects of ABL behaviour: (i) the presence of clouds within the ABL and (ii) the nature of the underlying surface. The presence of ABL clouds obviously affects climate in a very real and direct way. The ABL controls the evaporation and turbulent redistribution of water into the atmosphere, thus affecting significantly the global distributions of cumuliform and stratiform clouds. The clouds in turn influence the mean structure and turbulence of the ABL through cloud-induced circulations, radiation fields and precipitation. ABL stratus and stratocumulus cloud sheets are among the most common cloud types in the world. These clouds influence the distribution of both shortwave and longwave radiative fluxes, at the top of the atmosphere and also at the surface. The radiative forcing of the Earth's climate system is thus, in part, determined by the distribution of cloudiness. By reflecting solar radiation back to space, clouds exert a strong influence on the heating of the upper ocean and thereby on sea-surface temperatures.

The turbulent fluxes of sensible and latent heat at the surface are the most important source of moist available energy for the atmospheric circulation, and boundary-layer friction is the most important sink of atmospheric kinetic energy. The ABL, in fact, directly controls the interaction of the atmosphere with the oceans and land surface. If the nature of the land surface is changed, ABL fluxes will be affected, as will be the atmosphere or climate subsequently, at a scale dependent on the scale of the surface changes. This is a problem of great concern at present, in the context of man-made modifications to the land surface and to changes in land-use practice, particularly with reference to past deforestation (in the Sahel region, for example) and current deforestation in tropical regions.

Landscape changes (both natural and man-made) may affect climate in a number of ways, the most notable being through changes to the surface albedo. The sensitivity of climate to gross albedo changes can readily be demonstrated by considering a simple global energy balance, i.e.

$$\sigma T_e^{\,4} = (1 - \alpha)S_c/4 \qquad (9.1)$$

where σ is the Stefan-Boltzmann constant, S_c is the solar irradiance ($1367\,\mathrm{W\,m^{-2}}$) and T_e is an effective temperature. Note that the mean flux of energy into the system averaged over one year is exactly $S_c/4$ ($342\,\mathrm{W\,m^{-2}}$), since the solar energy is incident on a cross-sectional area of $\pi R_E^{\,2}$, where R_E is the radius of the Earth, and is distributed over a surface area of $4\pi R_E^{\,2}$. In the case of an Earth without an atmosphere, α can be interpreted as the surface albedo, giving $T_e = 264\,\mathrm{K}$ for a global mean albedo of about 0.2. In the presence of an atmosphere, and therefore clouds, α is the planetary albedo, giving $T_e = 255\,\mathrm{K}$ for a global mean albedo of 0.3. Differentiation of Eq. 9.1 gives a relation for the sensitivity of temperature to changes in albedo, i.e.

$$\partial T_e/\partial \alpha = -\,S_c/16\sigma T_e^{\,3} = -\,T_e/4(1 - \alpha) \qquad (9.2)$$

$\approx -\,0.8\,\mathrm{K}$ per 0.01 increase in albedo using values of the parameters quoted above. In the absence of clouds, and with changes in land albedo only, this value is reduced by a factor of four since land occupies only 25 per cent of the global surface area. Thus, $\Delta T_e \approx -\,20\Delta\alpha_\mathrm{land} \approx -\,0.2\,\mathrm{K}$ for an increase in α of 0.01. Realistic long-term increases in global mean land-surface albedo due to land-use practices may be an order of magnitude less (eg. Henderson-Sellers and Gornitz, 1984), implying temperature decreases of less than 0.02 K.

In the presence of an atmosphere, the response of the climate to changes in *surface* albedo will be more difficult to assess. This is the case because such changes will induce a complex response in the atmosphere, involving many feedback mechanisms that will result in changes to climatically important variables such as surface evaporation and cloud amount. For a simple calculation, we can consider an equilibrium atmosphere–surface system where the incoming shortwave flux is balanced by the outgoing longwave flux to space *and* the evaporative flux into the atmosphere. Differentiation of Eq. 5.1 with respect to albedo α_s produces

$$[4\sigma T^3 + (\rho\lambda/r_a)s]\partial T/\partial\alpha_s = -\,R_{s0}(1 - \alpha_s) \qquad (9.3)$$

where all other terms have been neglected in comparison with the dominant terms shown above, and we have assumed potential evaporation (given by Eq. 5.23). In Eq. 9.3, T is the surface temperature. Taking $T \approx 283$ K gives $s = 0.51$ (s is the slope of the saturation humidity–temperature curve), and with $r_a = 100 \, \mathrm{s \, m^{-1}}$ and $R_{s0} = 250 \, \mathrm{W \, m^{-2}}$, we get a reduced sensitivity of $\partial T / \partial \alpha_s \approx -15$ K or $\partial T / \partial \alpha_{\mathrm{land}} \approx -4$ K. Interestingly, model calculations have given somewhat greater sensitivities than those calculated above. One-dimensional radiative–convective models have yielded 1 K decreases in global surface temperature for 0.01 absolute increases in surface land albedo. Some estimates from the literature are shown in Table 9.1.

Many GCM simulations of the regional and global climate reveal a sensitivity of these climates to the land-surface evapotranspiration (Mintz, 1984), through prescribed changes in the available soil moisture or changes in the albedo (which, in model studies, tend to be large compared with those estimated from land-use changes). In the case of regional and large-scale surface-albedo increases, the major effects appear to be (i) decreased land evaporation and precipitation (regional changes) and (ii) increased precipitation over the sea (global changes). In the case of surface moisture changes, the major effects can be summarized as follows: (i) with zero surface moisture, continental rainfall is small except in the tropics where moisture convergence from the oceans is dominant; (ii) for saturated land, continental rainfall is significant.

At the smaller scale, there is both observational and modelling evidence that land-use changes at the regional scale can influence ABL structure and thus the local and regional climate. In the presence of soil and vegetation inhomogeneities of sufficiently large dimensions, horizontal gradients in surface albedo or surface moisture are likely to exist and be related to horizontal gradients (during the daytime) in the sensible heat flux. These, in turn, will induce thermally direct circulations, called non-classical mesoscale circulations or NCMCs, and often referred to as "inland sea breezes" (Mahfouf *et al.*, 1987; Segal *et al.*, 1988). Examples where NCMCs may exist include mesoscale regions of wetted soil following a storm in a generally semi-arid region; between dry soil and snow-covered ground following a winter storm; at discontinuities between soil of different thermal characteristics; between vegetation (particularly irrigated crops) and dry soil (extensive irrigation in semi-arid regions); and between forested and recently deforested areas. The generation of significant horizontal heat flux differences leads to differential boundary-layer evolution, and to horizontal temperature gradients throughout the lower atmosphere, resulting in a sea-breeze-like circulation. Sharp contrasts in surface temperature are readily observed from satellite imagery, and the intensity of the circulation is likely to be critically dependent upon the synoptic flow and on the presence of topographically induced mesoscale flows.

Theoretical, conceptual and numerical evaluations of the kinematic and thermodynamic processes associated with NCMCs have been made in recent years. One major study (Anthes, 1984) has hypothesized that bands of vegetation in semi-arid areas, of optimum spacing ~ 100 km, could, under favourable large-scale conditions, result in enhanced convective precipitation. The basis of this is related to (i) an increase in the low-level moist static energy (defined as

Table 9.1. *Estimates of global temperature changes* ΔT *produced by surface albedo changes* $\Delta \alpha_s$, *based on results from one-dimensional radiative–convective* (1D RC) *models and three-dimensional general circulation models* (3D GCMs)

Reference	Model	$\Delta \alpha_s$	$\Delta T(K)$
Sagan *et al.* (1971)	1D RC	0.01	− 2
Hansen *et al.* (1981)	1D RC	0.015	− 1.3
Henderson-Sellers and Gornitz (1984)	1D RC	0.0005*	− 0.05
Rowntree (1988) (quotes a priv. comm.)	3D GCM	0.0008	− 0.1

*This value of $\Delta \alpha_s$ was an estimated increase due to land-use change, with ΔT estimated from the results of Hansen *et al.* (1981)

$c_p T + gz + \lambda q$), (ii) an increase in boundary-layer water vapour due to increased evaporation and decreased runoff and (iii) the generation of NCMCs associated with the surface inhomogeneities created at this scale by the vegetation. These ideas were broadly confirmed in a later numerical study (Yan and Anthes, 1988) that showed that strips of dry and wet surface with width and separation ~ 100–$200\,\text{km}$ can initiate convective rainfall in a convectively unstable environment with weak synoptic flow and plentiful moisture.

9.1.2 The effect of climate change on the ABL

The second question posed above asks how climate changes may influence the ABL. Intuitively, there are two effects that would be expected to affect the mean climatological structure of the ABL over a period of time. Firstly, increased carbon dioxide concentrations in the atmosphere will affect the longwave radiative fluxes and thus the relative roles of radiative and turbulent cooling in the NBL. This is a fairly subtle and relatively minor effect. Secondly, higher mean temperatures are predicted in any climate change resulting from increased concentrations of "greenhouse gases". Surface sensible and latent heat fluxes (or the Bowen ratio) over saturated or very wet surfaces are affected considerably by the absolute surface temperature (see e.g. Fig. 5.7). For wet surfaces, under given atmospheric conditions, the surface evaporation increases as surface temperature increases, thus affecting directly the turbulent intensity in the ABL, and its depth. In association with the above effects, any changes to low-level cloudiness, or cloudiness in general, will impact directly on the surface energy balance, on energy partitioning and on the mean ABL structure.

There is also some indication from modelling studies of the likely response of the ABL to climate changes induced by the presence of a high-altitude smoke or dust layer, as might result from volcanic activity or a nuclear war. Near-surface temperature changes have been the focus of many studies (e.g. Ghan *et al.*, 1988; Schneider and Thompson, 1988), and are very much dependent on the changes in vertical ABL structure and its depth. Calculations have recently been reported (Garratt *et al.*, 1990) on the impact of smoke of varying absorption optical depth (AOD) on the ABL, over a time period of several days during which major smoke-induced cooling occurred. The presence of smoke reduces the amount of shortwave energy reaching the ground, and consequently the daytime mixed-layer depth is reduced. For large enough values of the AOD, a daytime surface inversion results, with large cooling confined to heights of less than a few hundred metres. Naturally, the magnitude of the impact of smoke is particularly sensitive to soil wetness and to the smoke AOD. Figure 9.1 demonstrates the change in the ABL thermal structure, over dry and wet soil, some five days after smoke of AOD equal to 0.2 is introduced. The smoke has produced a significant reduction in the amount of shortwave energy reaching the surface (though this has been partly balanced by an increase in downwards longwave radiation), and consequently a reduction in the ABL depth for both dry and wet surfaces, as well as significant ABL cooling.

9.2 Sensitivity of climate to the ABL and to land surface

Numerical models of the atmosphere can be used to study the sensitivity of climate to the choice of land surface and ABL parameterization schemes. Such modelling studies are particularly useful, since relationships between changes in surface parameters and gross climatic quantities are not always obvious, or easily deduced from intuitive arguments, because of the strongly non-linear nature of the climate system. These studies are necessary in order to evaluate (i) the level of sophistication of the physics needed in the model and (ii) climate response to land-use changes, and therefore to vegetation variations (e.g. the likely effects of tropical deforestation on climate).

Generally speaking, GCMs, as of 1991, use relatively crude representations of the surface and the ABL compared with the detailed formulations and parameterization schemes presented in this book. GCM sensitivity studies are, and have been, useful in evaluating the impact of refined or improved schemes on climate simulations. In addition, mesoscale or one-dimensional stand-alone models have been used to evaluate the impact of ABL schemes, in particular, on mesoscale circulations, and hence on regional climate. GCM studies have tended to concentrate on surface albedo, moisture and roughness effects, all of which are involved when vegetation cover is changed. A few have discussed the impacts of ABL cloud schemes and ABL depth representation. In smaller-scale models, the emphasis has been on turbulence closure schemes and their impact on one-dimensional ABL structure and mesoscale circulations.

Climate (whose primary indicators include near-surface air temperature, mean cloudiness and rainfall) responds to surface changes through the related changes in mean evaporation and heat flux into the atmosphere. From simple considerations, we might expect the following response to a mean change in evaporation: with increased evaporation, for example, ABL moisture increases, as does cloudiness and hence precipitation. But how does the evaporation depend upon changes in the surface parameters expected to affect climate? Let us take three relevant surface quantities whose effects on evaporation can be estimated.

(i) Firstly, the albedo: an increase in albedo reduces the shortwave energy absorbed at the surface, and hence the available energy and the evaporation. This is confirmed when we consider the relation between net radiation and net shortwave flux (Eq. 5.13), which suggests $\partial R_N / \partial \alpha_s < 0$.

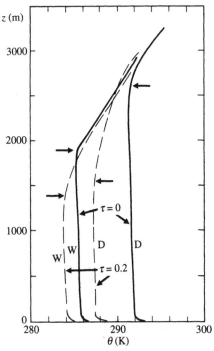

Fig. 9.1 One-dimensional numerical simulations of θ profiles, illustrating the impact of high-level smoke on ABL structure over both dry (D) and wet (W) soils. Profiles are for local noon, some five days after the introduction of the smoke. Horizontal arrows indicate the top of the convective mixed layer. Solid curves are for smokeless skies; pecked curves have smoke present of absorption optical depth (AOD) $\tau = 0.2$. After Garratt *et al.* (1990), *Journal of Applied Meteorology*, American Meteorological Society.

Now, taking Eqs. 5.26 and 5.39, it is readily shown that $\partial E/\partial R_N \approx \Gamma R_N (1 + r_s/r_a)^{-1} > 0$, where $\Gamma = s(s + \gamma)^{-1}$, so that $\partial E/\partial \alpha_s < 0$.

(ii) Secondly, the surface availability of moisture: a decrease in the availability of moisture through drying of the soil or through an increase in surface resistance produces a decrease in evaporation.

(iii) Thirdly, the surface roughness or aerodynamic resistance: according to the curves of Fig. 5.6, an increase in z_0 increases the potential evaporation and decreases sensible heat flux. However, in the presence of a moderate surface resistance, surface roughness changes may have a minimal effect on evaporation.

GCM studies describe climate responses to gross albedo and surface moisture changes that are broadly consistent with the above expectations. For albedo, some results are shown in Table 9.2, for illustration, showing that increases in mean albedo generally produce a decrease in evaporation and in precipitation. For soil moisture, model simulations have been carried out with the land surface saturated (giving potential evaporation) and dry (giving zero evaporation). A summary of results from one such study is given in Table 9.3(*a*). Overall, the

Table 9.2. *Sensitivity results of the response of three-dimensional GCM climate simulations to gross changes in surface albedo* $\delta\alpha_s$; *climate response is shown here in terms of average changes in evaporation* δE *and rainfall* δP *over land*

Reference	$\delta\alpha_s$	δE (mm day^{-1})	δP (mm day^{-1})
Charney et al. (1977)*	0.21	− 0.8	− 2.0
Chervin (1979)*	0.27	−	− 1.7
Carson & Sangster (1981)*	0.20	− 0.95	− 1.2
	0.10	−	− 0.4
Sud and Fenessey (1982)	0.16	− 0.4	− 0.6
Henderson-Sellers and Gornitz (1984)	0.06	− 0.45	− 0.6
Cunnington and Rowntree (1986)	0.06	−	− 0.75
	− 0.04	−	+ 0.6
Sud and Smith (1985)	0.06	−	− 1.6

*Discussed in Mintz (1984). This and other references are discussed in Rowntree (1988).

dry-soil case results in much less cloudiness over the continents than the wet-soil case, and in many parts much less precipitation. So far as the effects of roughness changes are concerned, Table 9.3(*b*) summarizes one GCM study in which a global change in z_0 *over the land* (from 0.45 m to 0.0002 m) produces little change in the mean land evaporation, but measurable regional changes in the land–sea precipitation distribution. These changes occur because of the large changes in the horizontal convergence of the water vapour transport in the boundary layer, related to increases in boundary-layer winds with decreased roughness.

The inclusion of vegetation through a suitable interactive canopy sub-model (see Chapter 8) ensures that the influences of albedo, surface moisture availability and roughness are included. Failing to incorporate such a model and accounting only for albedo and roughness changes may mean that surface-change effects are not properly represented. Land surfaces denuded of all vegetation, particularly forests, are likely to have higher albedo, smaller roughness and to be drier (through lack of access to the deep-soil moisture) than fully vegetated surfaces.

The inclusion of sub-models of the terrestrial biosphere in GCMs does have an impact on climate simulations, and this can be illustrated best by reference to several recent studies. In two of these, comparisons were made between simulations with a simple bucket hydrology and with a canopy sub-model ("simple biosphere model"). The results of the first study (Sato et al., 1989) are summarized in Fig. 8.10, discussed earlier in the context of canopy models, and show that the inclusion of a canopy produces a more realistic partitioning of energy at the land surface. In fact, the presence of a canopy results in more sensible heat and less evaporation over vegetated land compared with fluxes calculated with simple hydrology, with a deeper ABL and reduced precipitation over continents. The results from a second study, summarized in Table 9.3(*c*),

Table 9.3. (*a*) *Values of gross evaporation and precipitation evaluated in a GCM, for two cases of (prescribed) dry soil ($x = 0$) and saturated soil ($x = 1$). E and P represent monthly averages for the North American and Eurasian land masses. (b) Impact upon gross evaporation and precipitation of roughness changes in a GCM. (c) Impact upon mean evaporation and precipitation of introducing a canopy sub-model into a GCM. Values are shown for control simulations (bucket hydrology) and simulations having a canopy scheme (SiB)*

(*a*) Soil moisture	x	E (mm day^{-1})	P (mm day^{-1})
Shukla and Mintz (1982)	0	0	0
	1	4.3	4

(*b*) Surface roughness	δz_0	δE (mm day^{-1})	δP (mm day^{-1})
Sud *et al*. (1988)	0.0002–0.45	no change	large change in distribution

(*c*) Vegetation		E (mm day^{-1}) Jan	Jul	P (mm day^{-1}) Jan	Jul
Sud *et al*. (1990)	control	2.4	3.6	2.4	3.5
	canopy	1.6	2.3	1.9	2.4

confirm and extend these findings. In this study, the inclusion of a canopy formulation gives much smaller evaporation over land surfaces, and generates significantly different surface fluxes for both vegetated and bare-soil regions. These differences are accompanied by large and statistically significant changes in the simulated rainfall. The decrease in evaporation over vegetated land is the direct result of the canopy's control upon evapotranspiration, whilst the decrease over bare soil is the result of replacing the "bucket" hydrology scheme with the "force-restore" method. In the latter approach, surface drying of the soil and reduced evaporation replaces a partly filled "bucket" or moist thick soil layer with associated high evaporation rates. The overall impact of the canopy model on the simulation of the global hydrological cycle can be assessed by reference to the relative flux magnitudes shown in Fig. 9.2. Even though the overall global differences are small, the evaporation over land reduces by about 40–50 per cent for both January and July. Against this change, precipitation over land decreases by about 30 per cent, implying an increase in moisture convergence inland of about 10–20 per cent times the evaporation. Changes in the net radiation over land do indicate that the decrease in evaporation will be partly the result of reduced available energy.

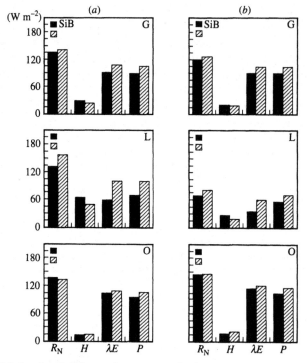

Fig. 9.2 (*a*) GCM-simulated July surface energy budget components, and precipitation, for the entire earth (G), the land (L), and the oceans (O). The components are net radiation R_N, sensible heat flux H, evaporation λE and precipitation P. The shaded regions represent the GCM with the SiB canopy and the standard GCM without a canopy formulation. (*b*) The same as (*a*) but for January simulations. From Sud *et al.* (1990), by permission of Elsevier Science Publishers and the authors.

The studies described above are concerned with the impact of an improved representation of the surface hydrology (i.e. replacing "bare soil" bucket hydrology with a relatively sophisticated canopy-soil hydrology) on the climate simulations. In contrast to this approach, GCM sensitivity studies have been reported that investigate the gross sensitivity of regional climate to canopy type – specifically, deforestation in Amazonia with forest replaced by degraded tropical pasture or short grass. Typically, long-term (several years) simulations of the climate are made in GCMs using canopy sub-models (Dickinson and Henderson-Sellers, 1988; Lean and Warrilow, 1989; Shukla *et al.*, 1990), initially with a forested surface across South America (and a realistic vegetation distribution elsewhere), and then with a deforested Amazonia. The main results can best be assessed by reference to the changes in precipitation, evaporation and near-surface air temperature (see Table 9.4). The impact of deforestation on the ABL is considerable, since reductions in long-term evaporation and precipitation occur, with increases in the near-surface air temperature. These changes are consistent with the tendency of the upper soil layers to dry with the removal of forest, since bulk moisture becomes less accessible to the shallow-rooted grass. The reduction in evaporation with removal of the forest is also consistent

Table 9.4. *Impact of the canopy type in a GCM upon regional climate (Amazonia); C, control simulation (forest); Ex, experimental simulation (deforested or tropical short grass). Long-term evaporation (E) and precipitation (P) rates are shown, together with runoff (R) and near-surface air temperature*

Reference		E	P (mm day^{-1})	R	T (°C)
Dickinson and	C	5.9	9	4	25
Henderson-Sellers	Ex	5.2	9	4	26.5
(1988)					
Lean and Warrilow (1989)	C	3.1	6.6	3.4	23.6
	Ex	2.3	5.3	3	26.0
Shukla *et al.* (1990)	C	4.5	6.7	–	23.5
	Ex	3.2	5.0	–	26.0

with the roughness effect seen in Fig. 5.6 (in terms of the combination relation for potential evaporation, the aerodynamic term decreases with decreased roughness), particularly since the vegetation is wet for much of the year.

ABL clouds are treated in a number of ways in the many models used around the world, but most models treat them very crudely, including little dependence on ABL mixing, on entrainment across the ABL or cloud-layer top, or on ABL depth. All these factors appear to be crucial to the development or dissipation of low-level stratiform clouds. Consequently, in most GCMs there is an excess of stratus and stratocumulus globally, particulary in high latitudes and in the central sub-tropical oceans, and a deficit of stratocumulus in the sub-tropical oceans west of the continents where maxima are observed. Unfortunately, too few model studies have been involved in evaluating the sensitivity of the amount of low cloud and its geographical distribution to ABL schemes and other effects. Reported improvements in low-cloud simulations have been found for quite different reasons, and seem to depend on the model used.

9.3 Research priorities

In terms of their relevance to climate simulations, several areas of ABL-related research are likely to expand over the coming years and provide improvements to physical parameterization schemes used in climate models. These priority areas can be summarized as follows:

(i) the problem of area averaging in heterogeneous terrain;
(ii) the application of ABL theory over complex terrain, and the influence of sub-grid orography;
(iii) canopy and surface hydrology representation in models;
(iv) ABL clouds.

9.3.1 Area-averaging problems
In large-scale numerical models of the atmosphere it is necessary to consider the properties of the ABL as averaged over the grid spacings in the model. Thus it

is important to be able to determine the averaged surface flux over an area of *heterogeneous* terrain and to modify, as appropriate, local formulations applicable to horizontally homogeneous surfaces in order to give the correct flux. A related problem arises in the interpretation of observations in heterogeneous terrain; not only is it necessary to form such averages, but the area over which the average must be made has to be deduced. In short, how can a set of randomly distributed observations over a region be collected together to give a meaningful (from the dynamical viewpoint) area average? The importance of this problem to modelling lies in the need to incorporate the effect of unresolved (sub-grid) real-world variations in roughness, surface temperature and moisture, and thus ABL thermal stability and winds.

Formulation for vertical fluxes
Let us consider the averaging process applied to the grid-area surface flux evaluated at a GCM grid point (Mahrt, 1987). Locally, for surface i, the surface flux of property s is given by

$$F_i = C_i u_i \Delta s_i \qquad (9.4)$$

with the areal flux given by

$$\langle F \rangle = n^{-1} \sum_i F_i = \langle C_i u_i \Delta s_i \rangle. \qquad (9.5)$$

Note that Eq. 9.4 is a parameterized form for the local flux at the ABL scale, so that the flux represented by Eq. 9.5 includes only those contributions from local scales. Where the grid size is much greater than the ABL scale (as in large-scale numerical models), the sub-grid scale flux parameterization does not explicitly include transport by motions that are larger than ABL scale but still unresolved. Such mesoscale flows include cloud-induced motions and flows responding to sub-grid terrain and differential heating (e.g. sea breezes). Parameterization of the mesoscale component of the area-averaged flux is usually ignored.

We can represent local variables as deviations from the area-averaged value, with $\phi_i = \langle \phi \rangle + \phi''$. The right-hand side of Eq. 9.5 can then be expanded to give

$$\langle F \rangle = \langle C_i \rangle \langle u \rangle \langle \Delta s \rangle + A_1$$
$$= \langle C_i(z_0, Ri_B) \rangle \langle u \rangle \langle \Delta s \rangle + A_1. \qquad (9.6)$$

Here, the term A_1 is meant to represent a set of sub-grid correlation terms:

$$A_1 = \langle C \rangle \langle u'' \Delta s'' \rangle + \langle u \rangle \langle C'' \Delta s'' \rangle + \langle \Delta s \rangle \langle C'' u'' \rangle + \langle C'' u'' \Delta s'' \rangle \quad (9.7)$$

where we have excluded any dependence of C upon scalar roughness length in Eq. 9.6. The functional dependence of C_i in Eq. 9.6 is shown to indicate that local values of z_0 and Ri_B are involved. However, modelling procedures use only grid-averaged variables (e.g. $\langle z_0 \rangle$ and $\langle Ri_B \rangle$) so that Eq. 9.6 needs to be rewritten as

$$\langle F \rangle = C(z_0^{\text{eff}}, \langle Ri_B \rangle) \langle u \rangle \langle \Delta s \rangle + A_1 + A_2, \qquad (9.8)$$

where

$$A_2 = \langle u \rangle \langle \Delta s \rangle [\langle C_i(z_0, Ri_B) \rangle - C(z_0^{\text{eff}}, \langle Ri_B \rangle)]. \tag{9.9}$$

The first term on the right-hand side of Eq. 9.8 is the form in which the surface flux is usually represented in numerical models, so that the sum of A_1 and A_2 represents the correction required to give the true flux in the presence of sub-grid variability. Note that we have introduced in place of $\langle z_0 \rangle$ an effective roughness length, which is the value of the roughness length required in heterogeneous terrain to give, e.g., the correct area-averaged stress when using bulk aerodynamic formulae (see later).

It is evident that the terms A_1 and A_2 need to be assessed and even parameterized in order to evaluate the effects of sub-grid variability in heterogeneous terrain. The set of correlation terms comprising A_1 is not readily determined from measurement and modelling seems to represent the best alternative. It is to be expected that some of these spatial correlation terms may be significant under certain conditions. The potential importance of the term A_2 is easier to assess, and can be illustrated in terms of specified statistical distributions of Ri_B across an averaging area, assuming constant z_0. We can expect differences in the two transfer coefficients in Eq. 9.9 because of the non-linear relation between C and stability (refer to Fig. 3.7). Consider the case where the area-averaged stratification is stable but varies within the grid area. The value of C predicted by the area-averaged stability may be quite small in stable conditions. However, due to strong non-linearity of the stability dependence, the area average of the local C may be significantly larger due to sub-grid areas where the stratification is near-neutral or unstable. In these sub-grid areas, the local C may be one or more orders of magnitude larger than that implied by the spatially averaged stability. Because of the non-linear dependence of C upon stability, small sub-grid regions could have a large influence on the grid-averaged C, and therefore flux, but little influence on the grid-averaged stability. This can be illustrated by considering sub-grid Gaussian distributions of Ri_B of varying widths (standard deviations). The variation of $\langle C(Ri_B) \rangle$ with surface bulk Richardson number for a range of Ri_B standard deviations is shown in Fig. 9.3, together with that of $C(\langle Ri_B \rangle)$; curve A is for zero standard deviation. Differences between any curve and curve A at specified values of Ri_B are directly proportional to A_2. In unstable conditions, such differences are not so large, but where the standard deviation exceeds about 1/2, differences in stable conditions become significant.

It should be noted that, although transfer coefficients in unstable conditions do not require systematic modification due to sub-grid variability, the flux itself could still be underestimated due to unaccounted mesoscale sub-grid motions. In contrast, in stable conditions, the systematic differences seen in Fig. 9.3 may also be supplemented by finite vertical-grid effects and the presence of internal gravity waves. A thick layer of air, for example, may be associated with a super-critical gradient Richardson number Ri, and so may be assumed to carry zero vertical flux. Realistically, values of sub-critical Ri are almost certainly present in thinner layers within the vertical grid interval, implying that a non-zero flux probably exists supported by the smaller-scale turbulence. In some modelling schemes, this effect is allowed for by increasing the magnitude of the critical Richardson number.

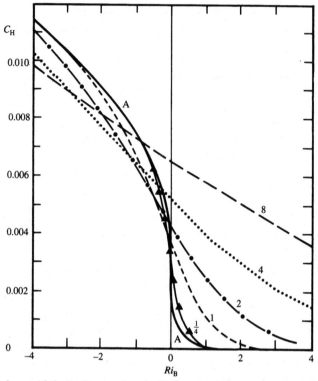

Fig. 9.3 Dependence of the bulk transfer coefficient for heat upon bulk Richardson (Ri_B) and assumed sub-grid distributions of Ri_B of varying widths or standard deviation (shown against each curve). Curve A represents the expected variation for zero standard deviation (refer to Eqs. 3.45 and 3.47, and Fig. 3.7) and $z/z_0 = 2000$. Differences between any other curve and curve A for a given Ri_B are directly proportional to A_2 in Eq. 9.8. After Mahrt (1987), *Monthly Weather Review*, American Meteorological Society.

Effective roughness length

We introduced above an effective roughness length z_0^{eff} required, in heterogeneous terrain, to account for sub-grid variations in roughness. This should be taken as the value of the roughness length which, in homogeneous terrain, gives a stress value equal to the area-averaged stress occurring in the heterogeneous terrain. In the past, it has been interpreted differently, in particular as the roughness length required to give the correct spatially-averaged velocity profile. Let us review the main approaches to determining the effective roughness length.

(i) The average geostrophic drag coefficient is used to define z_0^{eff}. In this approach (Smith and Carson, 1977), it is assumed that the wind scale most effectively constant over an area is the geostrophic or boundary-layer top wind. It then follows that averaging the surface stress is equivalent to a linear averaging of the local geostrophic drag coefficients, but only if it can be assumed that there is an equilibrium relation between the local stress and geostrophic wind. This may not be the case when horizontal variations in z_0 occur on scales less than a few tens of kilometres, since IBL considerations suggest a fetch of about 50–100 km to yield an equilibrium ABL of depth about 1 km. Taking Eq. 9.5 for stress, setting $u_i = \langle u \rangle = G$, and identifying C_{gi} with local surface i of fractional area f_i, then

$$C_g^{\text{eff}} = \sum_i f_i C_{gi} \tag{9.10}$$

with C_g defined by Eq. 3.21 in neutral conditions, for example, allowing Eq. 9.10 to relate z_0^{eff} (call it z_{0a}) to local roughness lengths.

(ii) The correct spatial average of the low-level wind profile (rather than the stress) is used to define z_0^{eff}. Thus (Taylor, 1987), Eq. 3.17 averaged across an area yields

$$\langle u \rangle = k^{-1}(\langle u_* \rangle \ln z - \langle u_* \ln z_0 \rangle) \tag{9.11}$$

indicating that the apparent friction velocity $\langle u_* \rangle$ is not equal to the square root of the average stress $\langle u_*^2 \rangle$, and implying an effective roughness length z_{0b} given by

$$\ln z_{0b} = \langle u_* \ln z_0 \rangle / \langle u_* \rangle \approx \langle \ln z_0 \rangle. \tag{9.12}$$

(iii) A value of the effective roughness length is required that gives the correct surface stress. For neutral conditions this amounts to averaging drag coefficients based on a blending height (Mason, 1988), i.e. the height at which flow changes from equilibrium with the local surface to independence of horizontal position. This blending height l_b is typically about $L_x/200$, where L_x is the horizontal scale of roughness change. For $z = l_b$ within the surface layer, the effective roughness length $z_0^{\text{eff}} = z_{0c}$ is then given by (see Eq. 9.5 for stress, with $u_i(l_b) = \langle u \rangle$),

$$[\ln (l_b/z_{0c})]^{-2} = \sum_i f_i [\ln (l_b/z_{0i})]^{-2}. \tag{9.13}$$

For sufficiently large L_x, say $L_x > 25$–50 km, Eq. 9.13 must tend towards the effective roughness length given by Eq. 9.10.

Typical values of effective roughness length based on several methods and results from a numerical model assumed to give the "true" value are given for comparison in Table 9.5 for the simple case of two adjacent local surfaces with quite different values of local z_0. Two scales are illustrated, showing that Mason's method is in reasonable agreement with the numerical results for both scales but Eq. 9.10 is acceptable at the larger scales only, as expected. Even so, all methods are within a factor of 2–5, a range of uncertainty that may not be critical in many GCM simulations of climate.

9.3.2 Complex terrain and sub-grid orography

We can take account of heterogeneous, flat terrain by suitable area averaging to give an effective roughness length or drag coefficient with which to calculate the area stress, heat flux and evaporation. In qualitative terms, the influence of orography on all scales must be to increase the stress, for a given geostrophic

Table 9.5. *Estimates of the effective roughness length for an area comprising two adjacent local surfaces, of roughness lengths 0.015 m and 1.5 m. Estimates are based on Eqs. 9.10 (z_{0a}), 9.12 (z_{0b}) and 9.13 (z_{0c}), for two values of the horizontal scale L_x (relevant to z_{0c} and the numerical z_0^{eff}). Also shown are numerical "true" values. For z_{0a}, we have taken the ABL depth equal to 1 km, and, for z_{0c}, the blending height $l_b = L_x/200$. The table is based partly on Mason (1988)*

L_x (km)	z_0^{eff} (m)	z_{0a} (m)	z_{0b} (m)	z_{0c} (m)
1	0.82	0.35	0.16	0.77
100	0.43	0.35	0.16	0.38

wind, and to be reflected in an increased roughness length or drag coefficient. The increased stress may be the result of form drag (pressure differences produced by orographic blocking of the low-level flow) in mainly neutral or unstable flow, or of gravity-wave drag in the ABL due to the forcing of internal gravity waves in stable conditions.

Form drag

The presence of form drag can be accounted for by an increase in the geostrophic drag coefficient or roughness length. This assumes that the form drag can be parameterized by direct analogy with the small-scale stress, so that, if stresses are additive, and related to the geostrophic or ABL wind scale,

$$C_g^{eff}(z_0^{eff}) = C_g(z_0) + C_{FD}. \tag{9.14}$$

Here, C_g is given by an equation like Eq. 3.21, and the effective drag coefficient C_g^{eff} is assumed to satisfy the same relation. Typically C_g^{eff} should apply over a distance greater than the horizontal scale of the topographical feature or mountain range in the downstream direction to incorporate the pressure drop across the feature, and should be associated with the immediate upwind or downwind geostrophic wind magnitude of the undisturbed flow. The problem reduces to evaluating C_{FD} and identifying the main variables upon which it depends. Having derived C_{FD} for neutral conditions, substitution into Eq. 9.14 then allows z_0^{eff} to be evaluated and the usual buoyancy influences incorporated through the coefficients A and B in Eq. 3.21.

For present purposes, let us assume the neutral drag coefficient is defined in terms of the geostrophic wind or mean ABL wind, and given by Eq. 3.21. Many GCMs utilize Cressman's (1960) values of C_g^{eff} based on a crude representation of C_{FD} given by

$$C_{FD}^1 = 0.125\sigma_h/\lambda_s \tag{9.15}$$

where σ_h is the root-mean-square variation in orography and λ_s is the mean spacing between the dominant orographic features. Equation 9.15 represents the simplest physical model for form drag that is consistent with the expected force exerted by a bluff body on the atmosphere in a given flow (some model parameterizations simplify even further, and incorporate an overall global value of λ_s, thus making $C_{FD}^1 \propto \sigma_h$). Equation 9.15 was derived specifically by Cressman for GCM grid scales of several hundreds of kilometres.

In contrast, the form drag can be parameterized in terms of a terrain roughness length z_{0h}, as was done by Smith and Carson (1977), who applied the results for grid scales of about 10 km. Their suggested relation is

$$z_{0h} = 0.2\sigma_h{}^2/\lambda_s \tag{9.16}$$

with the associated C_{FD} (call this C_{FD}^2) given by Eq. 3.21. There is of course no firm basis for believing that, for terrain roughness elements of horizontal scale much greater than the ABL height, z_{0h} should behave similarly to z_0 and appear within the logarithmic function in Eq. 3.21. A parameterization of form drag that seems to apply over a wide range of grid sizes has been developed by Deardorff and co-workers (e.g. Deardorff *et al.*, 1984), and is based on detailed numerical simulations of neutral and heated flow over complex terrain. Their results took into account only the contributions to form drag of terrain features of horizontal scale 20–250 km, being constrained by grid length and domain size. For the neutral case, the form drag coefficient is given by

$$C_{FD}^3 = 0.18\sigma_h{}^2 h_b{}^{-3/2}\lambda_s{}^{-1/2} \tag{9.17}$$

where h_b is the ABL depth averaged over the region of interest.

In all the formulations introduced above C_{FD} depends on orographic parameters σ_h and λ_s, whilst Eqs. 9.16 and 9.17 depend on neutral ABL height, implicitly through Eq. 3.21 in the former case. The inclusion of h_b in Eq. 9.17 simply reflects numerical model results, but physically the mixed-layer depth affects low-level winds, and thus form drag indirectly. Where heating of the terrain occurs, Deardorff *et al.* identified a thermal-anomaly form drag, and associated drag coefficient, resulting from the resistance of pockets of cool air to upslope motion and pockets of warm air to downslope motion. The drag coefficient has an additional dependence upon a horizontal root-mean-square temperature variable, producing a more complex formulation for the parameterized drag. Its details are beyond the scope of the present work, but it does highlight the problem of the realistic parameterization of sub-grid orography, and how to include the many important physical processes occurring for given flow situations.

The predictions of each of the above three methods can be readily compared for specified orographic parameter values. For example, we have chosen three values for each of σ_h (25, 100 and 250 m) and λ_s (1, 10 and 100 km) that reflect those found in the real world and would be appropriate for either mesoscale or large-scale numerical models. The values of h_b for a neutral ABL required in Eq. 9.17 cover the range 500 to 1500 m. Relevant values for comparison with the predictions of Eqs. 9.16 and 3.21 can be deduced by solving Eq. 3.21 for u_{*0} and then inferring the ABL depth from $h_b = 0.2u_{*0}/|f|$. In the calculations, we have taken $|f| = 10^{-4}$ s^{-1} and a geostrophic wind speed of 10 m s^{-1}. Table 9.6 summarizes values of C_{FD}, showing that Eq. 9.15 tends to overestimate at small λ_s and Eq. 9.16 tends to overestimate at large λ_s. This might have been anticipated, and relates to the grid-scale ranges for which each method was derived and tested (see remarks above). Otherwise, all three methods give

Table 9.6. *Values of the form-drag coefficient C_{FD} deduced from three specific formulations, for a range of values of the parameters σ_h and λ_s.*

Formulations are as follows: C60 is based on Eq. 9.15; S77 is based on Eqs. 9.16 and 3.21, for specified $G = 10 \text{ m s}^{-1}$ and $|f| = 10^{-4} \text{ s}^{-1}$; D84 is based on Eq. 9.17 with h_b based on the neutral relation $h_b = 0.2u_{*0}/|f|$ and u_{*0} inferred from the S77 C_{FD} values for $G = 10 \text{ m s}^{-1}$. C60, Cressman (1960); D84, Deardorff *et al.* (1984); S77, Smith and Carson (1977).

Formulation	σ_h (m)	λ_s (km)	$C_{FD} \times 10^3$		
			1	10	100
C60	25		3.1	0.31	0.03
S77			1.7	1.2	0.9
D84			0.2	0.08	0.03
C60	100		12.5	1.2	0.12
S77			2.9	1.9	1.4
D84			1.7	0.7	0.3
C60	250		31	3.1	0.3
S77			4.2	2.7	1.7
D84			7	3	1.3

comparable results at intermediate λ_s, particularly at larger σ_h. Typically, C_{FD} increases from about 0.0005 to 0.003 as σ_h increases from 25 to 250 m. For flat, vegetated terrain with $z_0 = 0.1$ m, the boundary-layer drag coefficient (Eq. 3.21) has a value of 0.0017, so the form drag is evidently significant.

All the above form-drag considerations apply to neutral conditions only, and the derived form-drag coefficients can be used to evaluate related effective roughness lengths using Eqs. 9.14 and 3.21. As with small-scale frictional effects, it is then to be expected that form-drag coefficients and the form drag itself decrease to small values even in rugged terrain, as conditions become strongly stable. With stable stratification, however, internal gravity waves are forced by flow over the terrain so account needs to be taken of the contribution of gravity waves to surface stress.

Gravity-wave drag

The drag on the lower atmosphere due to surface friction arises mainly from vertical momentum transfer by small-scale turbulent eddies. In stable conditions, the turbulence is suppressed and the drag is reduced towards zero. However, in stably stratified flow over orography, drag due to internal gravity waves is possible. For example, consider a fluid with buoyancy frequency $N (N^2 = (g/\theta) \partial\theta/\partial z)$ flowing at speed U over a roughness obstacle (e.g. hill) of characteristic width L_w. Then, if $L_w \gg U/N$, the time taken for a parcel to traverse the hill is much greater than the buoyancy time scale $(2\pi/N)$, so that the form drag across the obstacle is dominated by buoyancy effects, and thus by internal gravity-wave generation. In this case, the drag does not depend on any inertial effect (i.e. the wind speed) as with small-scale friction, and a drag force can be imposed on the fluid without any downstream boundary-layer separation.

Such a buoyancy-dependent drag depends on the stability N of the flow, its wind speed U and the roughness of the orography σ_h. A simple parameterization of this drag that is used in GCMs, and reflects these features, can be written as $\tau_0^{igw} = \rho C_g' U^2$, with (Miller *et al.*, 1989)

$$C_g' = k_0 N \sigma_h^2 / U. \tag{9.18}$$

A typical value of the "tuning" constant k_0 is $2.5 \times 10^{-5} \text{ m}^{-1}$, so for values of $N^2 = 0.00015 \text{ s}^{-2}$ and $U = 10 \text{ m s}^{-1}$ characteristic of the lower atmosphere this gives C_g' equal to 0.00002, 0.00031 and 0.0019 for σ_h equal to 25, 100 and 250 m respectively. Such values are comparable with small-scale friction and form-drag coefficients discussed earlier. Of course, such expressions as Eq. 9.18 are highly simplified representations of the actual complex physical processes occurring in stably stratified flow over complex orography.

9.3.3 Canopies and surface hydrology

Canopies

Equations 5.30 and 5.31 represent one of the simplest means of incorporating the effects of vegetation on evaporation. Such an approach treats the canopy as one layer, or "big leaf", with water availability described by a surface (canopy) resistance. Until recently this level of detail was as far as many GCMs went, with most having a very poor representation of the energy exchange processes at the Earth's surface. In Chapter 8, we described two improved canopy models for use in GCMs. Both are similar to the vegetation model described by Deardorff (1978) which was seen at the time as an ambitious attempt to introduce realistic interactions of vegetation with the atmosphere. In recent times, single-level canopy models that are somewhat more sophisticated, but that are still basically extensions of the Deardorff model, have appeared specifically for use in GCMs. These include the simple biosphere model (SiB), and the canopy component of the biosphere–atmosphere transfer system (BATS). These models each represent a particular level of sophistication and complexity. For example, SiB uses a canopy with an understorey and three soil layers; it is essentially two big-leaf models, one in direct contact with the soil and the other distributed as a tall porous upperstorey canopy. In addition, SiB contains nearly 50 adjustable constants for each vegetation and soil combination, so that the problem of assigning values to these "constants" over a heterogeneous grid area containing many surface types would seem to be daunting.

It is fair to ask whether this new level of complexity is the most appropriate for use in a GCM. One problem faced by all canopy models to date is how to deal properly with sparse vegetation cover, where ground fluxes are likely to be significant, or the averaging of fluxes over an area comprising sub-areas of bare soil and dense, but differing, vegetation types. The formulations described in Chapter 8, and found in the models above, do not distinguish between homogeneous foliage cover having $\sigma_f = 0.5$, or dense foliage over half the grid area and bare soil over the other half. From the plant physiologist's or

ecologist's point-of-view, such canopy models are seen as gross oversimplifica-
tions, yet many GCM modellers have quite the opposite viewpoint! In addition,
it is not clear that such added sophistication will lead to improvements in GCM
climate simulations without parallel improvements in other aspects of model
parameterization schemes, including clouds, rainfall, sub-grid orography and
ABL dynamics. Major requirements in this regard are very carefully designed
GCM sensitivity studies and comparison of simulations with compatible observations that take into account area averaging.

Surface hydrology
Interactive surface hydrology is now an accepted and necessary component of
land-surface schemes in GCMs. Such schemes allow for surface runoff only
when precipitation falls on an already saturated soil. The deep-soil percolation
can be set to a non-zero value. Improved interactive hydrological modelling is
necessary both for more realistic estimates of evaporation and to allow the
hydrology to respond more realistically to the evaporation and precipitation
rates generated in the GCM. In future, surface hydrological schemes should
account for deep-soil percolation in a physically realistic way, and should include
runoff occurring in response to intense precipitation events. To do this,
parameterizations must be developed for the representation of sub-grid hydrological processes and the spatially variable nature of precipitation in the real
world, particularly in convective conditions (e.g. Pitman *et al.*, 1990).

   For a physically realistic parameterization of surface runoff, two mechanisms
of overland flow must be modelled: firstly, runoff due to excess precipitation
over maximum soil infiltration and, secondly, runoff due to precipitation over
saturated or impermeable surfaces. For GCM grid-area scales, the area-averaged
precipitation intensity as normally calculated rarely assumes magnitudes in
excess of the area-averaged soil infiltration. Thus, large local runoff rates
following intense local storms cannot be represented with such schemes. What is
required is a fractional wetting parameterization within the GCM, where soil
hydraulic properties and precipitation intensity are assumed to be spatially
distributed across a grid area according to some specified probability distribution. The net effect should be to allow more runoff, and therefore to reduce the
soil moisture available for evaporation into the atmosphere. Experimentation
must determine whether this would result in a reduced value of globally
averaged rainfall (which tends to be excessive in most GCM climate simulations).

9.3.4 ABL clouds
Cloud-topped boundary-layer (CTBL) regimes are a dominant feature of the
climate over extensive areas of the globe and it is essential that the dominant
physical processes in such boundary layers should be studied so that eventually
they can be incorporated into weather prediction and climate models. Chapter 7
gives an introduction and overview of current understanding of the CTBL and,
though somewhat limited in terms of a detailed treatment, there is much there
that would serve as a basis for developing parameterization schemes in numerical models of the atmosphere. We reproduce below a number of recommended

actions proposed in a recent JSC/CAS workshop on modelling of the CTBL (World Meteorological Organization, 1985).

(i) Studies are required to determine the main processes affecting turbulence and entrainment and their interaction within the ABL, particularly the effect of horizontal cloud inhomogeneities on the spatial distribution of radiative properties and on evolution of the cloud structure.

(ii) Theoretical studies should be encouraged on organized mesoscale patterns of convective cells to identify the factors that determine the evolution and geometry of open and closed cells, and cloud streets, and to provide appropriate parameterization schemes.

(iii) Existing and planned CTBL observational data sets should be compared with large-eddy simulations and higher-order closure model results to clarify the generation and redistribution of TKE, the entrainment process and the separation between the cloud layer and the sub-cloud layer.

(iv) Future observational programmes should be carefully planned to ensure that all relevant physical processes are considered, using the results of past experiments and numerical modelling studies. The International Satellite Cloud Climatology Project data set should be used to develop procedures for identifying different CTBL regimes leading to the establishment of a global climatology of such regimes.

(v) Close attention should be paid to the strong direct coupling between radiation and turbulence in the design of GCM parameterizations of the CTBL. The most promising strategy for improving a given parameterization scheme is to study carefully the performance of that scheme in a particular GCM and to compare simulations both with observations and with the results from small-scale ABL models. Objective statistical methods for evaluating GCM performance against global, long-term data sets should be pursued. Sensitivity tests of GCMs should be conducted to help determine the accuracy required of CTBL data sets if they are to provide useful constraints on the parameterization schemes.

Notes and bibliography

Section 9.2

Two useful reference papers dealing with the sensitivity of numerically simulated climate to land-surface schemes are

Mintz, Y. (1984), The sensitivity of numerically simulated climates to land-surface boundary conditions, Chapter 6 in *Global Climate*, ed. J. T. Houghton, pp. 79–105. Cambridge University Press, and

Rowntree, P. R. (1988), Review of GCMs as a basis for predicting the effects of vegetation change on climate, in *Forests, Climate and Hydrology – Regional Impacts*, eds E. R. C. Reynolds and F. B. Thompson, pp. 162–96. The United Nations University, Tokyo.

The problems of including ABL clouds in GCMs, and the impact of their parameterization on the simulated climate, have not been discussed at length in this chapter. Two papers that deal with the ABL cloud parameterization problem are

Randall, D. A., Abeles, J. A. and T. G. Corsetti (1985), Seasonal simulations of the planetary boundary layer and boundary-layer stratocumulus clouds with a general circulation model, *J. Atmos. Sci.* **42**, 641–76, and

Slingo, A., Wilderspin, R. C. and R. N. B. Smith (1989), Effect of improved physical parameterizations on simulations of cloudiness and the Earth's radiation budget, *J. Geophys. Res.* **94**(D2), 2281–301.

On the subject of regional models and regional climate, far fewer studies on the importance of the ABL in climate simulations have been made with regional models than with GCMs. In one of these studies,

Mahfouf, J. F., Richard, E., Mascart, P., Nickerson, E. C. and R. Rosset (1987), A comparative study of various parameterizations of the planetary boundary layer in a numerical mesoscale model, *J. Clim. Appl. Meteor.* **26**, 1671–95,

a three-dimensional mesoscale model with a range of ABL schemes was used to simulate a sea breeze. The schemes included a TKE-based K (A); an O'Brien K with prognostic h (B); use of $h = 0.25u_*/|f|$ (C) and use of $h = 1$ km (D). In a diurnal integration, fields at 1200 local time were compared; for surface wind and ABL temperature, schemes gave similar results, but updraughts were weaker in A and B because of deeper mixing near the main updraughts.

Holt, T. and S. Raman (1988), A review and comparative evaluation of multilevel boundary layer parameterizations for first-order and turbulent kinetic energy closure schemes, *Rev. Geophys.* **26**, 761–80,

recently described an evaluation of multi-level ABL parameterization schemes for K, confining the study to first-order and TKE closures. Any of the closure schemes would be suitable for use in larger-scale models. Eleven schemes were assessed by comparison of one-dimensional ABL model simulations of mean and turbulent structure with MONEX 79 data: the two main conclusions showed that (i) the mean structure of the ABL is fairly insensitive to the type of closure scheme (assuming that the scheme properly accounts for turbulent mixing); (ii) TKE closure is preferable to first-order closure in predicting the overall turbulent structure of the boundary layer.

Section 9.3

Much of the material on vertical fluxes in Section 9.3.1 is based on the analysis of

Mahrt, L. (1987), Grid-averaged surface fluxes, *Mon. Wea. Rev.* **115**, 1550–60;

Mahrt has evaluated the function A_1 for the case of the surface heat flux, using a numerical model.

In the context of effective roughness lengths, also discussed in Section 9.3.1,

André, J. C. and C. Blondin (1986), On the effective roughness length for use in numerical 3-D models, *Bound. Layer Meteor.* **35**, 231–45,

assumed constant velocity at the lowest model level across the domain, and derived an expression for the effective z_0 that depended upon the height of the lowest level. There seems to be little physical support for their method.

For grid scales ~ 10 km,

Mason, P. J. (1986), On the parameterization of orographic drag, in *Observation, Theory and Modelling of Orographic Effects*, Vol. 1, pp. 167–94. European Centre for Medium-range Weather Forecasts Workshop, Reading, UK,

has utilized both linear theory (for shallow slopes) and bluff-body dynamics (for steep slopes) to provide a more physically based extension to Eq. 9.16, with the limit for large slopes of $z_0 < 0.1\sigma_h$.

Appendix 1
Cartesian tensor notation

In many fields of science, vector quantities are represented by three Cartesian coordinates. Thus, the three-dimensional velocity vector $\mathbf{V}$ represents $u\mathbf{i} + v\mathbf{j} + w\mathbf{k} = (u, v, w)$, where $\mathbf{i}$, $\mathbf{j}$ and $\mathbf{k}$ are unit vectors in the x-, y- and z-directions (also written as $\mathbf{e}_1$, $\mathbf{e}_2$ and $\mathbf{e}_3$ in tensor notation). However, it is often convenient so far as mathematical manipulation is concerned to adopt the formalism of tensor notation. I have done so in this book, and remind the reader here of the elementary properties of Cartesian tensors (for rectangular Cartesian coordinates). The three-dimensional velocity field is represented by the first-order tensor $u_i = (u_1, u_2, u_3)$. The stress σ_{ij} introduced in Chapter 2 is an example of a second-order tensor, with nine components (six of which are independent).

The introduction of Cartesian tensors requires the use of several simple rules, as follows.

(i) Repeated indices in a single term are summed. This is called the *summation convention*; thus, in three-dimensional space, $a_{ii} = a_{jj} = a_{11} + a_{22} + a_{33}$.

(ii) Non-repeated indices in a term indicate the order of that term, and are called free indices (e.g. u_i is a first-order tensor; σ_{ij} is a second-order tensor). The maximum value of the free index is equal to the number of spatial dimensions in the system. For the atmosphere, this is three.

(iii) Only tensors of the same order can be added together.

(iv) Addition and multiplication (dot and cross) of tensors can be performed as for vectors.

(v) Two special tensors exist:

    (a)   the Kronecker delta δ_{ij}, such that $\delta_{ij} = 1$ when $i = j$, and $\delta_{ij} = 0$ when $i \neq j$. This is relevant to the representation of the gravitational term in the equation of motion. Note that $\delta_{ij} = \mathbf{e}_i \cdot \mathbf{e}_j$.

    (b)   the alternating unit tensor ε_{ijk}, which is zero except when i, j and k are all different. In this case, the value is $+1$ or -1 according to whether i, j and k are, or are not, in cyclic order (e.g. $\varepsilon_{112} = 0$, $\varepsilon_{132} = -1$, $\varepsilon_{231} = 1$). Note that $\varepsilon_{ijk}\mathbf{e}_k = \mathbf{e}_i \times \mathbf{e}_j$ (the vector cross product).

The following identities exist between vector and tensor notation:

$$\mathbf{V} \equiv u_i \mathbf{e}_i \tag{A1}$$

$$\boldsymbol{\nabla} \equiv \mathbf{e}_j \, \partial/\partial x_j \tag{A2}$$

although the tensors are usually written without the $\mathbf{e}_i$, $\mathbf{e}_j$ so that the tensor is indicated by its ith component, e.g. $\mathbf{V}$ is indicated by u_i. We also have

$$\nabla^2 \equiv \partial^2/\partial x_j \partial x_j \tag{A3}$$

$$\partial/\partial t + \mathbf{V} \cdot \boldsymbol{\nabla} \equiv \partial/\partial t + u_j \, \partial/\partial x_j. \tag{A4}$$

Note that

$$\mathbf{V} \cdot \boldsymbol{\nabla} \equiv u_j \, \partial/\partial x_j \quad \text{and} \quad \mathbf{V}' \cdot \boldsymbol{\nabla} \equiv u_j' \, \partial/\partial x_j \tag{A5}$$

whereas

$$\boldsymbol{\nabla} \cdot \mathbf{V} \equiv \partial u_j/\partial x_j \quad \text{and} \quad \boldsymbol{\nabla} \cdot \mathbf{V}' \equiv \partial u_j'/\partial x_j. \tag{A6}$$

The gravitational and rotation terms appearing in the equation of motion (Eq. 2.40) are thus written in tensor and vector form as

$$- g\delta_{i3} - 2\Omega\varepsilon_{ijk}\eta_j u_k$$

and

$$- g\mathbf{k} - 2\Omega\boldsymbol{\eta} \times \mathbf{V}.$$

Here, $\boldsymbol{\eta}$ is a unit vector and η_j ($j = 1, 2, 3$) is the jth component, where

$$\boldsymbol{\eta} = (0, \cos\phi, \sin\phi), \tag{A7}$$

and ϕ is the latitude. Expansion of both the above forms gives, for the u-component ($i = 1$),

$$(2\Omega\sin\phi)v - (2\Omega\cos\phi)w \approx (2\Omega\sin\phi)v$$

for the v-component ($i = 2$),

$$- (2\Omega\sin\phi)u,$$

and for the w-component ($i = 3$),

$$- g + (2\Omega\cos\phi)u \approx - g.$$

As another example, consider the left-hand side of Eq. 2.40, i.e.

$$\partial\mathbf{V}/\partial t + \mathbf{V} \cdot \boldsymbol{\nabla}\mathbf{V} \quad \text{in vector form}$$

or

$$\partial u_i/\partial t + u_j \, \partial u_i/\partial x_j \quad \text{in component form.}$$

Setting $\mathbf{V} = \overline{\mathbf{V}} + \mathbf{V}'$ or $u_i = \overline{u}_i + u_i'$ and averaging gives Eq. 2.41, with the left-hand side of Eq. 2.40 becoming

$$\partial\overline{\mathbf{V}}/\partial t + \overline{\mathbf{V}} \cdot \boldsymbol{\nabla}\overline{\mathbf{V}} + \overline{\mathbf{V}' \cdot \boldsymbol{\nabla}\mathbf{V}'} \quad \text{in vector form}$$

or

$\partial \overline{u_i}/\partial t + \overline{u_j}\, \partial \overline{u_i}/\partial x_j + \overline{u_j'\partial u_i'/\partial x_j}$ in component form,

where we note that, by continuity, $\nabla \cdot \mathbf{V} = 0$ or $\partial u_j/\partial x_j = 0$ and

$$\overline{\mathbf{V}' \cdot \nabla \mathbf{V}'} \equiv e_i \overline{u_j'\partial u_i'/\partial x_j} = e_i \overline{\partial u_i' u_j'/\partial x_j} \tag{A8}$$

where e_i is the magnitude (unity) of $\mathbf{e}_i$.

The TKE equation (Eq. 2.72) is derived by multiplying Eq. 2.43 (for u_i') by u_i', summing for $i = 1, 2$ and 3 and averaging. In vector notation, the analogous equation for $\mathbf{V}'$ is dot multiplied by $\mathbf{V}'$ and averaged. Thus, the left-hand side terms in the fluctuation equations are

$$\partial \mathbf{V}'/\partial t + \overline{\mathbf{V}} \cdot \nabla \mathbf{V}' + \mathbf{V}' \cdot \nabla \overline{\mathbf{V}} + \mathbf{V}' \cdot \nabla \mathbf{V}' - \overline{\mathbf{V}' \cdot \nabla \mathbf{V}'} \quad \text{in vector form}$$

and

$$\partial u_i'/\partial t + \overline{u_j}\partial u_i'/\partial x_j + u_j'\partial \overline{u_i}/\partial x_j + u_j'\partial u_i'/\partial x_j - \overline{u_j'\partial u_i'/\partial x_j}$$

in component form,

and the resulting terms in the TKE equation are

$$\partial \bar{e}/\partial t + \overline{\mathbf{V}} \cdot \nabla \bar{e} + \overline{\mathbf{V}' \cdot (\mathbf{V}' \cdot \nabla \overline{\mathbf{V}})} + \nabla \cdot (\overline{\mathbf{V}'e})$$

and

$$\partial \bar{e}/\partial t + \overline{u_j}\partial \bar{e}/\partial x_j + \overline{u_i'u_j'}\,\partial \overline{u_i}/\partial x_j + \partial \overline{u_j'e}/\partial x_j.$$

Note that $\bar{e}$ represents the TKE and should not be confused with the unit vector $\mathbf{e}_i$. For the transport term, we have used

$$\partial \overline{u_j'e}/\partial x_j = \overline{u_j'\partial e/\partial x_j} + \overline{e\partial u_j'/\partial x_j} = \overline{u_j'\partial e/\partial x_j} \tag{A9}$$

and

$$\nabla \cdot (\overline{\mathbf{V}'e}) = \overline{\mathbf{V}' \cdot \nabla e} + \overline{e\nabla \cdot \mathbf{V}'} = \overline{\mathbf{V}' \cdot \nabla e}. \tag{A10}$$

The molecular term in the TKE equation, $\varepsilon = - \overline{vu_i'\partial^2 u_i'/\partial x_j\partial x_j}$, can also be written in vector notation as $\varepsilon = - v\overline{\mathbf{V}' \cdot \nabla^2 \mathbf{V}'}$.

Appendix 2
Geophysical and radiation quantities, and thermodynamic properties of dry air and water vapour

Geophysical and radiation properties

The Smithsonian Physical Tables (Forsythe, 1959) and Meteorological Tables (List, 1949) are widely used reference sources for atmospheric scientists and those working in associated fields. Recently, Beer (1989, 1990) has published a set of Oceanographic and Meteorological Tables which are comparable in many respects with the Smithsonian Tables, and have the additional advantages of using up-to-date values of numerous physical constants and being available in a PC-compatible form. We have used Beer's Tables as the source of many of the data given in these Appendices.

Royal Society (1975) gives a list of recommended values of many physical constants, some of which have been updated. The publication itself is strongly recommended for its statement on the use of SI units, and its discussion on quantities, units and symbols used in science in general. Throughout the book, I have endeavoured to use SI units.

Geophysical and radiation parameters used throughout the text are listed, with some elaboration, below.

Angular velocity of rotation of the Earth, $\Omega = 7.292116 \times 10^{-5}\,\text{rad}\,\text{s}^{-1}$, where $\Omega = 2\pi/T$, and T is the length of the sidereal day, 23 h 56 m 4.09 s.

Radius of the Earth: at the equator, $R_E = 6378.137$ km; at the poles, $R_E = 6356.750$ km corresponding to a flattening factor of 1/298.257.

Acceleration due to gravity, g: this varies with location on the Earth's surface and with altitude. Modern formulae give it as a smoothly varying function of latitude ϕ and altitude above mean sea level z. Then, with g_0 representing the value at $z = 0$ for any latitude,

$$g = g_0(1 + z/R_E)^{-2} \tag{A11}$$

where

$$g_0 = 9.78033(1 + G_1) \tag{A12}$$

and

$$G_1 = (0.0053 \sin^2 \phi - 5.8 \times 10^{-6} \sin^2 2\phi). \qquad (A13)$$

This gives $g = 9.7803$ and $9.8322 \, \mathrm{m\,s}^{-2}$ at mean sea level at the equator and the poles respectively, with the standard value of $9.80665 \, \mathrm{m\,s}^{-2}$ corresponding to latitude approximately $45.5°$.

Solar constant is $1367 \, \mathrm{W\,m}^{-2}$.

Stefan–Boltzmann constant $\sigma = 5.67051 \times 10^{-8} \, \mathrm{W\,m}^{-2}\,\mathrm{K}^{-4}$.

1 atmosphere is $1013.25 \, \mathrm{hPa}$ (or mbar).

The freezing point of water at 1 atmosphere is $0 \, °\mathrm{C}$ or $273.15 \, \mathrm{K}$.

Thermodynamic properties of dry air and water vapour

We set out below the values of some basic quantities relevant to the thermodynamical properties of dry air, water vapour, and a mixture of both (moist air).

Universal gas constant $R = 8.3145 \, \mathrm{J\,mol}^{-1}\,\mathrm{K}^{-1}$.

Mean molecular weight for dry air $M_d = 0.028965 \, \mathrm{kg\,mol}^{-1}$.

Gas constant for dry air $R_d = 287.05 \, \mathrm{J\,kg}^{-1}\,\mathrm{K}^{-1}$.

Molecular weight for water vapour $M_v = 0.018015 \, \mathrm{kg\,mol}^{-1}$.

Gas constant for water vapour $R_v = 461.53 \, \mathrm{J\,kg}^{-1}\,\mathrm{K}^{-1}$.

The densities of air and of water vapour at temperature T (in kelvins) can be deduced from the ideal gas equation:

$$\rho_d = p/R_d T, \qquad \rho_v = p/R_v T, \qquad (A14)$$

where p is the pressure. The gas equation for moist air is given by

$$\rho = p/R_d T_v \qquad (A15)$$

where $T_v = T(1 + 0.61q)$ is the virtual temperature. For dry air, density values are given in Table A1.

For use in the context of evaporation, the saturation vapour pressure (svp) and slope of the svp *versus* temperature curve are needed; both are strong functions of temperature.

Let e be the vapour pressure above a plane surface of pure water and e^* the saturation vapour pressure. For ideal gases,

Mixing ratio $r = \rho_v/\rho_d = 0.622e/(p - e)$ (A16)

Specific humidity $q = \rho_v/\rho = 0.622e/(p - 0.378e)$ (A17)

where $\rho = \rho_v + \rho_d$, $0.622 = M_v/M_d$ and $0.378 = 1 - M_v/M_d$.

The variation of e^*, and thus of q^*, the saturation specific humidity, with T is given by the Clausius–Clapeyron equation

$$de^*/dT = \lambda e^*/(R_v T^2) \qquad (A18)$$

with the dependence of λ, the latent heat of vaporization of water, upon T given by

$$d\lambda/dT = c_{pv} - c_l \qquad (A19)$$

where c_{pv} is the specific heat for water vapour at constant pressure and c_l is the specific heat of pure water.

Table A1. *Values of the density of dry air in* kg m^{-3} *based on the ideal gas equation*

T (°C)	Pressure (hPa)							
	1013.25	1000	975	950	925	900	875	850
0	1.292	1.275	1.243	1.212	1.180	1.148	1.116	1.084
5	1.269	1.252	1.221	1.190	1.159	1.127	1.096	1.065
10	1.247	1.230	1.200	1.169	1.138	1.107	1.077	1.046
15	1.225	1.209	1.179	1.149	1.118	1.088	1.058	1.028
20	1.204	1.188	1.159	1.129	1.099	1.070	1.040	1.010
25	1.184	1.168	1.139	1.110	1.081	1.052	1.022	0.993
30	1.164	1.149	1.120	1.092	1.063	1.034	1.006	0.977
35	1.145	1.131	1.102	1.074	1.046	1.017	0.989	0.961
40	1.127	1.112	1.085	1.057	1.029	1.001	0.973	0.946

For dry air, the isobaric specific heat c_p has a weak temperature dependence given by

$$c_p = 1005 + (T - 250)^2/3364 \text{ J kg}^{-1}\text{ K}^{-1}. \tag{A20}$$

In the range of temperature of interest, say $-30\,°C$ to $50\,°C$, the *specific heats* have the following values.

> *Specific heat of dry air,* c_p varies from 1005.0 at $-30\,°C$ to 1006.6 J kg^{-1} K^{-1} at 50 °C.
> *Specific heat of water vapour in dry air at* 0 °C, $c_{pv} = 1859$ J kg^{-1} K^{-1}.
> *Specific heat of water at* 0 °C, $c_l = 4217$ J kg^{-1} K^{-1}.

Latent heats at 0 °C are as follows.

> *Latent heat of vaporization,* $\lambda = 2.502 \times 10^6$ J kg^{-1}.
> *Latent heat of fusion of ice,* $\lambda_f = 0.334 \times 10^6$ J kg^{-1}.

The *saturation specific humidity,* q^* is pressure dependent, so the values given in Table A2 are referred to $p = 1000$ hPa. Values of e^* are based on Bolton's (1980) fit to Wexler (1976),

$$e^* = 6.112 \exp[17.67(T - 273.15)/(T - 29.65)], \tag{A21}$$

with T in kelvins, though other alternatives that give values that are nearly as accurate in the range of temperature 0–40 °C are given by Richards (1971), Wexler (1976) and the Goff–Gratch formulae (see Beer, 1990). Values of de^*/dT are obtained from Eq. A18, with λ given by the integrated form of Eq. A19 (assuming the specific heats are constant with temperature and given by their values at 0 °C). These values are very close to those evaluated from the derivative of the Bolton formula. Values of dq^*/dT are evaluated from the derivative of Eq. A17, i.e.

$$dq^*/dT = [0.622p/(p - 0.378e^*)^2]de^*/dT. \tag{A22}$$

Table A2. *Variation with temperature of a range of thermodynamic quantities*

Here, $\gamma = c_p/\lambda$, $s_1 = de*/dT$ (from Eq. A18) and $s_2 = dq*/dT$; both are accurate determinations at the specified temperatures. For intermediate temperatures, linear interpolation will suffice for most purposes, otherwise the above Bolton formula should be used to evaluate $e*$ and to evaluate s_1 from Eq. A18. Values of the specific heat for dry air (c_p) and pure water (c_1) are also given. Calculations are for a pressure of 1000 hPa.

T (°C)	$e*$ (hPa)	$q*$ 3.81×10^{-3}	λ (J kg^{-1}) 2.50×10^{6}	γ (K^{-1}) 0.402×10^{-3}	s_1 (hPa K^{-1})	s_2 (K^{-1}) 0.278×10^{-3}	c_p (J kg^{-1} K^{-1})	c_1 (J kg^{-1} K^{-1})
0	6.11				0.444		1005.16	4217
2.5	7.31	4.56	2.50	0.402	0.521	0.326	1005.20	
5	8.72	5.44	2.49	0.404	0.608	0.381	1005.24	4202
7.5	10.36	6.47	2.48	0.405	0.708	0.440	1005.28	
10	12.27	7.67	2.48	0.405	0.822	0.516	1005.33	4192
12.5	14.48	9.06	2.47	0.407	0.951	0.598	1005.38	
15	17.04	10.7	2.47	0.407	1.097	0.691	1005.44	4185
17.5	19.99	12.5	2.46	0.409	1.261	0.797	1005.49	
20	23.37	14.7	2.45	0.410	1.447	0.916	1005.55	4182
22.5	27.25	17.1	2.45	0.410	1.654	1.050	1005.62	
25	31.67	19.9	2.44	0.412	1.886	1.202	1005.69	4179
27.5	36.72	23.2	2.44	0.412	2.145	1.372	1005.76	
30	42.46	26.8	2.43	0.414	2.434	1.564	1005.84	4178
32.5	48.96	31.0	2.43	0.414	2.754	1.778	1005.92	
35	56.31	35.8	2.42	0.416	3.109	2.019	1006.01	4178
37.5	64.61	41.2	2.41	0.417	3.501	2.288	1006.10	
40	73.95	47.3	2.41	0.418	3.934	2.590	1006.19	4178

Table A3. *Variation with temperature of the kinematic viscosity (v) and the thermal diffusivity (κ_T) of dry air, and the molecular diffusivity of water vapour (κ_V) in dry air, at $p = 1000$ hPa.*

T (°C)	v	κ_T (m^2 s^{-1})	κ_V
0	1.35×10^{-5}	1.90×10^{-5}	2.26×10^{-5}
5	1.39	1.96	2.34
10	1.44	2.03	2.41
15	1.48	2.09	2.49
20	1.53	2.16	2.57
25	1.58	2.23	2.65
30	1.62	2.30	2.73
35	1.67	2.37	2.81
40	1.72	2.44	2.89

Values of e^* calculated from the Goff–Gratch and Bolton formulae, and of de^*/dT calculated from Eq. A18 and the differential of Eq. A21, differ very little over the temperature range of greatest interest, 0–40 °C, as illustrated below.

T (°C)	e^* (hPa)		de^*/dT (hPa K^{-1})	
	Goff–Gratch	Bolton	Eq. A18	d/dT of Eq. A21
0	6.11	6.11	0.444	0.444
10	12.27	12.27	0.822	0.822
20	23.37	23.37	1.447	1.448
30	42.43	42.46	2.434	2.442
40	73.78	73.95	3.934	3.959

Molecular transport properties of dry air

The dynamic viscosity of dry air, μ, is a function of temperature but not pressure. The kinematic viscosity v by definition varies with pressure, as do the thermal diffusivity of dry air, κ_T, and the molecular diffusivity of water vapour in dry air, κ_V. Table A3 gives the temperature variation of v, κ_T and κ_V at a pressure of 1000 hPa. Between $T = 0$ °C and 40 °C, the Prandtl number ($Pr = v/\kappa_T$) remains constant at about 0.71, and the Schmidt number ($Sc = v/\kappa_V$) at about 0.60.

Appendix 3
Depth of the neutral boundary layer

The neutral ABL in barotropic, horizontally homogeneous conditions is also referred to as the Ekman layer, following the work of Ekman (1905) on the influence of the Earth's rotation on ocean currents. Asymptotic similarity theory for the neutral, barotropic ABL introduces a scale height $u_{*0}/|f|$ (e.g. Csanady, 1967; Blackadar and Tennekes, 1968) or $ku_{*0}/|f|$ (Zilitinkevich, 1975; Zilitinkevich and Monin, 1974) that does not apply at low latitudes where $f \rightarrow 0$.

The depth of the neutral ABL is proportional to $u_{*0}/|f|$, where the constant of proportionality depends on the exact definition of the ABL depth. In many examples in the literature this definition is omitted. Otherwise, the depth is defined variously by the height where the wind direction first reaches that of the geostrophic wind (Clarke and Hess, 1973), where the wind magnitude first reaches the geostrophic value (Wyngaard, 1973) or where the stress approaches zero (Plate, 1971). The neutral ABL depth h is given by

$$h = cu_{*0}/|f| \tag{A23}$$

where the value of c depends on the exact definition of h. Equation A23 is sometimes referred to as the Rossby–Montgomery formula. From a practical viewpoint, the value of c is needed

   (i) to allow comparison of neutral resistance or drag formulations (Eqs. 3.20 to 3.22) with the non-neutral analogues (Eqs. 3.80, 3.81), as used in the data of Fig. 3.12 and
   (ii) to allow comparison of typical ABL depths in non-neutral and neutral conditions.

Table A4 shows values of c taken from the literature, with values tending to lie close to 0.2, the value used in Fig. 3.12. In Chapter 8, the CBL growth equation (Eq. 8.46, 8.50) implies a value of $c = 0.33$ since this is derived from the formulation of Deardorff (1974). Note that, in the drag relations (Eqs. 3.20, 3.71 and 3.80), a value for c of 0.2 or 0.4 introduces an absolute error in the normalized velocities (ku_g/u_{*0} or $k\hat{u}_g/u_{*0}$) of about 0.8 only. This compares

Table A4. *Values of the constant c taken from the literature*

In many cases, the value is quoted, without reference, and is not derived. Often, there is no clear definition of the neutral ABL depth; exceptions are indicated as follows: [a] the height at which the actual and geostrophic wind directions first coincide; [b] the height at which the stress tends to zero. In the derivation of Plate (1971), the value of c was actually given as $2k/B_0$, with $B_0 = 4.3$; [c] the height at which the actual and geostrophic wind magnitudes first coincide.

Reference	c	Comments
Csanady (1967)		Introduces a scaling depth ($\propto u_*/f$) but gives no value for c
Blackadar and Tennekes (1968)	0.25	Quote, but do not derive
Hanna (1969)	0.2[a]	Quotes three references that utilize numerical simulations
Clarke (1970)	0.2	Evaluates from observations, and interpolation to neutral ABL
Plate (1971)	0.185[b]	Analytical derivation
Wyngaard (1973)	0.25[c]	Quotes, but does not derive
Tennekes (1973)	0.3	Quotes, but does not derive
Clarke and Hess (1973)	0.3[a]	Based on an Ekman spiral assumption and observations
Deardorff (1974)	0.33	Neutral limit, from an ABL growth formula
Zilitinkevich and Monin (1974)	0.4	Thickness of neutral Ekman layer, with $c = k$
Zilitinkevich and Deardorff (1974)	0.3	Quote, but do not derive
Yamada (1976)	0.3	Quotes, but does not derive
Brutsaert (1982)	0.15 to 0.3	Quotes a typical range
Panofsky & Dutton (1984)	≈ 0.2	Quote, but do not derive

with the expected range in the normalized velocity of about 5 in light winds over rough surfaces to about 20 in strong winds over smooth surfaces.

The value of $h = 0.2u_{*0}/|f|$ can be contrasted readily with the heights of the CBL and the NBL. In the unstable case, Fig. 3.11 shows an observed range of the scale–height ratio $R = |f|h/u_{*0}$ of about 0.04 to 0.2 for a low-latitude site and about 0.25 to 0.4 for two mid-latitude sites. These latter values are thus larger then the neutral value of 0.2, demonstrating the expected deeper CBL for a given u_{*0}. In the stable case, the equilibrium height is given by Eq. 6.55 with $\gamma_c = 0.4$; this can be rewritten as $h = c'u_{*0}/|f|$, with $c' = \gamma_c^2 L/h = 0.16(h/L)^{-1}$. In strongly stable conditions, $h/L = 10$, whence $c' = 0.016$, and in only slightly stable conditions, $h/L = 1$, whence $c' = 0.16$. These smaller c' values, together with the smaller u_{*0} values that tend to occur in stable conditions, emphasize the shallower nature of the stable ABL compared with the neutral layer.

Appendix 4
Tabulations of physical data

Surface-layer constants

Values of the surface-layer "constants" used in Eqs. 3.23–3.25, 3.29 and 3.33 are given in Table A5, based on atmospheric observations and the review of Dyer (1974). The Wieringa (1980) results are based on a re-analysis of the Kansas observations used by Businger *et al.* (1971). For comments on the Hogstrom (1985) results, Telford and Businger (1986) should be consulted.

Table A5. *Values of surface-layer constants, where k is the von Karman constant, P_{tN} is the neutral turbulent Prandtl number and β and γ are experimental constants appearing in the expressions for the Monin–Obukhov stability functions Φ_M and Φ_H*

	k	P_{tN}	γ_1	γ_2	β_{1M}	β_{1H}
Observations						
W70	—	—	—	—	5.2	5.2
DH70	0.41	1	16	—	—	—
B71	0.35	0.74	15	9	4.7	4.7
G77	0.41	—	—	—	—	—
W80	0.41	1	22	13	6.9	9.2
DB82	0.40	1	28	14	—	—
W82	—	1	20.3	12.2	—	—
H85	0.40	1	—	—	4	—
H88	0.40	0.95	19	11.6	6.0	7.8
Z88	0.40	—	—	—	—	—
Review						
D74	0.41	1	16	16	5	5

Sources: W70, Webb (1970); DH70, Dyer and Hicks (1970); B71, Businger *et al.* (1971); G77, Garratt (1977a); W80, Wieringa (1980); DB82, Dyer and Bradley (1982); W82, Webb (1982); H85, Hogstrom (1985); H88, Hogstrom (1988); Z88, Zhang *et al.* (1988); D74, Dyer (1974).

The aerodynamic roughness length and the zero-plane displacement

Values of z_0 and d/h_c are given in Table A6 for a range of natural surfaces and values of h_c. Additional values of z_0 and d/h_c, usually in tabular form, can be found in e.g. Sutton (1953), Brutsaert (1982), Pielke (1984) and Stull (1988) though many values are based on the same original source. The wind dependence for flexible crops and grasses is not included here; that for the sea can be deduced from Eqs. 4.5 and 4.23.

Table A6. *Values of aerodynamic roughness length and zero-plane displacement for a range of natural surfaces*

Surface	Reference	h_c (m)	z_0 (m)	d/h_c
soils			0.001–0.01	
grass				
thick	Sutton (1953)	0.1	0.023	
thin	Sutton (1953)	0.5	0.05	
sparse	Clarke *et al.* (1971)	0.025	0.0012	
	Deacon (1953)	0.015	0.002	
		0.45	0.018	
		0.65	0.039	
crops				
wheat stubble	Izumi (1971)	0.18	0.025	
wheat	Garratt (1977b)	0.25	0.005	
		0.4	0.015	
		1.0	0.05	
corn	Kung (1961)	0.8	0.064	
beans	Thom (1971)	1.18	0.077	
vines	Hicks (1973)	0.9	0.023[a]	
		1.4	0.12[b]	
vegetation	Fichtl and McVehil (1970)	1–2	0.2	
woodland				
trees	Fichtl and McVehil (1970)	10–15	0.4	
savannah	Garratt (1980)	8	0.4	0.6
		9.5	0.9	0.75
forests				
pine	Hicks *et al.* (1975)	12.4	0.32	
pine	Thom *et al.* (1975)	13.3	0.55	
		15.8	0.92	
coniferous	Jarvis *et al.* (1976)	10.4[c]–27.5	0.28–3.9	0.61–0.92
tropical	Thomson and Pinker (1975)	32	4.8	
tropical	Shuttleworth (1989)	35	2.2	0.85

[a] Flow parallel to rows.
[b] Flow normal to rows.
[c] Range in h_c for 11 sites; the mean z_0/h_c is 0.076 and the mean d/h_c is 0.78.

Thermal properties of natural surfaces

Variations of thermal conductivity or thermal diffusivity with soil moisture content for a range of soils can be found in a number of texts, including Philip (1957, Fig. 4), Sellers (1965, Figs. 37 and 39), Marshall and Holmes (1979, Fig. 11.5), Hillel (1982, Fig. 9.1) and Pielke (1984; Fig. 11-17). Empirical relations describing these dependences are not well established, but in the case of the volumetric heat capacity, its dependence on soil moisture can be expressed as (Sellers, 1965, Chapter 9)

$$C_s = (1 - \eta_s)C_{si} + \eta C_w \qquad (A24)$$

where η is the volumetric moisture content, η_s is the saturation value (see Section 5.3.4), C_{si} is the volumetric heat capacity of the dry soil type i, and C_w is the volumetric heat capacity for water. Equation A24 is in fact a weighting of the contributions to C_s from the dry soil and the water present in the soil, with the contribution from air neglected.

Representative values of a range of properties are given in Table A7.

Table A7. *Representative values of the thermal conductivity* k_s, *specific heat* c_s, *density* ρ_s *and thermal diffusivity* κ_s *for various types of surface based mainly on Table* 11-3 *in Pielke* (1984)

Data for clay and sand are approximately consistent with Eq. A24, in which C_{si} is equal to 2.7×10^6 and 2.2×10^6 J m^{-3} K^{-1} for clay and sand respectively; C_w is equal to $\rho_w c_1$, with $\rho_w = 1000$ kg m^{-3} and $c_1 = 4186$ J kg^{-1} K^{-1}; and η_s is taken from Table A9. The reader should also consult e.g. Geiger (1965, Table 10), Hillel (1982, Table 9.3) and Oke (1987, Table 2.1).

Surface	k_s (W m^{-1} K^{-1})	c_s (J kg^{-1} K^{-1})	ρ_s (kg m^{-3})	κ_s (10^{-6} m^2 s^{-1})
Sand soil				
dry	0.3	800	1600	0.23
$\eta = 0.2$	1.9	1260	1800	0.84
$\eta = 0.4$	2.2	1480	2000	0.74
Clay soil				
dry	0.25	890	1600	0.18
$\eta = 0.2$	1.1	1170	1800	0.52
$\eta = 0.4$	1.6	1550	2000	0.52
rock	2.9	750	2700	1.4
ice	2.5	2100	910	1.3
snow				
old	1.0	2090	640	0.7
new	0.1	2090	150	0.3
water	0.6	4186	1000	0.14

Appendices

Radiative properties of natural surfaces

Values of approximate mean shortwave albedo α_s and longwave emissivity ε_s for a range of natural surfaces, including ABL clouds, are given in Table A8. Data have been taken from many sources in the literature.

Relations exist in the literature describing the dependence of soil albedo upon moisture status; thus (e.g. see Idso *et al.*, 1975)

$$\alpha_s = 0.31 - 0.34(\eta/\eta_s) \qquad \text{for } \eta/\eta_s \leqslant 0.5 \qquad (A25)$$

$$= 0.14 \qquad \text{for } \eta/\eta_s > 0.5 \qquad (A26)$$

is a typical empirical relation expressing soil albedo as a function of the fractional wetness, η/η_s.

Moisture properties of soils

Values of soil moisture quantities for 11 soil types are given in Table A9.

Table A8. *Representative values of shortwave albedo (α_s) and longwave emissivity (ε_s) for a range of natural surfaces*

Surface type	Comments	α_s	ε_s
ocean	high sun	0.05	0.95
	low sun	0.1–0.5	0.95
forest	tropical rain	0.07–0.15	0.98
	coniferous	0.1–0.19	0.98
	deciduous	0.14–0.2	0.96
crops		0.15–0.25	0.96
grasses		0.15–0.30	0.96
soils[a]	dark, wet	0.1	
	wet sandy	0.1–0.25	0.98
	wet clay	0.1–0.2	0.97
	dry sandy	0.2–0.4	0.9[b]–0.95
	dry clay	0.2–0.35	0.95
snow	fresh	0.65–0.95	0.95
	old	0.45–0.65	0.9
clouds	thick stratus, stratocumulus, nimbostratus	0.6–0.8	
	large cumulus	0.2–0.5	
	small cumulus	< 0.2	

[a] Values of albedo quoted here are consistent with Eqs. A25 and A26 giving the dependence of albedo on soil moisture.

[b] This low value is given by Sutherland (1986).

Sources: data are based on values given in Sellers (1965, Table 4), Paltridge and Platt (1976, Fig. 6.12 and Table 6.2), Pielke (1984, Tables 11-2, 11-4 and 11-8), Oke (1987, Table 1.1) and Arya (1988, Table 3.1).

Table A9. *Soil moisture quantities for a range of soil types, based on Clapp and Hornberger* (1978)

Quantities shown are as follows: η_s is the saturation moisture content (volume per volume), η_w is the wilting value of the moisture constant which assumes 150 m suction (i.e. the value of η when $\psi = -150$ m), ψ_s is the saturation moisture potential and $K_{\eta s}$ is the saturation hydraulic conductivity; b is an index parameter (see Eqs. 5.46–5.48).

Soil type	η_s $(m^3\ m^{-3})$	ψ_s (m)	$K_{\eta s}$ $(10^{-6}\ m\ s^{-1})$	b	η_w $(m^3\ m^{-3})$
1. sand	0.395	− 0.121	176	4.05	0.0677
2. loamy sand	0.410	− 0.090	156.3	4.38	0.075
3. sandy loam	0.435	− 0.218	34.1	4.90	0.1142
4. silt loam	0.485	− 0.786	7.2	5.30	0.1794
5. loam	0.451	− 0.478	7.0	5.39	0.1547
6. sandy clay loam	0.420	− 0.299	6.3	7.12	0.1749
7. silty clay loam	0.477	− 0.356	1.7	7.75	0.2181
8. clay loam	0.476	− 0.630	2.5	8.52	0.2498
9. sandy clay	0.426	− 0.153	2.2	10.40	0.2193
10. silty clay	0.492	− 0.490	1.0	10.40	0.2832
11. clay	0.482	− 0.405	1.3	11.40	0.2864

References

Albrecht, B. A., Penc, R. S. and W. H. Schubert (1985), An observational study of cloud-topped mixed layers, *J. Atmos. Sci.* **42**, 800–22.

American Meteorological Society (1990), *Symposium on FIFE* (preprints). American Meteorological Society, Boston, MA, 180 pp.

André, J. C. and L. Mahrt (1982), The nocturnal surface inversion and influence of clear-air radiative cooling, *J. Atmos. Sci.* **39**, 864–78.

André, J. C., Goutorbe, J. P. and A. Perrier (1986), HAPEX-MOBILHY: A hydrologic atmospheric experiment for the study of water budget and evaporation flux at the climatic scale, *Bull. Amer. Met. Soc.* **67**, 138–44.

Anthes, R. A. (1984), Enhancement of convective precipitation by mesoscale variations in vegetative covering in semiarid regions, *J. Clim. Appl. Meteor.* 23, 541–54.

Arya, S. P. S. (1977), Suggested revisions to certain boundary layer parameterization schemes used in atmospheric circulation models, *Mon. Wea. Rev.* **105**, 215–27.

Arya, S. P. S. (1988), *Introduction to Micrometeorology*. Academic Press, New York, 303 pp.

Arya, S. P. S. and J. C. Wyngaard (1975), Effect of baroclinity on wind profiles and the geostophic drag law for the convective planetary boundary layer, *J. Atmos. Sci.* **32**, 767–78.

Atlas, D., Walter, B., Chou, S.-H. and P. J. Sheu (1986), The structure of the unstable marine boundary layer viewed by lidar and aircraft observations, *J. Atmos. Sci.* **43**, 1301–18.

Augstein, E., Riehl, H., Ostapoff, F. and V. Wagner (1973), Mass and energy transports in an undisturbed Atlantic trade wind flow, *Mon. Wea. Rev.* **101**, 101–11.

Ball, F. K. (1960), Control of inversion height by surface heating, *Quart. J. Roy. Met. Soc.* **86**, 483–94.

Batchelor, G. K. (1953), *The Theory of Homogeneous Turbulence*. Cambridge University Press, 197 pp.

Batchelor, G. K. (1967), *An Introduction to Fluid Mechanics*. Cambridge University Press, 615 pp.

Beer, T. (1989), *Applied Environmetrics Oceanographic Tables*. Applied Environmetrics, Victoria, Australia, 38 pp.

Beer, T. (1990), *Applied Environmetrics Meteorological Tables*. Applied Environmetrics, Victoria, Australia, 56 pp.

Betts, A. K. (1973), Non-precipitating cumulus convection and its parameterization, *Quart. J. Roy. Met. Soc.* **99**, 178–96.

Betts, A. K. (1974), Reply to comment on the paper "Non-precipitating cumulus convection and its parameterization", *Quart. J. Roy. Met. Soc.* **100**, 469–71.

Betts, A. K. (1990), Diurnal variation of California coastal stratocumulus from two days of boundary layer soundings, *Tellus* **42A**, 302–4.

Betts, A. K. and W. Ridgway (1989), Climatic equilibrium of the atmospheric convective boundary layer over a tropical ocean, *J. Atmos. Sci.* **46**, 2621–41.

Bhumralkar, C. M. (1975), Numerical experiments on the computation of ground surface temperature in an atmospheric general circulation model, *J. Appl. Meteor.* **14**, 1246–58.

Blackadar, A. K. (1957), Boundary layer wind maxima and their significance for the growth of nocturnal inversions, *Bull. Amer. Met. Soc.* **38**, 283–90.

Blackadar, A. K. (1962), The vertical distribution of wind and turbulent exchange in a neutral atmosphere, *J. Geophys. Res.* **67**, 3095–102.

Blackadar, A. K. (1979), High-resolution models of the planetary boundary layer, in *Advances in Environmental Science and Engineering*, Vol. 1, eds J. R. Pfafflin and E. N. Zeigler, pp. 50–85. Gordon and Breach, New York.

Blackadar, A. K. and H. Tennekes (1968), Asymptotic similarity in neutral barotropic planetary boundary layers, *J. Atmos. Sci.* **25**, 1015–20.

Bolton, D. (1980), The computation of equivalent potential temperature, *Mon. Wea. Rev.* **108**, 1046–53.

Bougeault, P. (1985), The diurnal cycle of the marine stratocumulus layer: A higher-order model study, *J. Atmos. Sci.* **42**, 2826–43.

Bradley, E. F. (1968), A micrometeorological study of velocity profiles and surface drag in the region modified by a change in surface roughness, *Quart. J. Roy. Met. Soc.* **94**, 361–79.

Bradley, E. F., Coppin, P. A. and J. S. Godfrey (1991), Measurements of sensible and latent heat flux in the western equatorial Pacific Ocean, *J. Geophys. Res.* **96** (Supplement), 3375–89.

Brost, R. A. and J. C. Wyngaard (1978), A model study of the stably stratified planetary boundary layer, *J. Atmos. Sci.* **35**, 1427–40.

Brutsaert, W. (1975), A theory for local evaporation (or heat transfer) from rough and smooth surfaces at ground level, *Water Resour. Res.* **11**, 543–50.

Brutsaert, W. (1982), *Evaporation into the Atmosphere*. Reidel, Dordrecht, 299 pp.

Businger, J. A. (1982), Equations and concepts, in *Atmospheric Turbulence and Air Pollution Modelling*, eds F. T. M. Nieuwstadt and H. van Dop, pp. 1–36. Reidel, Dordrecht.

Businger, J. A. and H. Charnock (1983), Boundary layer structure in relation to larger-scale flow: Some remarks on the JASIN observations, *Phil. Trans. Roy. Soc. London* **A308**, 445–9.

Businger, J. A., Wyngaard, J. C., Izumi, Y. and E. F. Bradley (1971), Flux profile relationships in the atmospheric surface layer, *J. Atmos. Sci.* **28**, 181–9.

Carson, D. J. (1982), Current parameterizations of land-surface processes in atmospheric general circulation models, in *Land Surface Processes in Atmospheric General Circulation Models*, ed. P. S. Eagleson, pp 67–108. Cambridge University Press, Cambridge.

Carson, D. J. and A. B. Sangster (1981), The influence of land-surface albedo and soil moisture on general circulation model simulations, in *GARP/WCRP: Research*

Activities in Atmospheric and Oceanic Modelling, ed. I. D. Rutherford, pp. 5.14–5.21. Numerical Experimentation Programme Report No. 2, Geneva.

Carson, D. J. and F. B. Smith (1974), Thermodynamic model for the development of a convectively unstable boundary layer, in *Turbulent Diffusion in Environmental Pollution*, eds F. N. Frenkiel and R. E. Munn, *Advances in Geophysics*, Vol. 18A, pp. 111–24. Academic Press, New York.

Caughey, S. J. and S. G. Palmer (1979), Some aspects of turbulence structure through the depth of the convective boundary layer, *Quart. J. Roy. Met. Soc.* **105**, 811–27.

Caughey, S. J., Wyngaard, J. C. and J. C. Kaimal (1979), Turbulence in the evolving stable boundary layer, *J. Atmos. Sci.* **36**, 1041–52.

Chamberlain, A. C. (1983), Roughness length of sea, sand and snow, *Bound. Layer Meteor.* **25**, 405–9.

Charney, J., Quirk, W. J., Chow, S.-H. and J. Kornfield (1977), A comparative study of the effects of albedo change on drought in semi-arid regions, *J. Atmos. Sci.* **34**, 1366–85.

Charnock, H. (1955), Wind stress on a water surface, *Quart. J. Roy. Met. Soc.* **81**, 639.

Charnock, H. (1958), A note on empirical wind-wave formulae, *Quart. J. Roy. Met. Soc.* **84**, 443–7.

Charnock, H. and R. T. Pollard, eds (1983), *Results of the Royal Society Joint Air–Sea Interaction Project* (JASIN). The Royal Society, London, 229 pp.

Chervin, R. M. (1979), Response of the NCAR GCM to changed land surface albedo, in *Performance, Intercomparison and Sensitivity Studies Report of the JOC Study Conference on climate models*, ed. W. L. Gates, GARP Publication Series No. 22, Vol. 1, pp. 563–81. World Meteorological Organization, Geneva.

Clapp, R. B. and G. M. Hornberger (1978), Empirical equations for some soil hydraulic properties, *Water Resour. Res.* **14**, 601–4.

Clarke, R. H. (1970), Observational studies in the atmospheric boundary layer, *Quart. J. Roy. Met. Soc.* **96**, 91–114.

Clarke, R. H. and R. R. Brook (1979), *The Koorin Expedition – Atmospheric Boundary Layer Data over Tropical Savannah Land*, Australian Government Publishing Service, Canberra, 359 pp.

Clarke, R. H. and G. D. Hess (1973), On the appropriate scaling for velocity and temperature in the planetary boundary layer, *J. Atmos. Sci.* **30**, 1346–53.

Clarke, R. H., Dyer, A. J., Brook, R. R., Reid, D. G. and A. J. Toup (1971), *The Wangara Experiment: Boundary-layer Data*, Technical Paper No.19, Division of Meteorological Physics, CSIRO, Australia, 21 pp.

Corrsin, S. (1951), On the spectrum of isotropic temperature fluctuations in isotropic turbulence, *J. Appl. Phys.* **22**, 469–73.

Cressman, G. P. (1960), Improved terrain effects in barotropic forecasts, *Mon. Wea. Rev.* **88**, 327–342.

Csanady, G. T. (1967), On the "resistance law" of a turbulent Ekman layer, *J. Atmos. Sci.* **24**, 467–71.

Cunnington, W. M. and P. R. Rowntree (1986), Simulations of the Saharan atmosphere – Dependence on moisture and albedo, *Quart. J. Roy. Met. Soc.* **112**, 971–99.

Deacon, E. L. (1953), *Vertical Profiles of Mean Wind in the Surface Layers of the Atmosphere*, Geophysical Memoirs No. 91, The Meteorological Office, Bracknell, UK, 68 pp.

Deacon, E. L. (1977), Gas transfer to and across an air-water interface, *Tellus* **29**, 363–74.

Deacon, E. L. (1988), The streamwise Kolmogoroff constant, *Bound. Layer Meteor.* **42**, 9–17.

Deacon, E. L. and E. K. Webb (1962), Interchange of properties between sea and air, in *The Sea*, ed. M. N. Hill, Vol. 1, pp. 43–87. Interscience, New York.

Deardorff, J. W. (1972), Numerical investigation of neutral and unstable planetary boundary layers, *J. Atmos. Sci.* **29**, 91–115.

Deardorff, J. W. (1973), An explanation of anomalously large Reynolds stresses within the convective planetary boundary layer, *J. Atmos. Sci.* **30**, 1070–6.

Deardorff, J. W. (1974), Three-dimensional numerical study of the height and mean structure of a heated planetary boundary layer, *Bound. Layer Meteor.* **7**, 81–106.

Deardorff, J. W. (1976), On the entrainment rate of a stratocumulus-topped mixed layer, *Quart. J. Roy. Met. Soc.* **102**, 563–82.

Deardorff, J. W. (1977), A parameterization of ground-surface moisture content for use in atmospheric prediction models, *J. Appl. Meteor.* **16**, 1182–5.

Deardorff, J. W. (1978), Efficient prediction of ground surface temperature and moisture, with inclusion of a layer of vegetation, *J. Geophys. Res.* **83** (C4), 1889–903.

Deardorff, J. W. (1979), Prediction of convective mixed layer entrainment for realistic capping inversion structure, *J. Atmos. Sci.* **36**, 424–36.

Deardorff, J. W. (1980), Stratocumulus-capped mixed layers derived from a three-dimensional model, *Bound. Layer Meteor.* **18**, 495–527.

Deardorff, J. W. (1981), On the distribution of mean radiative cooling at the top of a stratocumulus-capped mixed layer, *Quart. J. Roy. Met. Soc.* **107**, 191–202.

Deardorff, J. W., Ueyoshi, K. and Y.-J. Han (1984), Numerical study of terrain-induced mesoscale motions and hydrostatic form drag in a heated, growing mixed layer, *J. Atmos. Sci.* **41**, 1420–41.

Denmead, O. T. and E. F. Bradley (1985), Flux-gradient relationships in a forest canopy, in *The Forest–Atmosphere Interaction*, eds B. A. Hutchinson and B. B. Hicks, pp. 421–42. Reidel, Dordrecht.

Derbyshire, S. H. (1990), Nieuwstadt's stable boundary layer revisited, *Quart. J. Roy. Met. Soc.* **116**, 127–58.

Dickinson, R. E. (1984), Modeling evapotranspiration for three-dimensional global climate models, in *Climate Processes and Climate Sensitivity*, eds J. E. Hansen and T. Takahashi, Geophysical Monographs, No. 29, pp. 58–72. American Geophysical Union, Washington DC.

Dickinson, R. E. (1988), The force-restore model for surface temperature and its generalizations, *J. Climate* **1**, 1086–97.

Dickinson, R. E. and A. Henderson-Sellers (1988), Modelling tropical deforestation: A study of GCM land-surface parametrizations, *Quart. J. Roy. Met. Soc.* **114**, 439–62.

Dickinson, R. E., Henderson-Sellers, A., Kennedy, P. J. and M. F. Wilson (1986), *Biosphere–atmosphere Transfer Scheme (BATS) for the NCAR Community Climate Model*, NCAR Technical Note NCAR/TN-275+STR, 69 pp.

Dorman, J. L. and P. J. Sellers (1989), A global climatology of albedo, roughness length and stomatal resistance for atmospheric general circulation models as represented by the simple biosphere model (SiB), *J. Appl. Meteor.* **28**, 833–55.

Driedonks, A. G. M. and P. G. Duynkerke (1989), Current problems in the stratocumulus-topped atmospheric boundary layer, *Bound. Layer Meteor.* **46**, 275–303.

Durand, P., Brière, S. and A. Druilhet (1989), A sea–land transition observed during the COAST experiment, *J. Atmos. Sci.* **46**, 96–116.

Duynkerke, P. G. (1989), The diurnal variation of a marine stratocumulus layer: A model sensitivity study, *Mon. Wea. Rev.* **117**, 1710–25.

Dyer, A. J. (1974), A review of flux-profile relationships, *Bound. Layer Meteor.* **7**, 363–72.

Dyer, A. J. and E. F. Bradley (1982), An alternative analysis of flux-gradient relationships at the 1976 ITCE, *Bound. Layer Meteor.* **22**, 3–19.

Dyer, A. J. and B. B. Hicks (1970), Flux-gradient relationships in the constant flux layer, *Quart. J. Roy. Met. Soc.* **96**, 715–21.

Dyer, A. J. and B. B. Hicks (1982), Kolmogoroff constants at the 1976 ITCE, *Bound. Layer Meteor.* **22**, 137–50.

Ebert, E. E., Schumann, U. and R. B. Stull (1989), Nonlocal turbulent mixing in the convective boundary layer evaluated from large-eddy simulation, *J. Atmos. Sci.* **46**, 2178–207.

Ekman, V. W. (1905), On the influence of the Earth's rotation on ocean currents, *Arkiv. Mat. Astron. Fysik.* **2**(11), 1–53.

Fichtl, G. H. and G. E. McVehil (1970), Longitudinal and lateral spectra of turbulence in the atmospheric boundary layer at the Kennedy Space Center, *J. Appl. Meteor.* **9**, 51–63.

Fiedler, B. H. (1984), An integral closure model for the vertical turbulent flux of a scalar in a mixed layer, *J. Atmos. Sci.* **41**, 674–80.

Finnigan, J. J. and P. J. Mulhearn (1978), A simple mathematical model of airflow in waving plant canopies, *Bound. Layer Meteor.* **14**, 415–31.

Finnigan, J. J. and M. R. Raupach (1987), Transfer processes in plant canopies in relation to stomatal characteristics, in *Stomatal Function*, eds E. Zeiger, G. Farquhar and I. Cowan, pp. 385–429. Stanford University Press, Stanford, CA.

Firestone, J. K. and B. A. Albrecht (1986), The structure of the atmospheric boundary layer in the central equatorial Pacific during January and February of FGGE, *Mon. Wea. Rev.* **114**, 2219–31.

Forsythe, W. E. (1959), *Smithsonian Physical Tables*, 9th revised edition, Smithsonian Institution, Washington DC, 827 pp.

Fravalo, C., Fouquart, Y. and R. Rosset (1981). The sensitivity of a model of low stratiform clouds to radiation, *J. Atmos. Sci.* **38**, 1049–62.

Friedlander, S. K. and L. Topper, eds (1961), *Turbulence: Classic Papers on Statistical Theory*. Interscience, New York, 187 pp.

Garratt, J. R. (1977a), Review of drag coefficients over oceans and continents, *Mon. Wea. Rev.* **105**, 915–29.

Garratt, J. R. (1977b), Aerodynamic roughness and mean monthly surface stress over Australia, CSIRO Division of Atmospheric Physics Technical Paper No. 29, 19 pp.

Garratt, J. R. (1978), Transfer characteristics for a heterogeneous surface of large aerodynamic roughness, *Quart. J. Roy. Met. Soc.* **104**, 491–502.

Garratt, J. R. (1980), Surface influence upon vertical profiles in the atmospheric near-surface layer, *Quart. J. Roy. Met. Soc.* **106**, 803–19.

Garratt, J. R. (1982), Observations in the nocturnal boundary layer, *Bound. Layer Meteor.* **22**, 21–48.

Garratt, J. R. (1983), Surface influence upon vertical profiles in the nocturnal boundary layer, *Bound. Layer Meteor.* **26**, 69–80.

Garratt, J. R. (1987), The stably stratified internal boundary layer for steady and diurnally varying offshore flow, *Bound. Layer Meteor.* **38**, 369–94.

Garratt, J. R. (1990), The internal boundary layer – A review, *Bound. Layer Meteor.* **50**, 171–203.

Garratt, J. R. and R. A. Brost (1981), Radiative cooling effects within and above the nocturnal boundary layer, *J. Atmos. Sci.* **38**, 2730–46.

Garratt, J. R. and R. J. Francey (1978), Bulk characteristics of heat transfer in the unstable, baroclinic atmosphere boundary layer, *Bound. Layer Meteor.* **15**, 399–421.

Garratt, J. R. and B. B. Hicks (1973), Momentum, heat and water vapour transfer to and from natural and artificial surfaces, *Quart. J. Roy. Met. Soc.* **99**, 680–7.

Garratt, J. R. and P. Hyson (1975), Vertical fluxes of momentum, sensible heat and water vapour during the air mass transformation experiment (AMTEX) 1974, *J. Met. Soc. Japan* **53**, 149–60.

Garratt, J. R. and B. F. Ryan (1989), The structure of the stably stratified internal boundary layer in offshore flow over the sea, *Bound. Layer Meteor.* **47**, 17–40.

Garratt, J. R. and M. Segal (1988), On the contribution of atmospheric moisture to dew formation, *Bound. Layer Meteor.* **45**, 209–236.

Garratt, J. R., Pittock, A. B. and K. Walsh (1990), Response of the atmospheric boundary layer and soil layer to a high altitude, dense aerosol cover, *J. Appl. Meteor.* **29**, 35–52.

Garratt, J. R., Wyngaard, J. C. and R. J. Francey (1982), Winds in the atmospheric boundary layer – Prediction and observation, *J. Atmos. Sci.* **39**, 1307–16.

Gates, D. M. (1980), *Biophysical Ecology*. Springer-Verlag, New York, 611 pp.

Geernaert, G. L., Katsaros, K. B. and K. Richter (1986), Variation of the drag coefficient and its dependence on sea state, *J. Geophys. Res.* **91**, 7667–79.

Geiger, R. (1965), *The Climate Near the Ground*. Harvard University Press, Cambridge, MA, 611 pp.

Ghan, S. J., MacCracken, M. C. and J. J. Walton (1988), Climatic response to large atmospheric smoke injections: Sensitivity studies with a tropospheric general circulation model, *J. Geophys. Res.* **93** (D7), 8315–37.

Hanna, S. R. (1969), The thickness of the planetary boundary layer, *Atmos. Environ.* **3**, 519–36.

Hansen, J., Johnson, D., Lacis, A., Lebedeff, S., Lee, P., Rind, D. and G. Russell (1981), Climate impact of increasing atmospheric carbon dioxide, *Science* **213**, 957–66.

Haren, L. Van and F. T. M. Nieuwstadt (1989), The behaviour of passive and buoyant plumes in a convective boundary layer, as simulated with a large-eddy model, *J. Appl. Meteor.* **28**, 818–32.

Hasse, L. (1971), The sea surface temperature deviation and the heat flow at the sea–air interface, *Bound. Layer Meteor.* **1**, 368–379.

Hess, G. D. (1973), On Rossby-number similarity theory for a baroclinic planetary boundary layer, *J. Atmos. Sci.* **30**, 1722–3.

Henderson-Sellers, A. and V. Gornitz (1984), Possible climatic impacts of land cover transformations, with particular emphasis on tropical deforestation, *Climatic Change* **6**, 231–57.

Hicks, B. B. (1973), Eddy fluxes over a vineyard, *Agric. Meteor.* **12**, 203–15.

Hicks, B. B., Hyson, P. and C. J. Moore (1975), A study of eddy fluxes over a forest, *J. Appl. Meteor.* **14**, 58–66.

Hillel, D. (1971), *Soil and Water: Physical Principles and Processes*. Academic Press, New York, 288 pp.

Hillel, D. (1982), *Introduction to Soil Physics*. Academic Press, New York, 364 pp.

Hinze, J. O. (1975), *Turbulence: An Introduction to Its Mechanism and Theory*, 2nd edition. McGraw-Hill, New York, 790 pp.

Hogstrom, U. (1985), Von Karman's constant in atmospheric boundary layer flow: Reevaluated, *J. Atmos. Sci.* **42**, 263–70.

Hogstrom, U. (1988), Non-dimensional wind and temperature profiles in the atmospheric surface layer: A re-evaluation, *Bound. Layer Meteor.* **42**, 55–78.

Hogstrom, U. (1990), Analysis of turbulence structure in the surface layer with a modified similarity formulation for near neutral conditions, *J. Atmos. Sci.* **47**, 1949–72.

Holtslag, A. A. M. and C.-H. Moeng (1991), Eddy diffusivity and countergradient transport in the convective atmospheric boundary layer, *J. Atmos. Sci.* **48**, 1690–8.

Hsu, S. A. (1983), On the growth of a thermally modified boundary layer by advection of warm air over a cooler sea, *J. Geophys. Res.* **88**(C1), 771–4.

Hsu, S. A. (1986), A note on estimating the height of the convective internal boundary layer near shore, *Bound. Layer Meteor.* **35**, 311–16.

Idso, S. B., Aase, J. K. and R. D. Jackson (1975), Net radiation – Soil heat flux relations as influenced by soil water content variations, *Bound. Layer Meteor.* **9**, 113–22.

Inoue, E. (1963), On the turbulent structure of airflow within crop canopies, *J. Met. Soc. Japan* **41**, 317–26.

Iribarne, J. V. and W. L. Godson (1981), *Atmospheric Thermodynamics*, 2nd edition, Reidel, Dordrecht, 259 pp.

Izumi, Y. (1971), *Kansas 1968 Field Program Data Report*, Air Force Cambridge Research Papers No. 379, AFCRL-72-0041, Air Force Cambridge Research Laboratory, Bedford, MA, 79 pp.

Izumi, Y, and S. J. Caughey (1976), *Minnesota 1973 Atmospheric Boundary Layer Experimental Data Report*, AFCRL Res. Rep. No. AFCRL-TR-76-0038. Air Force Cambridge Research Laboratory, Bedford, MA. 28 pp.

Jackson, N. A. (1976), The propagation of modified flow downstream of a change in roughness, *Quart. J. Roy. Met. Soc.* **102**, 924–33.

Jacobs, C. A. and P. S. Brown (1973), An investigation of the numerical properties of the surface heat-balance equation, *J. Appl. Meteor.* **12**, 1069–72.

Jarvis, P. G. (1976), The interpretation of the variations in leaf water potential and stomatal conductance found in canopies in the field, *Phil. Trans. Roy. Soc. London* **B273**, 593–610.

Jarvis, P. G., James, G. B. and J. J. Landsberg (1976), Coniferous forest, in *Vegetation and the Atmosphere*, Vol. 2, ed. J. L. Monteith, pp. 171–240. Academic Press, New York.

Kader, B. A. and A. M. Yaglom (1990), Mean fields and fluctuation moments in unstably stratified turbulent boundary layers, *J. Fluid Mech.* **212**, 637–62.

Kaimal, J. C., Wyngaard, J. C. Izumi, Y. and O. R. Coté (1972), Spectral characteristics of surface layer turbulence, *Quart. J. Roy. Met. Soc.* **98**, 563–89.

Kaimal, J. C., Wyngaard, J. C., Haugen, D. A., Coté, O. R., Izumi, Y., Caughey, S. J. and C. J. Readings (1976), Turbulence structure in the convective boundary layer, *J. Atmos. Sci.* **33**, 2152–69.

Kazanski, A. B. and A. S. Monin (1960), A turbulent regime above the ground atmosphere layer, *Bull. (Izv.) Acad. Sci. USSR, Geophys. Ser.* 110–12.

Kazanski, A. B. and A. S. Monin (1961), On the dynamic interaction between the atmosphere and Earth's surface, *Bull. (Izv.) Acad. Sci. USSR, Geophys. Ser.* 514–15.

Kitaigorodskii, S. A. and Y. A. Volkov (1965), On the roughness parameter of the sea surface and the calculation of momentum flux in the near-water layer of the atmosphere, *Akad. Nauk SSSR, Izv. Atmos. Oceanic Phys.* **1**, 566–74.

Kitchen, M., Leighton, J. R. and S. J. Caughey (1983), Three case studies of shallow convection using a tethered balloon, *Bound. Layer Meteor.* **27**, 281–308.

Kondo, J., Fujinawa, Y. and G. Naito (1973), High frequency components of ocean waves and their relation to aerodynamic roughness, *J. Phys. Oceanogr.* **3**, 197–202.

Kuettner, J. P. and J. Holland (1969), The BOMEX project, *Bull. Amer. Met. Soc.* **50**, 394–402.

Kuettner, J. P. and D. E. Parker (1976), GATE: Report on the field phase, *Bull. Amer. Met. Soc.* **57**, 11–30.

Kung, E. C. (1961), Derivation of roughness parameters from wind profile data above tall vegetation, in *Annual Report of the Meteorology Department*, University of Wisconsin, DA-36-039-SC-80282.

Kuo, H.-C. and W. H. Schubert (1988), Stability of cloud-topped boundary layers, *Quart. J. Roy. Met. Soc.* **114**, 887–916.

Kutzbach, J. E. (1961), Investigations of the modification of wind profiles by artificially controlled surface roughness, in *Annual Report of the Meteorology Department*, University of Wisconsin, DA-36-039-SC-80282, pp. 71–113.

Large, W. G. and S. Pond (1982), Sensible and latent heat flux measurements over the ocean, *J. Phys. Oceanogr.* **12**, 464–82.

Lean, J. and D. A. Warrilow (1989), Simulation of the regional climatic impact of Amazon deforestation, *Nature* **342**, 411–13.

LeMone, M. A. (1980), The marine boundary layer, in *Workshop of the Planetary Boundary Layer*, pp. 182–234. American Meteorological Society, Boston, MA.

Lenschow, D. H. and E. M. Agee (1976), Preliminary results from the air mass transformation experiment (AMTEX), *Bull. Amer. Met. Soc.* **57**, 1346–55.

Lenschow, D. H., Wyngaard, J. C. and W. T. Pennell (1980), Mean-field and second moment budgets in a baroclinic convective boundary layer, *J. Atmos. Sci.* **37**, 1313–26.

Lenschow, D. H., Li, X. S., Zhu, C. J. and B. B. Stankov (1988), The stably stratified boundary layer over the Great Plains. I. Mean and turbulence structure, *Bound. Layer Meteor.* **42**, 95–121.

Lettau, H. H. (1967), Small to large-scale features of boundary layer structure over mountain slopes, in *Proceedings of the Symposium on Mountain Meteorology*, Atmos. Sci. Paper 122, pp. 1–73, Colorado State University.

Lettau, H. H. (1969), Note on aerodynamic roughness-parameter estimation on the basis of roughness element description, *J. Appl. Meteor.* **8**, 828–32.

Lettau, H. H. and B. Davidson (1957), *Exploring the Atmosphere's First Mile*, Vols 1–2. Pergamon Press, New York, 578 pp.

Lilly, D. K. (1968), Models of cloud-topped mixed layers under a strong inversion, *Quart. J. Roy. Met. Soc.* **94**, 292–309.

List, R. J. (1949), *Smithsonian Meteorological Tables*, 6th revised edition. Smithsonian Institution, Washington, 527 pp.

Liu, W. T. and J. A. Businger (1975), Temperature profile in the molecular sublayer near the interface of a fluid in turbulent motion, *Geophys. Res. Lett.* **2**, 403–4.

Liu, W. T., Katsaros, K. B. and J. A. Businger (1979), Bulk parameterization of air–sea exchanges of heat and water vapor including the molecular constraints at the interface, *J. Atmos. Sci.* **36**, 1722–35.

Louis, J. F. (1979), A parametric model of vertical eddy fluxes in the atmosphere, *Bound. Layer Meteor.* **17**, 187–202.

Lumley, J. L. (1978), Computational modeling of turbulent flows, *Adv. Appl. Mech.* **18**, 123–76.

Lumley, J. L. and B. Khajeh-Nouri (1974), Computational modeling of turbulent transport, in *Turbulent Diffusion in Environmental Pollution*, eds F. N. Frenkiel and R. E. Munn, *Advances in Geophysics*, Vol. 18A, pp. 169–92. Academic Press, New York.

Lumley, J. L. and H. A. Panofsky (1964), *The Structure of Atmospheric Turbulence*. Wiley Interscience, New York, 239 pp.

Lundgren, T. S. and F. C. Wang (1973), Eddy viscosity models for free turbulent flows, *Phys. Fluids* **16**, 174–8.

Mahfouf, J. F., Richard, E. and P. Mascart (1987), The influence of soil and vegetation on the development of mesoscale circulations, *J. Clim. Appl. Meteor.* **26**, 1483–95.

Mahrt, L. (1976), Mixed-layer moisture structure, *Mon. Wea. Rev.* **104**, 1403–7.

Mahrt, L. (1987), Grid-averaged surface fluxes, *Mon. Wea. Rev.* **115**, 1550–60.

Manins, P. C. and B. L. Sawford (1979), A model of katabatic winds, *J. Atmos. Sci.* **36**, 619–30.

Manins, P. C. and J. S. Turner (1978), The relation between the flux ratio and energy ratio in convectively mixed layers, *Quart. J. Roy. Met. Soc.* **104**, 39–44.

Marshall, T. J. and J. W. Holmes (1979), *Soil Physics*. Cambridge University Press, 345 pp.

Mason, P. J. (1988), The formation of areally-averaged roughness lengths, *Quart. J. Roy. Met. Soc.* **114**, 399–420.

Mason, P. J. (1989), Large-eddy simulation of the convective atmospheric boundary layer, *J. Atmos. Sci.* **46**, 1492–1516.

Mason, P. J. and N. S. Callen (1986), On the magnitude of the subgrid-scale eddy coefficient in large-eddy simulations of turbulent channel flow, *J. Fluid Mech.* **162**, 439–62.

Mason, P. J. and D. J. Thomson (1987), Large-eddy simulations of the neutral-static-stability planetary boundary layer, *Quart. J. Roy. Met. Soc.* **113**, 413–43.

McIlroy, I. C. (1984), Terminology and concepts in natural evaporation, *Agric. Water Management*, **8**, 77–98.

Mellor, G. L. and H. J. Herring (1973), A survey of the mean turbulent field closure models, *AIAA J.* **11**, 590–9.

Mellor, G. L. and T. Yamada (1974), A hierarchy of turbulence closure models for planetary boundary layers, *J. Atmos. Sci.* **31**, 1791–806. (*Corrigenda* (1977) *J. Atmos. Sci.* **34**, 1482.)

Mellor, G. L. and T. Yamada (1982), Development of a turbulence closure model for geophysical fluid problems, *Rev. Geophys, Space Phys.* **20**, 851–75.

Melville, W. K. (1977), Wind stress and roughness length over breaking waves, *J. Phys. Oceanogr.* **7**, 702–10.

Merry, M. and H. A. Panofsky (1976), Statistics of vertical motion over land and water, *Quart. J. Roy. Met. Soc.* **102**, 255–60.

Miles, J. W. (1957), On the generation of surface waves by shear flows, *J. Fluid Mech.* **3**, 185–204.

Miles, J. W. (1961), On the stability of heterogeneous shear flows, *J. Fluid Mech.* **10**, 496–508.

Miller, M. J., Palmer, T. N. and R. Swinbank (1989), Parametrization and influence of subgridscale orography in general circulation and numerical weather prediction models, *Meteor. Atmos. Phys.* **40**, 84–109.

Mintz, Y. (1984), The sensitivity of numerically simulated climates to land-surface boundary conditions, Chapter 6 in *Global Climate*, ed. J. T. Houghton, pp. 79–105. Cambridge University Press.

Miyake, M. (1965), Transformation of the atmospheric boundary layer over inhomogeneous surfaces, Science Report 5R-6, University of Washington, Seattle.

Moeng, C.-H. (1984), A large-eddy-simulation model for the study of planetary boundary-layer turbulence, *J. Atmos. Sci.* **41**, 2052–62.

Moeng, C.-H. (1986), Large-eddy simulation of a stratus-topped boundary layer. Part I: Structure and budgets, *J. Atmos. Sci.* **43**, 2886–900.

Moeng, C.-H. and J. C. Wyngaard (1984), Statistics of conservative scalars in the convective boundary layer, *J. Atmos. Sci.* **41**, 3161–9.

Moeng, C.-H. and J. C. Wyngaard (1989), Evaluation of turbulent transport and dissipation closures in second-order modeling, *J. Atmos. Sci.* **46**, 2311–30.

Monin, A. S. and A. M. Obukhov (1954), Basic laws of turbulent mixing in the atmosphere near the ground, *Tr. Akad. Nauk SSSR Geofiz. Inst.* **24**(151), 163–87.

Monin, A. S. and A. M. Yaglom (1971), *Statistical Fluid Mechanics: Mechanics of Turbulence.* Vol. 1, English translation, ed. J. L. Lumley. MIT Press, Cambridge, MA, 769 pp.

Monteith, J. L. (1957), Dew, *Quart. J. Roy. Met. Soc.* **83**, 322–41.

Monteith, J. L. (1981), Evaporation and surface temperature, *Quart. J. Roy. Met. Soc.* **107**, 1–27.

Mulhearn, P. J. (1978), A wind-tunnel boundary-layer study of the effects of a surface roughness change: rough to smooth, *Bound. Layer Meteor.* **15**, 3–30.

Mulhearn, P. J. (1981), On the formation of a stably stratified internal boundary layer by advection of warm air over a cooler sea, *Bound. Layer Meteor.* **21**, 247–54.

Nicholls, S. (1985), Aircraft observations of the Ekman layer during the joint air–sea interaction experiment, *Quart. J. Roy. Met. Soc.* **111**, 391–426.

Nicholls, S. (1989), The structure of radiatively driven convection in stratocumulus, *Quart. J. Roy. Met. Soc.* **115**, 487–511.

Nicholls, S. and J. Leighton (1986), An observational study of the structure of stratiform cloud sheets, Part I: Structure, *Quart. J. Roy. Met. Soc.* **112**, 431–60.

Nicholls, S., LeMone, M. A. and G. Sommeria (1982), The simulation of a fair weather marine boundary layer in GATE using a three dimensional model, *Quart. J. Roy. Met. Soc.* **108**, 167–90.

Nicholls, S. and J. D. Turton (1986), An observational study of the structure of stratiform cloud shects, Part II: Entrainment, *Quart. J. Roy. Met. Soc.* **112**, 461–80.

Nieuwstadt, F. T. M. (1984), The turbulent structure of the stable nocturnal boundary layer, *J. Atmos. Sci.* **41**, 2202–16.

Nieuwstadt, F. T. M. (1985), A model for the stationary, stable boundary layer, in *Turbulence and Diffusion in Stable Environments*, ed. J. C. R. Hunt, pp. 149–79. Clarendon Press, Oxford.

Nieuwstadt, F. T. M. and R. A. Brost (1986), The decay of convective turbulence, *J. Atmos. Sci.* **43**, 532–46.

Nieuwstadt, F. T. M. and J. P. J. M. M. de Valk (1987), A large-eddy simulation of buoyant and non-buoyant plume dispersion in the atmospheric boundary layer, *Atmos. Environ.* **21**, 2573–87.

Nieuwstadt, F. T. M. and H. Tennekes (1981), A rate equation for the nocturnal boundary-layer height, *J. Atmos. Sci.* **38**, 1418–28.

Noilhan, J. and S. Planton (1989), A simple parameterization of land surface processes for meteorological models, *Mon. Wea. Rev.* **117**, 536–49.

O'Brien, J. J. (1970), A note on the vertical structure of the eddy exchange coefficient in the planetary boundary layer, *J. Atmos. Sci.* **27**, 1213–15.

Oke, T. R. (1987), *Boundary Layer Climates*, 2nd edition. Methuen, New York, 435 pp.

Owen, P. R. and W. R. Thomson (1963), Heat transfer across rough surfaces, *J. Fluid Mech.* **15**, 321–34.

Paeschke, W. (1937), Experimentelle Untersuchungen zum Rauhigkeits- und Stabilitats-problem in der Bodennahen Luftschicht, *Beit. Phys. Freien Atmos.* **24**, 163–89.

Palmén, E. and C. W. Newton (1969), *Atmospheric Circulation Systems*. Academic Press, New York, 603 pp.

Paltridge, G. W. and C. M. R. Platt (1976), *Radiative Processes in Meteorology and Climatology*. Elsevier, Amsterdam, 318 pp.

Panofsky, H. A. (1973), Tower Micrometeorology, Chapter 4 in *Workshop on Microme-teorology*, ed. D. A. Haugen, pp. 151–76. American Meteorological Society, Boston, MA.

Panofsky, H. A. and J. A. Dutton (1984), *Atmospheric Turbulence – Models and Methods for Engineering Applications*. John Wiley and Sons, New York, 397 pp.

Panofsky, H. A. and A. A. Townsend (1964), Change of terrain roughness and the wind profile, *Quart. J. Roy. Met. Soc.* **90**, 147–55.

Panofsky, H. A., Tennekes, H., Lenschow, D. H. and J. C. Wyngaard (1977), The characteristics of turbulent velocity components in the surface layer under convec-tive conditions, *Bound. Layer Meteor.* **11**, 355–61.

Pasquill, F. and F. B. Smith (1983), *Atmospheric Diffusion*, 3rd edition. Wiley and Sons, New York, 437 pp.

Penman, H. L. (1948), Natural evaporation from open water, bare soil and grass, *Proc. Roy. Soc. London* **A193**, 120–46.

Perrier, A. (1982), Land-surface processes: Vegetation, in *Land Surface Processes in Atmospheric General Circulation Models* ed. P. S. Eagleson, pp. 395–448. Cam-bridge University Press.

Philip, J. R. (1957), Evaporation, and moisture and heat fields in the soil, *J. Meteor.* **14**, 354–66.

Philip, J. R. (1959), The theory of local advection, *J. Meteor.* **16**, 535–47.

Phillips, O. M. (1957), On the generation of waves by turbulent wind, *J. Fluid Mech.* **2**, 417–45.

Phillips, O. M. (1958), The equilibrium range in the spectrum of wind-generated waves, *J. Fluid Mech.* **4**, 426–34.

Pielke, R. A. (1984), *Mesoscale Meteorological Modeling*. Academic Press, New York, 612 pp.

Pitman, A. J., Henderson-Sellers, A. and Z.-L. Yang (1990), Sensitivity of regional climates to localized precipitation in global models, *Nature*, **346**, 734–7.

Plate, E. J. (1971), *Aerodynamic Characteristics of Atmospheric Boundary Layers*. US Atomic Energy Commission, Division of Technical Information, Oak Ridge, TN. 190 pp.

Prandtl, L. (1905), Ueber Flussigkeitsbewegung bei Sehr Kleiner Reibung, Verh. III, *Proceedings of the 3rd International Congress of Mathematicians* (Heidelberg 1904), pp. 484–91. Trans. (1928), Tech. Mem. No. 452, National Advisory Committee for Aeronautics, Washington DC.

Priestley, C. H. B. (1954), Convection from a large horizontal surface, *Austr. J. Phys.* **7**, 176–201.

Priestley, C. H. B. and R. J. Taylor (1972), On the assessment of surface heat flux and evaporation using large-scale parameters, *Mon. Wea. Rev.* **100**, 81–92.

Randall, D. A. (1980a), Entrainment into a stratocumulus layer with distributed radiative cooling, *J. Atmos. Sci.* **37**, 148–59.

Randall, D. A. (1980b), Conditional instability of the first kind upside-down, *J. Atmos. Sci.* **37**, 125–30.

Randall, D. A. (1984), Stratocumulus cloud deepening through entrainment, *Tellus* **36A**, 446–57.

Rao, K. S. (1975), Effect of thermal stratification on the growth of the internal boundary layer, *Bound. Layer Meteor.* **8**, 227–34.

Rao, K. S., Wyngaard, J. C. and O. R. Coté (1974), The structure of the two-dimensional internal boundary layer over a sudden change of surface roughness, *J. Atmos. Sci.* **31**, 738–46.

Raupach, M. R. (1988), Canopy transport processes, in *Flow and Transport in the Natural Environment: Advances and Applications*, eds W. L. Steffen and O. T. Denmead, pp. 95–127. Springer-Verlag, New York.

Raupach, M. R. (1989), Stand overstorey processes, *Phil. Trans. Roy. Soc. London* **B324**, 175–90.

Raupach, M. R., Antonia, R. A. and S. Rajagopalan (1991), Rough-wall turbulent boundary layers, *Appl. Mech. Rev.* **44**, 1–25.

Raupach, M. R., Thom, A. S. and I. Edwards (1980), A wind-tunnel study of turbulent flow close to regularly arrayed rough surfaces, *Bound. Layer Meteor.* **18**, 373–97.

Reynolds, O. (1883), An experimental investigation of the circumstances which determine whether the motion of water shall be direct or sinuous, and of the law of resistance in parallel channels, *Phil. Trans. Roy. Soc. London* **A174**, 935–82.

Reynolds, O. (1895), On the dynamical theory of incompressible viscous fluids and the determination of the criterion, *Phil. Trans. Roy. Soc. London* **A186**, 123–64.

Richards, J. M. (1971), A simple expression for the saturation vapour pressure of water in the range $-50°$ to $140°C$, *J. Phys. D: Appl. Phys.* **4**, L15–L18.

Roach, W. T. and A. Slingo (1979), A high resolution infrared radiative transfer scheme to study the interaction of radiation with cloud, *Quart. J. Roy. Met. Soc.* **105**, 603–14.

Rotta, J. C. (1951), Statistische Theorie Nichthomogener Turbulenz, *Z. Phys.* **129**, 547–72.

Rowland, J. R. and A. Arnold (1975), Vertical velocity structure and geometry of clear air convective elements, in *Preprints of the 16th Radar Meteorology Conference*, Houston, Texas, pp. 296–303. American Meteorological Society, Boston, MA.

Rowntree, P. R. (1988), Review of GCMs as a basis for predicting the effects of vegetation change on climate, in *Forests, Climate and Hydrology – Regional Impacts*, eds E. R. C. Reynolds and F. B. Thompson, pp. 162–96. The United Nations University, Tokyo.

Royal Society (1975), *Quantities, Units, and Symbols*. Royal Society, London, 54 pp.

Russell, G. (1980), Crop evaporation, surface resistance and soil water status, *Agric. Meteor.* **21**, 213–26.

Sagan, C., Toon, O. B. and J. B. Pollack (1971), Anthropogenic albedo changes and the earth's climate, *Science* **206**, 1363–8.

Sato, N., Sellers, P. J., Randall, D. A., Schneider, E. K., Shukla, J., Kinter, J. L. III, Hou, Y.-T. and E. Albertazzi (1989), Effects of implementing the simple biosphere model in a general circulation model, *J. Atmos. Sci.* **46**, 2757–82.

Saunders, P. M. (1967), The temperature at the ocean–air interface, *J. Atmos. Sci.*, **24**, 269–73.

Sawford, B. L. and F. M. Guest (1987), Lagrangian stochastic analysis of flux-gradient relationships in the convective boundary layer, *J. Atmos. Sci.* **44**, 1152–65.

Schlichting, H. (1979), *Boundary-layer Theory* (translated by J. Kestin), 7th edition. McGraw-Hill, Hamburg, 817 pp.

Schmetz, J. and E. Raschke (1981), An approximate computation of infrared radiative fluxes in a cloudy atmosphere, *Pure Appl. Geophys.* **119**, 248–58.

Schmetz, J., Raschke, E. and H. Fimpel (1981), Solar and thermal radiation in maritime stratocumulus clouds, *Contrib. Atmos. Phys.* **54**, 442–52.

Schmidt, H. and U. Schumann (1989), Coherent structure of the convective boundary layer derived from large-eddy simulations, *J. Fluid Mech.* **200**, 511–62.

Schneider, S. H. and S. L. Thompson (1988), Simulating the climatic effects of nuclear war, *Nature*, **333**, 221–7.

Schumann, U. (1989), Large-eddy simulation of turbulent diffusion with chemical reactions in the convective boundary layer, *Atmos Environ.* **23**(8), 1713–27.

Schumann, U. (1990), Large-eddy simulation of the up-slope boundary layer, *Quart. J. Roy. Met. Soc.* **116**, 637–70.

Segal, M. (1990), On the impact of thermal stability on some rough flow effects over mobile surfaces, *Bound. Layer Meteor.* **52**, 193–8.

Segal, M., Avissar, R., McCumber, M. C. and R. A. Pielke (1988), Evaluation of vegetation effects on the generation and modification of mesoscale circulations, *J. Atmos. Sci.* **45**, 2268–92.

Seginer, I. (1974), Aerodynamic roughness of vegetated surfaces, *Bound. Layer Meteor.* **5**, 383–93.

Sellers, W. D. (1965) *Physical Climatology*. University of Chicago Press, Chicago, 272 pp.

Sellers, P. J. and J. L. Dorman (1987), Testing the simple biosphere model (SiB) using point micrometeorological and biophysical data, *J. Clim. Appl. Meteor.* **26**, 622–51.

Sellers, P. J., Mintz, Y., Sud, Y. C. and A. Dalcher (1986), A simple biosphere model (SiB) for use within general circulation models, *J. Atmos. Sci.* **43**, 505–31.

Sheppard, P. A. (1968), The atmospheric boundary layer in relation to large-scale dynamics, in *The Global Circulation of the Atmosphere*, ed. G. A. Corby, pp. 91–112. Royal Meteorological Society, London.

Sheppard, P. A., Charnock, H. and J. R. D. Francis (1952), Observations of the westerlies over the sea, *Quart. J. Roy. Met. Soc.* **78**, 563–82.

Shir, C. C. (1972), A numerical computation of air flow over a sudden change of surface roughness, *J. Atmos. Sci.* **29**, 304–10.

Shukla, J. and Y. Mintz (1982), Influence of land-surface evapotranspiration on the earth's climate, *Science* **215**, 1498–501.

Shukla, J., Nobre, C. and P. Sellers (1990), Amazon deforestation and climate change, *Science* **247**, 1322–5.

Shuttleworth, W. J. (1989), Micrometeorology of temperate and tropical forest, *Phil. Trans. Roy. Soc. London* **B324**, 299–334.

Shuttleworth, W. J. and R. J. Gurney (1990), The theoretical relationship between foliage temperature and canopy resistance in sparse crops, *Quart. J. Roy. Met. Soc.* **116**, 497–519.

Shuttleworth, W. J. and J. S. Wallace (1985), Evaporation from sparse crops – An energy combination theory, *Quart. J. Roy. Met. Soc.* **111**, 839–55.

Shuttleworth, W. J. *et al.* (1984), Eddy correlation of energy partition for Amazonian forest, *Quart. J. Roy. Met. Soc.* **110**, 1143–62.

Slatyer, R. O. (1967), *Plant–water Relationships*. Academic Press, London, 366 pp.

Slingo, A. and H. M. Schrecker (1982), On the shortwave radiative properties of stratiform water clouds, *Quart. J. Roy. Met. Soc.* **108**, 407–26.

Slingo, A., Nicholls, S. and J. Schmetz (1982a), Aircraft observations of marine stratocumulus during JASIN, *Quart. J. Roy. Met. Soc.* **108**, 833–56.

Slingo, A., Brown, R. and C. L. Wrench (1982b), A field study of nocturnal strato-cumulus: III. High resolution radiative and microphysical observations, *Quart. J. Roy. Met. Soc.* **108**, 145–65.

Smith, S. D. and E.G. Banke (1975), Variation of the sea surface drag coefficient with wind speed, *Quart. J. Roy. Met. Soc.* **101**, 665–73.

Smith, F. B. and D. J. Carson (1977), Some thoughts on the specification of the boundary layer relevant to numerical modelling, *Bound. Layer Meteor.* **12**, 307–30.

Sommeria, G. (1976), Three-dimensional simulation of turbulent processes in an undis-turbed trade wind boundary layer. *J. Atmos. Sci.* **33**, 216–41.

Sorbjan, Z. (1989), *Structure of the Atmospheric Boundary Layer*. Prentice Hall, NJ, 317 pp.

Stewart, J. B. and A. S. Thom (1973), Energy budgets in pine forest, *Quart. J. Roy. Met. Soc.* **99**, 154–70.

Stewart, R. W. (1979), *The Atmospheric Boundary Layer*. WMO No. 523, World Meteorological Organization, Geneva, 44 pp.

Stull, R. B. (1985), A fair-weather cumulus cloud classification scheme for mixed layer studies, *J. Clim. Appl. Meteor.* **24**, 49–56.

Stull, R. B. (1988), *An Introduction to Boundary Layer Meteorology*, Kluwer Academic Publishers, Dordrecht, 666 pp.

Sud, Y. C. and M. Fennessey (1982), A study of the influence of surface albedo on July circulation in semi-arid regions using the GLAS GCM, *J. Climatol.* **2**, 105–25.

Sud, Y. C. and W. E. Smith (1985), Influence of local land-surface processes on the Indian monsoon: A numerical study, *J. Clim. Appl. Meteor.* **24**, 1015–36.

Sud, Y. C., Shukla, J. and Y. Mintz (1988), Influence of land surface roughness on atmospheric circulation and precipitation: A sensitivity study with a GCM, *J. Appl. Meteor.* **27**, 1036–54.

Sud, Y. C., Sellers, P. J., Mintz, Y., Chou, M. D., Walker, G. K. and W. E. Smith (1990), Influence of the biosphere on the global circulation and hydrologic cycle – A GCM simulation experiment, *Agric. Forest Meteor.* **52**, 133–80.

Sutherland, R. A. (1986), Broadband and spectral emissivities (2–18 μm) of some natural soils and vegetation, *J. Atmos. Oceanic Tech.* **3**, 199–202.

Sutton, O. G. (1953), *Micrometeorology*. McGraw-Hill, London, 333 pp.

Swinbank, W. C. (1963), Long-wave radiation from clear skies, *Quart. J. Roy. Met. Soc.* **89**, 339–48.

Swinbank, W. C. (1968), A comparison between predictions of dimensional analysis for the constant flux layer and observations in unstable conditions, *Quart. J. Roy. Met. Soc.* **94**, 460–7.

Sykes, R. I. and D. S. Henn (1989), Large-eddy simulation of turbulent sheared convection, *J. Atmos. Sci.* **46**, 1106–18.

Takeuchi, K. (1961), On the structure of the turbulent field in the surface boundary layer, *J. Met. Soc. Japan* **39** (II), 346–67.

Taylor, P. A. (1971), Airflow above changes in surface heat flux, temperature and roughness: An extension to include the stable case, *Bound. Layer Meteor.* **1**, 474–97.

Taylor, P. A. (1987), Comments and further analysis on effective roughness lengths for use in numerical three-dimensional models, *Bound. Layer Meteor.* **39**, 403–18.

Telford, J. W. and J. A. Businger (1986), Comments on "Von Karman's constant in atmospheric boundary layer flow: Reevaluated", *J. Atmos. Sci.* **43**, 2127–34.

Tennekes, H. (1970), Free convection in the turbulent Ekman layer of the atmosphere, *J. Atmos. Sci.* **27**, 1027–34.

Tennekes, H. (1973), Similarity laws and scale relations in planetary boundary layers, Chapter 5 in *Workshop on Micrometeorology*, ed. D. A. Haugen, pp. 177–216. American Meteorological Society, Boston, MA.

Tennekes, H. (1982), Similarity relations, scaling laws and spectral dynamics, in *Atmospheric Turbulence and Air Pollution Modelling*, eds F. T. M. Nieuwstadt and H. van Dop, pp. 37–68. Reidel, Dordrecht.

Tennekes, H. and A. G. M. Driedonks (1981), Basic entrainment equations for the atmospheric boundary layer, *Bound. Layer Meteor.* **20**, 515–31.

Tennekes, H. and J. L. Lumley (1972), *A First Course in Turbulence*. MIT Press, Cambridge, MA, 300 pp.

Thom, A. S. (1971), Momentum absorption by vegetation, *Quart. J. Roy. Met. Soc.* **97**, 414–28.

Thom, A. S. (1972), Momentum, mass and heat exchange of vegetation, *Quart. J. Roy. Met. Soc.* **98**, 124–34.

Thom, A. S. (1975), Momentum, mass and heat exchange of plant communities, in *Vegetation and the Atmosphere*, Vol. 1, *Principles*, ed. J. L. Monteith, pp. 57–109. Academic Press, London.

Thom, A. S. and H. R. Oliver (1977), On Penman's equation for estimating regional evaporation, *Quart. J. Roy. Met. Soc.* **103**, 345–57.

Thom, A. S., Stewart, J. B., Oliver, H. R. and J. H. C. Gash (1975), Comparison of aerodynamic and energy budget estimates of fluxes over a pine forest, *Quart. J. Roy. Met. Soc.* **101**, 93–105.

Thompson, O. E. and R. T. Pinker (1975), Wind and temperature profile characteristics in a tropical evergreen forest in Thailand, *Tellus* **27**, 562–73.

Thorpe, A. J. and T. H. Guymer (1977), The nocturnal jet, *Quart. J. Roy. Met. Soc.* **103**, 633–53.

Turner, J. S. (1973), *Buoyancy Effects in Fluids*. Cambridge University Press, 367 pp.

Turton, J. D. and S. Nicholls (1987), A study of the diurnal variation of stratocumulus using a multiple mixed layer model, *Quart. J. Roy. Met. Soc.* **113**, 969–1009.

Van Dyke, M. (1975), *Perturbation Methods in Fluid Mechanics* (annotated edition). The Parabolic Press, Stanford, CA, 271 pp.

Venkatram, A. (1977), A model of internal boundary-layer development, *Bound. Layer Meteor.* **11**, 419–37.

Verma, S. B., Baldocchi, D. D., Anderson, D. E., Matt, D. R. and R. J. Clement (1986), Eddy fluxes of CO_2, water vapour, and sensible heat over a deciduous forest, *Bound. Layer Meteor.* **36**, 71–91.

Webb, E. K. (1970), Profile relationships: The log-linear range and extension to strong stability, *Quart. J. Roy. Met. Soc.* **96**, 67–90.

Webb, E. K. (1975), Evaporation from catchments, in *Prediction in Catchment Hydrology*, eds T. G. Chapman and F. X. Dunin, pp. 203–36. Australian Academy of Science, Canberra.

Webb, E. K. (1977), Convection mechanisms of atmospheric heat transfer from surface to global scales, in *Proceedings of the Second Australasian Conference on Heat and Mass Transfer*, ed. R. W. Bilger, pp. 523–539. University of Sydney, Sydney.

Webb, E. K. (1982), Profile relationships in the superadiabatic surface layer, *Quart. J. Roy. Met. Soc.* **108**, 661–88.

Weil, J. C. (1990), A diagnosis of the asymmetry in top-down and bottom-up diffusion using a Lagrangian stochastic model, *J. Atmos. Sci.* **47**, 501–15.

Weill, A., Baudon, F., Resbraux, G., Mazaudier, C., Klapisz, C. and A. G. M. Driedonks (1985), A mesoscale shear-convective organization and boundary-layer modification: An experimental study performed with acoustic Doppler sounders during the COAST experiment, in *Proceedings of the 2nd Conference on Mesoscale Processes*. American Meteorological Society, Boston, MA.

Wexler, A. (1976), Vapor pressure formulation for water in range 0 to 100 °C. A revision, *J. Res. Nat. Bur. Stand.* **80A**, 775–85.

Wichmann, M. and E. Schaller (1986), On the determination of the closure parameters in higher-order closure models, *Bound. Layer Meteor.* **37**, 323–41.

Wieringa, J. (1980), A revaluation of the Kansas mast influence on measurements of stress and cup anemometer overspeeding, *Bound. Layer Meteor.* **18**, 411–30.

Wipperman, F. (1973), *The Planetary Boundary Layer of the Atmosphere*. Deutschen Wetterdienst, Offenbach, 346 pp.

Wood, D. H. (1982), Internal boundary layer growth following a step change in surface roughness, *Bound. Layer Meteor.* **22**, 241–4.

Wooding, R. A., Bradley, E. F. and J. K. Marshall (1973), Drag due to regular arrays of roughness elements of varying geometry, *Bound. Layer Meteor.* **5**, 285–308.

World Meteorological Organization (1985), *Report of the JSC/CAS Workshop on Modelling of Cloud Topped Boundary Layer*, WCP-106 (WMO/TD No. 75), Unofficial WMO Document (World Climate Programme Series), pp. 25–6. World Meteorological Organization, Geneva.

Wu, Jin (1969), Froude number scaling of wind-stress coefficients, *J. Atmos. Sci.* **26**, 408–13.

Wu, Jin (1975), Wind-induced drift currents, *J. Fluid Mech.* **68**, 49–70.

Wu, Jin (1980), Wind-stress coefficients over sea surface near neutral conditions – A revisit, *J. Phys. Oceanogr.* **10**, 727–40.

Wu, Jin (1982), Wind-stress coefficients over sea surface from breeze to hurricane, *J. Geophys. Res.* **87**, 9704–6.

Wyngaard, J. C. (1973), On surface layer turbulence, Chapter 3 in *Workshop on Micrometeorology*, ed. D. A. Haugen, pp. 101–49. American Meteorological Society, Boston, MA.

Wyngaard, J. C. (1982), Boundary-layer modeling, in *Atmospheric Turbulence and Air Pollution Modelling*, eds F. T. M. Nieuwstadt and H. van Dop, pp. 69–158. Reidel, Dordrecht.

Wyngaard, J. C. (1988), Structure of the PBL, in *Lectures on Air Pollution Modeling*, eds A. Venkatram and J. C. Wyngaard, pp. 9–61. American Meteorological Society, Boston, MA.

Wyngaard, J. C. and R. A. Brost (1984), Top-down and bottom-up diffusion of a scalar in the convective boundary layer, *J. Atmos. Sci.* **41**, 102–12.

Wyngaard, J. C. and O. R. Coté (1972), Cospectral similarity in the atmospheric surface layer, *Quart. J. Roy. Met. Soc.* **98**, 590–603.

Wyngaard, J. C., Coté, O. R. and Y. Izumi (1971), Local free convection, similarity, and the budgets of shear stress and heat flux, *J. Atmos. Sci.* **28**, 1171–82.

Yaglom, A. M. (1977), Comments on wind and temperature flux-profile relationships, *Bound. Layer Meteor.* **11**, 89–102.

Yamada, T. (1976), On the similarity functions A, B and C of the planetary boundary layer, *J. Atmos. Sci.* **33**, 781–93.

Yamada, T. (1979), Prediction of the nocturnal surface inversion height, *J. Appl. Meteor.* **18**, 526–31.

Yamada, T. and G. Mellor (1975), A simulation of the Wangara atmospheric boundary layer data, *J. Atmos. Sci.* **32**, 2309–29.

Yan, H. and R. A. Anthes (1988), The effect of variations in surface moisture on mesoscale circulations, *Mon. Wea. Rev.* **116**, 192–208.·

Zeman, O. (1979), Parameterization of the dynamics of stable boundary layers and nocturnal jets, *J. Atmos. Sci.* **36**, 792–804.

Zeman, O. (1981), Progress in the modeling of planetary boundary layers, *Ann. Rev. Fluid Mech.* **13**, 253–72.

Zhang, S. F., Oncley, S. P. and J. A. Businger (1988), A critical evaluation of the Von Karman constant from a new atmospheric surface layer experiment, in *Proceedings of the 8th Symposium on Turbulence and Diffusion*, pp. 148–50. American Meteorological Society, Boston, MA.

Zilitinkevich, S. S. (1972), On the determination of the height of the Ekman boundary layer, *Bound. Layer Meteor.* **3**, 141–5.

Zilitinkevich, S. S. (1975), Resistance laws and prediction equations for the depth of the planetary boundary layer, *J. Atmos. Sci.* **32**, 741–52.

Zilitinkevich, S. S. (1989), Velocity profiles, the resistance law and the dissipation rate of mean flow kinetic energy in a neutrally and stably stratified planetary boundary layer, *Bound. Layer Meteor.* **46**, 367–87.

Zilitinkevich, S. S. and J. W. Deardorff (1974), Similarity theory for the planetary boundary layer of time-dependent height, *J. Atmos. Sci.* **31**, 1449–52.

Zilitinkevich, S. S. and A. S. Monin (1974), Similarity theory for the planetary boundary layer of the atmosphere, *Akad. Nauk SSSR Izv. Atmos. Oceanic Phys.* **10**, 353–59.

Index

Printed in the United States
By Bookmasters